Prokaryotic Genetics

GENOME ORGANIZATION,

TRANSFER AND PLASTICITY

STUDIES IN MICROBIOLOGY

EDITOR

N. G. CARR
Department of Biological Sciences
University of Warwick

STUDIES IN MICROBIOLOGY

Prokaryotic Genetics

GENOME ORGANIZATION, TRANSFER AND PLASTICITY

FRANÇOISE JOSET
Professor, CNRS and Université Aix-Marseille II

JANINE GUESPIN-MICHEL
Professor, CNRS and Université de Rouen

WITH THE COLLABORATION OF
LESLIE O. BUTLER
Professor, St George's Hospital Medical School,
London

Blackwell
Science

© 1993 by
Blackwell Science Ltd
Editorial Offices:
Osney Mead, Oxford OX2 0EL
25 John Street, London WC1N 2BL
23 Ainslie Place, Edinburgh EH3 6AJ
238 Main Street, Cambridge
  Massachusetts 02142, USA
54 University Street, Carlton
  Victoria 3053, Australia

Other Editorial Offices:
Arnette Blackwell SA
1, rue de Lille, 75007 Paris
France

Blackwell Wissenschafts-Verlag GmbH
Kurfürstendamm 57
10707 Berlin, Germany

Blackwell MZV
Feldgasse 13, A-1238 Wien
Austria

First published 1993
Reprinted 1994

Set by Excel Typesetters Company Ltd,
Hong Kong
Printed and bound in Great Britain
at the University Press, Cambridge

DISTRIBUTORS

Marston Book Services Ltd
PO Box 87
Oxford OX2 0DT
(Orders: Tel: 01865 791155
       Fax: 01865 791927
       Telex: 837515)

USA
  Blackwell Science, Inc.
  238 Main Street
  Cambridge, MA 02142
  (Orders: Tel: 800 759-6102
           617 876-7000)

Canada
  Oxford University Press
  70 Wynford Drive
  Don Mills
  Ontario M3C 1J9
  (Orders: Tel: 416 441 2941)

Australia
  Blackwell Science Pty Ltd
  54 University Street
  Carlton, Victoria 3053
  (Orders: Tel: 03 347-5552)

A catalogue record for this title is available from
the British Library

ISBN 0-632-027282

Library of Congress
Cataloging-in-Publication Data

Joset, François.
   Prokaryotic genetics:
   Genome organization, transfer,
   and plasticity /
   François Joset, Janine Guespin-Michel
       p.   cm.
   (Studies in microbiology)
       Includes bibliographical references
   and index.
       ISBN 0-632-02728-2
       1. Bacterial genetics.
       2. Prokaryotes – Genetics.
   I. Guespin – Michel, Janine.
   II. Title.   III. Series.
       [DNLM: 1. Bacteria – genetics.
   2. Cytogenetics.   3. Prokaryotic Cells.
   ST924/QW 51 J836p]
   QH434.J67   1993
   589.9'015 – dc20

Contents

Preface

Prokaryotic genetics? While the meaning of 'genetics' is nowadays clear, the word 'prokaryote' may raise some questions. It is used here in place of the better-known word 'bacterium'. Its usage emphasizes the central fact that the component organisms characteristically lack a distinct nucleus and nuclear membrane, in sharp contrast with all other living cells. Prokaryotic cell division does not involve the degree of structural elaboration associated with eukaryotic cell division. The essential difference between these two groups is not that of unicellular versus multicellular structures but lies in the organization of the cells and genomes, and consequently of the genetic exchange processes involved.

Although viruses may at one level be considered as a borderline between living organisms and macromolecular complexes, at the genetic level they share a fundamental characteristic of all living organisms in having an autonomously reproducing genome. Furthermore, the understanding of their genetics and reproduction processes has played an important role in the unravelling of genetics, especially that of bacteria. For this reason, the viruses which infect bacteria, the bacteriophages, lie within the scope of this book, to the extent that they have contributed to the study of bacterial genetics.

Bacteria are usually assumed to have been the earliest living organisms on earth. Although they had already been observed in the seventeenth century, their importance in everyday life as pathogenic agents or as activators in chemical transformations, in particular in industrial processes, was recognized only during the nineteenth century, after Pasteur's work. The existence of bacterial viruses, although postulated for some time, was established almost a century later in the works by Twort and d'Hérelle. The history of work with prokaryotes has been one of identifying and separating an ever-increasing number of individual species, the classification of which has been based on our evolving knowledge of morphology, physiology, biochemistry and latterly molecular sequences, the more recent stages of this process having been increasingly underpinned by genetic knowledge. The first report of a study on bacterial genetics as such was in 1942, with the first recognition that DNA was the agent of inheritance. Successive landmarks in bacterial genetics which opened the way to molecular genetics of both prokaryotes and eukaryotes, were revealed over a period of about 20 years, during the 1940s to 1960s. These included: the establishment of the role of DNA as carrier of hereditary information, the deciphering of the chemical structure of DNA and the elucidation of the molecular organization of the DNA double helix and of the genetic code. Since then, bacterial genetics has developed at such an accelerating rate that it has, in fact, brought us most of our present knowledge of molecular genetics. The accompanying creation of *in vivo* and *in vitro* genetic tools has proved powerful not only for the improvement of our understanding of fundamental biological processes, but also for the utilization of genetic capabilities in biotechnology.

The diversity of prokaryotic species present in today's world is a measure of the varied genetic potentialities which yield stable, viable, individuals. These genetic characters are organized in coded structures, or genetic units, to form the bacterial chromosomes, which must, in a manner not yet understood, have been the precursors of those of eukaryotes. The complex fine structure of the bacterial chromosomes had long been assumed to be a token of their stability until recent discoveries indicated a degree of flexibility, a feature which may have had an important evolutionary role.

This text on prokaryotic genetics adopts the following sequence. Part 1 describes the genetic features by which individual organisms are determined and identified (i.e. the characteristics of the elements forming their genetic material), and discusses the perpetuation of the strains and species thus determined (i.e. the transmission and protection of their genetic content). The second part discusses the available means or opportunities of the strains or species to modify their genetic information, while maintaining limits compatible with their survival. The applications of these genetic capabilities as tools for further understanding of bacterial metabolic processes and for molecular genetic technology are discussed in Part 3.

Most of our knowledge in prokaryotic genetics has been acquired from only a small number of bacterial species and, even more limiting, from only particular strains of these species. The ease of manipulation of these 'model' strains in the laboratory, especially *Escherichia coli* K12, is responsible for this situation. Interest in the genetics of a greater range of species has arisen mostly since microbiology has revealed the large variety of resources existing in these organisms and the possibility of their exploitation by molecular genetics. Such techniques have allowed the jump from descriptive accounts of whole cells to precise molecular analysis of particular genetic functions, often with much basic genetic information remaining unknown.

While it is obviously impossible to describe the genetics of 'non-model' prokaryotes without reference to 'model' bacteria, we will attempt to emphasize the similarities and diversities of genetic processes in a range of species. Such comparisons, while illustrating the variety of biological processes, should encourage readers to apply existing knowledge to a greater range of organisms.

The first book on bacterial genetics was written by William Hayes in 1964 (*The Genetics of Bacteria and their Viruses*, published by Blackwell Scientific Publications) and was unique in presenting the totality of available knowledge in this field. It was a 'Bible' for both students and researchers thanks to its clarity, enthusiasm and thorough presentation of the existing knowledge, and it had a crucial role in converting microbiologists into geneticists. The difficulty at that time, remarkably overcome by Hayes, probably lay in the lack of general models linking together diverse information. The difficulty at the present time, in contrast, is that information and models are now so abundant that difficult choices must be made as to what to present.

This book is intended for readers with some knowledge of biochemistry, general genetics and cellular physiology and metabolism. It is aimed at a main readership of undergraduate students in the biological sciences, but it should also be of value to technicians and research workers in biology or biotechnology.

The authors would like to offer particular thanks to Professor Leslie Butler for his contribution to the entire manuscript. Professor Butler is the sole author of Chapter 9, and coauthor of Chapters 7 and 10. We would also like to express our appreciation to Dr Pam McAthey of the Polytechnic of North London for contributing Chapter 8 on DNA repair, to acknowledge the helpful discussions with many colleagues and to thank those who have provided or given permission for the reproduction of figures.

F.J., J.G.-M.

PART 1
STRUCTURE OF THE
GENETIC MATERIAL

1: The Prokaryotes:
a Brief Introduction

1 Identity and structure: a review

Prokaryotes (a synonym of Bacteria) are the simplest autonomous organisms known, i.e. organisms which possess all the functions (and hence the genetic information) allowing them to reproduce, although some need various forms of association, such as syntrophy, symbiosis and parasitism, either with other prokaryotes or with eukaryotes, to do so permanently or in certain conditions. They are also the smallest cells described, although their sizes vary widely, the largest ones reaching dimensions similar to those of some eukaryotic protists. Their morphological types are limited, but various external structures (such as flagella, pili, fimbriae, sheath, capsule, etc.) with different functions are often present. Most prokaryotes exist as unicellular organisms, although multicellular associations of cells within the same strain, endowed with particular physiological functions, are encountered; among the more complex, some form truly multicellular networks of branching hyphae or pluricellular fruiting bodies.

All prokaryotes share a number of common traits, the most specific of which is their lack of internal cellular organization into physically separated and functionally specialized compartments as found in eukaryotes, which are thought to have developed later. Present trends based on ribosomal RNA sequence comparisons divide living beings into three domains or kingdoms, the Bacteria, the Archaea and the Eukarya. The relation between these three domains is still debated, although a recent proposal by Woese (1987, 1990) accepts that a common ancestor, a progenote, led first to two branches, the Bacteria on one side and on the other a second branch which later divided into the Archaea and the Eukarya (Fig. 1.1). Bacteria (currently referred to as Eubacteria) are composed of Gram-positive and Gram-negative families, reflecting the presence of either a thick peptidoglycan cell wall (the former) or a thin peptidoglycan cell wall surrounded by an outer membrane (the latter), external to the cytoplasmic membrane and protecting the cell. The cell walls of Archaea show no structural uniformity and have completely different compositions from those of Eubacteria. Many other features, particularly at the molecular level, distinguish organisms from these two domains.

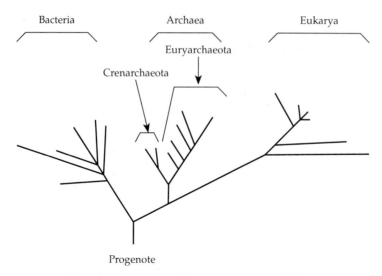

Fig. 1.1 Proposed phylogenetic tree of living organisms. This tree, showing three domains as presently known, is based particularly on ribosomal RNA comparisons. A putative original Progenote cell would have evolved into two branches, one generating the Bacteria and the other later dividing into two other domains, the Archaea and the Eukarya. The lengths of the different branches are proportional to the 'evolutionary distance' separating the various groups indicated, as known today. Each domain, or kingdom, subdivides into several 'groups'.

One feature important for the geneticist is the particular state of the genetic material of prokaryotes (Section 5). As might be expected, they contain less genetic information than even the least elaborate eukaryotes.

Although they have a fairly minimal structural complexity, prokaryotes display an amazing variety of adaptive capacities allowing them to populate a very large range of environmental conditions. These capacities are reflected both in the variety of 'nutritional' groups that can be defined, and in the intrinsic capacities of a given species or strain to adapt to varying environments. These features, together with morphological and other biochemical characteristics, have been the initial criteria used to define taxonomical groups. Accumulation of molecular data has led to modifications or confirmation of the original classifications. The nutritional criteria have, however, proved in most instances convenient and fruitful.

These extremely varied metabolic capacities, the frequent synthesis of industrially interesting secondary metabolites (Section 4) and the ease of growing large populations (Section 3) have promoted the use of bacteria as biological tools for applied purposes. This field of biotechnology has developed rapidly in the past decade or so, mostly because of the parallel increase of our knowledge in bacterial genetics. Molecular genetics and genetic engineering make use of most of the genetic processes described (Part 3).

2 Nutrition and nutritional modes

The life of a unicellular organism such as most prokaryotes, apart from possible differentiation processes, consists mainly in nutritional reactions leading to cell doubling and division, the result of the co-ordinated functioning of energetic pathways (transformation of external energy sources into energy-rich metabolites such as ATP) and assimilatory pathways (the making up of cellular material). These biosynthetic pathways, in general very similar among prokaryotes and eukaryotes, will not be described here. The complex molecular composition of the bacterium *E. coli* (Table 1.1) reflects the intense metabolic activity that must take place in such cells, considering that in this particular case generation times range from around 20 min to 2 h.

Bacteria can fulfil their needs for energy and for the basic constituent carbon through the use of several processes, characterized by the nature of the energy and carbon sources and that of the electron source and acceptor.

Table 1.1 Average molecular composition of an *E. coli* cell.

Constituents	Number of copies/cell	Proportion in % of dry weight
Cell wall and outer layers including exopolysaccharides	1	≈10
Membrane including lipids	1	10
DNA		
chromosome	1 to 3–4	2 to 6–8
plasmids	0 to 8–n	
RNA		
mRNA	>3 000	2
tRNA	>150 000	3
rRNA	20 000	21
Proteins (soluble and ribosomal)	>10^6	51
Small organic molecules	$6–7 \times 10^6$	1
Inorganic molecules	?	Negligible
Reserves (carbonic, nitrogenous, etc.)		Dependent on physiological state

E. coli is a rod-shaped bacterium, approximately 1.2 μ long × 0.5 μ diameter. This represents a volume of approximately 2.5×10^{-7} cm³, or a weight of 2.5×10^{-7} g/cell (supposing an average density of 1), of which 70% is water, the rest comprising the constituents listed in the table.

Table 1.2 Prokaryotes grouped according to their energy source.

| Class | Energy source | Electron | | Energetic pathway |
		Donor	Acceptor	
Chemoorganotrophs	Reduced carbon molecules		O_2 NO_3^-, SO_4^{2-} Organic substances	Aerobic respiration (oxidative phosphorylation) Anaerobic respiration Fermentation (substrate phosphorylation)
Photolithotrophs	Light	H_2O, H_2S, S	NAD, NADP	Photosynthesis (photophosphorylation)
Photoorganotrophs	Light	Organic substances (e.g. acids, sugars)	NAD, NADP	Anoxygenic photosynthesis (photophosphorylation)
Chemolithotrophs	Mineral reduced substances (e.g. H_2, NH_4, H_2S, S, Fe^{2+}, Cu^{2+})		O_2, NO_3^-, CO_2, H_2SO_4, etc.	Sulphatoreducing, methanogenic, nitrifying, etc.

NAD, nicotinamide adenine dinucleotide; NADP, nicotinamide adenine dinucleotide phosphate.

Table 1.3 Nutritional groups.

Groups	Carbon source	Energetic groups
Heterotrophs	Organic molecules such as sugars, amino acids, phenols, lignins, etc.	Chemoorganotrophs Some photo- or chemolithotrophs
Autotrophs	CO_2, $-CH_3$	Photoorganotrophs Photolithotrophs Chemolithotrophs
Methylotrophs	CO_2	Chemoorganotrophs

Energy can be derived from three main sources: light, oxidation of minerals or oxidation of organic compounds. Most bacteria can use one or two of these energy sources, e.g. light through photosynthesis or a sugar through aerobic or anaerobic respiration. The transformation of these energy sources into an assimilable form of energy is coupled to oxidation–reduction reactions, the transfer of electrons and protons involving various electron donors and acceptors (Table 1.2). The names given to the different classes refer to both energy sources and electron donor criteria.

The nature (organic/mineral) of the carbon source used provides a helpful classification (Table 1.3). Chemoorganotrophs are heterotrophic organisms which usually use the same molecule as both energy and carbon sources. For instance, glucose is respired, yielding ATP, CO_2, electrons and H^+, and assimilated into other organic compounds via the so-called intermediary metabolism.

Members of the three other energetic groups, photoorganotrophs, photolithotrophs and chemolithotrophs are most often autotrophs, i.e. use a mineral carbon source, CO_2, although some may show heterotrophic capacities.

One group, the methylotrophs, at the frontier between organo- and lithotrophs, utilizes $-CH_3$ as electron donor and CO_2 as carbon source.

3 Cell division and population growth

3.1 Cell division

With few exceptions, prokaryotes multiply by binary fission, one cell yielding two daughter cells, equivalent and genetically identical to the parental one and derived from a grossly equal partition of all constituents. It is thus normally impossible to distinguish the parent from the progeny in the two cells of the following generation. These cells will, in time, give rise to two new progeny cells each, through the same process. Thus the material of the original parent will pass to all successive descendants (of course, accidents may lead to the death of a cell, and thus to the loss of this original material). Except for some species displaying complex multicellular structures and certain levels of differentiation, no ageing or death of the original organism, as defined for multicellular beings, thus occurs.

A prokaryotic cell includes two categories of constituents (Table 1.1). Firstly, there are those present in numerous copies, such as ribosomes, metabolites, etc., of which more can always be made provided the genetic information is conserved. The second category covers only two types of complexes, each present in only one copy per cell, the envelope (taken as the sum of all the external layers) and the genetic material. While an approximate distribution of constituents of the first category in the two daughter cells is sufficient, it is of vital importance that the genetic material be copied and shared equally, and that the envelope be continuously synthesized so as to ensure cell integrity upon division.

Molecules and macromolecules of the first category are synthesized permanently. So is the synthesis of membrane and cell wall materials, although a particular process of synthesis starting along a diameter of the cell, but different for Gram-positive and Gram-negative bacteria, takes place at the initiation of division. At cell division, the newly made layers proceed towards the inside of the cell, making a septum which separates the two daughter cells.

Reproduction of the chromosome, which may require the whole span of time separating two divisions, is initiated only once per cell cycle, under usual growth conditions. It is coordinated in both space and time with the formation of the septum, so as to ensure the presence of one copy in each progeny cell, except in very fast-growing cultures for which chromosome reproduction will take longer than the division time, resulting in the necessity for simultaneous chromosome replication cycles (Chapter 2).

3.2 Cell and population growth cycles

Bacteria do not display a cell cycle in the strict sense, with a succession of well-defined phases leading to duplication and equal partitioning of the unique cell constituents, as takes place during mitosis in eukaryotic cells. Under theoretical conditions in which no limitation or modification of the available nutrients occurred, cell growth and divisions could continue indefinitely. In practice, however, be it in natural habitats or under artificial laboratory conditions, the amount of nutrients is limited. This, together with secondary effects (changes in pH, excretion of toxins, etc.), always sets a limit to proliferation. Thus, in a given set of physically and nutritionally determined conditions, a population growth cycle can be defined.

3.3 A typical growth curve

A characteristic curve, schematically illustrated in Fig. 1.2, represents the population growth cycle as obtained under strictly defined conditions in laboratory assays. Several phases are apparent:

1 The latent phase is not, strictly speaking, part of the cycle since it is not a compulsory phase. It is present only when the organisms need to adapt to different, less favourable, conditions, or when starved cells are used as inoculum.

2 The exponential phase is the actual growth phase. It derives its name from the binary mode of division, which yields an exponential increase in cell number as divisions take place. The mean time necessary for one such division (a doubling of the number of cells per unit volume of culture medium), called the generation time (usually denoted *G*), can easily be determined from the semilogarithmic representation of the growth curve. It is a characteristic parameter of the growth of a given strain under a given set of conditions (physical and chemical). The mathematical equation of exponential growth utilizes another term, μ, the growth rate, which is strictly related to *G*:

$$\mu(h^{-1}) = \ln 2/G,$$

and is determined as $dX/dt = \mu X$, where *X* is the biomass at time *t*.

3 The stationary phase indicates the end of growth. It takes place

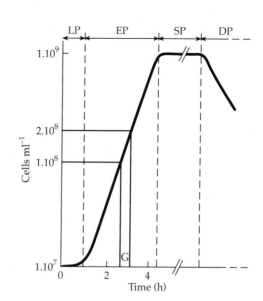

Fig. 1.2 Standard schematic growth cycle. Schematic growth cycle of *E. coli* at 37 °C in a rich medium under aeration. DP, death phase; EP, exponential phase; LP, latent phase; SP, stationary phase.

when an unfavourable factor prevents continuation of growth, e.g. exhaustion of one (or several) nutrient(s) (referred to as limiting factor(s)), or an unacceptable modification of the physical conditions (change in pH for instance). This may represent an equilibrium between the number of cells dividing and dying.

Since growth takes place at the expense of nutrients, a yield can be defined for each nutrient concentration, N:

$$Y_F(X/N) = dX/dN.$$

The general equation for the determination of yield can be written:

$$Y_F(X/N_1) = \frac{X_{max} - X_0}{N_0 - N_{max}}$$

where X_{max} = maximal biomass obtained at the stationary phase, X_0 = biomass of the inoculum (at t_0), N_0 = concentration of nutrient N at time t_0, and N_{max} = concentration of nutrient N at stationary phase. When growth ceases because of exhaustion of a nutrient, i.e. this nutrient is limiting, the yield for this nutrient can easily be evaluated, since the equation becomes

$$Y_F(X/N_1) = \frac{X_{max} - X_0}{N_0}.$$

Note that Y is usually larger than 1, except if referring to the energy source.

The yield constitutes the second parameter, with the growth rate, which characterizes the growth of organisms of a given strain under given conditions.

4   The last phase depicted in Fig. 1.2, the death phase, although a normal outcome of the end of growth, can follow very variable kinetics depending on the conditions and the organism. As noted above, death takes place statistically in the whole population and not in particular cells sorted out through a physiological process of ageing. The kinetics of death are also exponential. It follows that, unless unfavourable conditions last long enough, a fraction of the population will usually survive. This is important to remember when one is setting up sterilization conditions or mutant selection procedures.

4 Secondary metabolism

The functions grouped under the designation secondary metabolism are very heterogeneous, their common characteristic being their apparent dispensability for the producing cells. The molecules produced through secondary metabolic processes belong to a wide spectrum of chemical families, e.g. alkaloids, peptides, epoxides, pyrrols, hydrocarbons. Their physiological function is often unknown, although some are thought to be useful as self-protecting agents. They display extremely varied biological activities, which are very important for human welfare, e.g. enzyme inhibition (useful for therapeutic purposes), bactericidal activity (most of the natural antibiotics, representing at least eight different chemical families), and even insect-killing substances and herbicides.

These substances are often excreted by the cells. This observation may, however, be biased, since it is much easier to detect such a product when it is present in a culture medium than when it is accumulated inside cells. There is frequently a period of production at the end of fast growth, or at the beginning of the stationary phase, generally termed the idiophase.

The chemical heterogeneity of the known secondary metabolites is accompanied by a large variation in profiles of production among bacterial genera, or even among species or strains inside a genus. However, newly discovered metabolites are often produced by the same or related strains that are already known as producers of other such substances. This is particularly true for the production of antibiotics.

Secondary metabolites are generally produced in low amounts by the wild-type isolates. A large part of the efforts made in industrial microbiology has been orientated toward increasing these yields. After a long period during which random mutagenesis and screening was the only method available to isolate high-producing clones, an era has lately begun in which extremely efficient *in vivo* and *in vitro* genetic methods should produce industrially improved strains, although it is still too early to have actually led to such applications.

5 Organization of prokaryotic genetic information

All the genetic information considered necessary for the life of prokaryotic cells is usually assembled into a unique molecule, rendering these organisms monochromosomic (Chapters 2 and 14), although an exception has recently been revealed. The structural organization of these DNA molecules inside the cell is much simpler than that of eukaryotic chromosomes, which explains why the name 'chromosome' applied to these structures has sometimes been questioned. However, it will be used throughout this book. The absence of a true nucleus, as implicit in the word prokaryote ('primitive nucleus'), allows chromosome reproduction and partition to be simpler than the mitotic process of eukaryotes.

Other accessory, stable, autonomously duplicating, DNA molecules, the plasmids, are also often present (Chapter 3), although the distinction between some large plasmids and actual chromosomes might prove to be difficult or spurious as our knowledge of these units increases.

Many viruses that are specific for prokaryotes have been described (Chapter 4). The study of these bacteriophages has been extremely fruitful not only as models for viruses infecting eukaryote organisms, but also as tools in deciphering bacterial physiology and genetics themselves.

Mobile genetic units, known as transposable elements, the existence of which in bacteria was detected some 25 years ago, have since proved to be extremely widespread. Their biological role is still unclear (Chapter 5), but they have become a marvellous tool in present-day genetics (Chapters 13 and 16).

As in all organisms, the genetic content of bacteria as we currently observe it results from an equilibrium between processes tending to modify it, i.e. mutations (Chapters 5 and 7) and inter- and intraspecies exchanges (Chapters 9–12), and protection mechanisms aimed at maintaining the initial status (Chapters 6 and 8).

The natural modes of transfer of genetic information between prokaryote cells are distinctive (Chapters 10–12), although the events leading to molecular exchange between the parental DNA molecules, i.e. recombination, follow the same general rules as in eukaryotes (Chapter 9).

As just outlined, it is the aim of this book to analyse the various genetic constituents known in prokaryotes, highlighting diversities among overall common features. Unfortunately our knowledge concerning archaebacterial genetics is still very limited, explaining their frequent absence in the body of the book, which is a pity since the uniqueness of many of their known characteristics suggests that they exhibit some original genetic traits. Another bias to a presentation reflecting the diversity of the prokaryotic world results from the fact that only a few bacteria have served as models for geneticists for many years, the enteric bacterium *E. coli* outrunning all others. While considered to be a faithful representative for a long time, it now appears that it is unusual in many of the physiological traits which have in fact favoured its choice as working material. This situation explains why many descriptions will apply to this bacterium, but, depending on individual cases, can either be taken as paradigms, as is probably true for bacterial viruses, or should be generalized with caution.

This book should also be a demonstration of how accumulation of understanding of bacterial genetic processes has fed more discoveries and provided extremely powerful tools for both extending our knowledge of these systems and opening ways to modify them artificially almost at will (Part 3).

References

Woese C.R. (1987) Bacterial evolution. *Microbiological Reviews*, 51, 221–71.

Woese C.R., Kandl O., & Wheelis M.L. (1990) Towards a natural system of organisms: proposal for the domains Archae, Bacteria and Eucarya. *Proceedings of the National Academy of Science, USA*, 87, 4576–9.

2: Bacterial Chromosomes: Structure and Reproduction

As in eukaryotes, the hereditary information of prokaryotes is organized in molecules of double-stranded deoxyribonucleic acid (DNA). Two classes of DNA molecules are found in most prokaryotes. One of them, usually called the chromosome through analogy with the eukaryotic equivalent structures, carries all (or almost all, in the vast majority of prokaryotes) of the genetic information necessary for the cell to live and reproduce. The second class of DNA molecules, termed the plasmids, are dispensable for normal growth. Their biological significance will be described in the next chapter.

The absolute importance of chromosomal DNA for both the life of the cell and the propagation of the species accounts for the complex biochemical processes elaborated to ensure the accuracy of its reproduction and its protection against all kinds of potential damage, such as might be caused by destructive agents (e.g. mutagens, Chapter 7) or by invasion of the cell by other DNA molecules (Chapters 6 and 9–12).

1 Structure of prokaryotic chromosomes

A number of authors have recommended that the name 'chromosome' should not be used to describe the bacterial genetic material, because of its structural differences from eukaryotic chromosomes. The words 'chromonema' or 'genophore' have been proposed, but have never gained widespread usage. Another term, the nucleoid, specifically refers to the structure of the chromosome inside the cell (Section 1.5).

The objections are based on the reference of the word chromosome to the capacity of the corresponding eukaryotic structure to retain coloration by several dyes, a property of the nucleic acid part of the complex. However, although initially describing the whole highly organized complex made of DNA associated with specific proteins, since the particular roles of these constituents have been elucidated the term is now used also to describe the chemical material *per se* holding the genetic information. This now extends to all living organisms (including viruses), whatever the chemical nature and structure of their genetic material (DNA, RNA, single- or double-stranded). We shall thus conform to the general custom and use it in this sense.

1.1 Chemical composition

Eukaryotic chromosomes form a complex and highly organized association of linear, double-stranded DNA and specific proteins, the histones. Prokaryotic chromosomes contain a much smaller proportion of proteins, of which most do not seem to belong to a discrete number of easily defined classes. It is thus considered that the chromosome does not have an overall highly regular protein–nucleic acid-based structure. Some, at least, of these proteins make only transient associations with the DNA, and are known to play specific roles in the regulation of gene expression. Their role will be discussed in more detail later (Chapters 15 and 16). A few proteins, the histone-like proteins, have now been described in several bacteria belonging to very different groups (Table 2.1). These small (8–10 kD), most frequently basic, proteins are abundant (for instance, the HU protein in *E. coli* is present in $3–5 \times 10^4$ copies per cell) and show very conserved amino acid compositions and sequences. Their possible roles will be discussed below.

The HTa protein of the archaebacterium *Thermoplasma acidophilum* appears to be intermediate between prokaryote and eukaryote histones; weakly homologous to the bacterial ones, it has a particularly strong

Table 2.1 Histone-like proteins in several bacterial species.

Species	Number of known H-like proteins	% Homologies with	
		E. coli HU	Eukaryotic histones
EUBACTERIA			
Enterobacteria			
E. coli	4–5	100 (HU)	Weak (H2A, H2B)
S. typhimurium	2 ?	100	
Rhizobium meliloti	1	50	Weak (H2A, H2B)
Bacillus stearothermophilus	1	58	Weak (H2A, H2B)
Pseudomonas aeruginosa	1	76 (?)	Weak (H2A, H2B)
Clostridium pasteurianum	1		
Cyanobacteria			
Synechocystis sp.	2 ?	40	
Anabaena sp.	1	45	
CHLOROPLAST (spinach)	1	51	
MITOCHONDRIA (yeast)	1	None	
ARCHAEBACTERIA			
Thermoplasma acidophilum	1	27	25 (H2A, H3)
Sulfolobus brierlevi	5		
S. acidocaldarius	2		
S. solfataricus	2		
Methanosarcina barkeri	1		
Methanobacterium thermoautotrophicum	1		

similarity to the eukaryotic H2A and H3 histones. Its stabilizing effect allows a shift of 40 °C in the critical temperature normally leading to thermal denaturation of the DNA, an important fact considering that the optimal growth temperature of this strain is around 55–60 °C.

1.1.1 *Short review of the chemical composition and structure of a DNA molecule* (Fig. 2.1)

Each DNA strand is a polymer of four types of monomeric units, the monophosphate deoxyribonucleotides adenosine, guanosine, thymidine and cytidine. The molecule is built of alternating phosphates and deoxyriboses, each phosphate being covalently linked to the carbons 3′ and 5′ (the primes indicate the sugar carbons, whereas the carbons of the bases are denoted with plain numerals) of each adjacent sugar molecule. The organization of this 'sugar–phosphate backbone' confers an orientation to the chain.

The nitrogenous bases, adenine (Ade or A), guanine (Gua or G), thymine (Thy or T) and cytosine (Cyt or C), are joined to the carbon 1′ of the deoxyribose. The sequence of bases determines the genetic information carried. The four bases belong to two chemical groups, the purines (A and G) and the pyrimidines (T and C). Another pyrimidine, uracil (U) replaces thymine in RNA molecules.

Two important rules control the formation of a double-stranded structure from such single-stranded molecules: (i) the two chains must pair lengthwise through specific and exclusive associations (hydrogen bonding) of A with T and of G with C; (ii) the sugar–phosphate backbones must show antiparallel orientations. One consequence of the base-pairing specificity is of utmost importance: the base sequence of either chain can be deduced from that of the other. They are said to be complementary.

Double-stranded DNA stabilizes in a helical structure with right-hand coiling, known as the double helix. Each turn of the helix includes approximately 10 pairs of nucleotides. Such double-stranded helices frequently exist as fully closed circles (covalently closed circles, CCC) among prokaryote plasmids and viruses. A variety of constraints (resulting from a balance between enzymes that either increase or decrease the tension on the helix), lead to the organization of the CCC molecules into supercoils, i.e. the formation of a second level of twisting of the double helix about itself (Fig. 2.1d). The detailed structure of the double helix has been extensively studied, and is beyond the scope of this book. Other helical coiling structures have been observed, at least on limited regions of the molecules. They might be transient structures related to the functioning of the molecule (accessibility of particular genes or sequences).

(a)

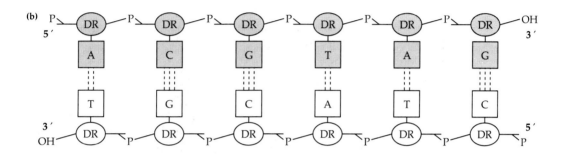

GUANINE   CYTOSINE   ADENINE   THYMINE

(b)

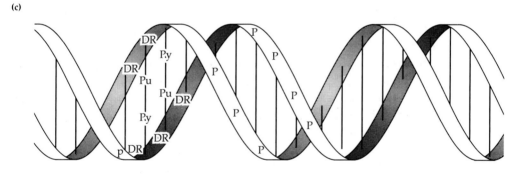

(c)

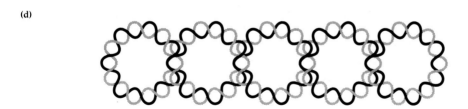

(d)

Fig. 2.1 Composition and structure of DNA. (a) Constituents: the deoxyribonucleotide pairs. (b) Structure of the double-strand. Base complementarity. A, adenine; C, cytosine; DR, deoxyribose; G, guanine; P, phosphate; T, thymine. (c) The double-helix, with usual right-hand coiling. Pu, purine; Py, pyrimidine. (d) Super-coiling of a covalently closed circular DNA.

The elucidation of the structural organization of double-stranded DNA resulted from the discoveries, among others, of the equal proportions in DNA of A+G and T+C, of A and T and of G and C (1950s), of the non-overlapping three-letter code (1960s) and of the characteristic X-ray pictures made by crystalline DNA molecules (1950s).

1.1.2 *The bacterial chromosome*

All known prokaryote chromosomes are formed of double-stranded DNA molecules containing the standard bases. A few unusual bases, 5-methyl- and 4-methylcytosine and N6-methyladenine, have been found in a low proportion in many disparate species of eubacteria and archaebacteria. These methylations may be important as a protection of the DNA against specific nucleases or as signals for the control of certain functions (Chapter 6).

The average G+C content of its DNA is characteristic of a species. A survey of the G+C values among all prokaryotes for which the information is available shows that, as might be expected, the majority of the values are around 50%. However, the range of values is very wide, since extremes from 24% (e.g. *Mycoplasma* spp., 24%, *Clostridium* spp., 27%) to 76% (e.g. *Micrococcus*, 76%) have been recorded. Whether the extreme values (between 24% and 35%) found among mycoplasmas, the smallest known prokaryotes, devoid of a rigid wall, are related to their particular taxonomical position is an open question.

It has been calculated that the minimum percentage of G+C required to ensure a normal capacity to include all amino acids in the cell proteins is 26%. The rationale for this estimation is as follows. Among the 64 possible codons, a number of them are synonymous (including the three nonsense, chain-terminating, ones). It is possible to imagine an organism in which only one of each of these different synonyms is exclusively utilized, the choice being dictated for instance as the G+C-poorest, or the G+C-richest, one. Considering that most of the DNA codes for proteins, theoretical minimal and maximal G+C contents of 24 and 76% are thus obtained. The extreme figures known in actual organisms fit these calculations.

Of course, these calculations do not take into account the extent of DNA which would not participate in the coding of proteins, and which might show a more extreme bias in base composition. This includes the ribosomal and transfer RNA genes, which represent a very small fraction of the whole genome, and other regions which might correspond to the structures described in eukaryotes as representing intergenic regions (spacers). This is indeed the case for mycoplasmas. It is established that the *E. coli* chromosome contains particular regions amounting to approximately 1% of its DNA. They appear as short stretches of repeated sequences, which probably play a role in the chromosomal structure or in the control of gene expression.

The G+C value was the first taxonomical criterion based on molecular arguments to be used, although the information it bears is limited in this regard, since this average value can hide important individual deviations when codon sequences of individual genes are considered. There may exist large variations of average base composition even among closely related groups. Similarly, little correlation can be drawn between this criterion and the membership to a group or a family. G+C contents in different families may cover similar overlapping values. Thus members of the archaebacteria show a similarly wide range of base compositions to that found in eubacteria. This criterion, however, has served to separate into two independent orders the Myxobacteriales (70% G+C) and the cytophagales (30% G+C), which up to then had been classified as two families on the ground that they share strikingly similar metabolic and physiological characteristics.

Nowadays, more accurate analyses of DNA base composition, taking into account differences in actual base sequences, play a large part in taxonomic determinations. On the one hand, the delicate problem of defining a given bacterial species is greatly facilitated by the use of total DNA hybridization. The percentage of hybridization between the DNAs from two strains helps to determine whether the two strains belong to the same species (if they show around 80% hybridization) or to different ones (if less than 20% hybridization is observed). On the other hand, a true revolution in taxonomic determinations was initiated when Woese proposed to use the ribosomal RNA (rRNA) sequences as a timer for evolution. While it is beyond the scope of this book to describe modern phylogeny, the underlying principle stems from the fact that the rRNAs from most living entities share some extremely conserved regions, whereas other regions of these molecules, although they have diverged more widely through mutation, have done so at a much slower pace than 'ordinary' genes. Thus all sequenced rRNA genes share about 50% homology (i.e. 50% similarity) overall. Comparisons are thus rendered much more accurate and permit the measurement of evolutionary time. They have led Woese to define three domains, the Bacteria, the Archaea and the Eukarya, all probably originating from the same common ancestor (Chapter 1). These analyses have thus destroyed the former 'unity of the prokaryotes' dogma. The phylogenic trees obtained often correlate well with numerical taxonomy, but are sometimes in complete contradiction with it. The tremendous advantage of this approach is that it relies on non-ambiguous quantitative data (the percentage of similarity between very well-defined RNA sequences), even though their interpretation may be subject to further improvement. With the present rapid development of molecular biology techniques, it may be speculated that molecular phylogeny will soon completely replace classical taxonomy.

1.2 Haploidy

With one exception known at present (Chapter 14), all prokaryotic cells are considered to possess all their necessary genetic information on a single chromosome. No eukaryotic organism is known with only one chromosome.

This single chromosome may be present as one copy per cell, or, depending on the species or the growth conditions, as two or more, e.g. *E. coli* dividing in a good growth medium will have up to four copies. A probably extreme case is that of *Azotobacter* species, in which 20–40 chromosome equivalents (calculated from the ratio of the amount of DNA per cell to the molecular mass of the chromosome as determined by gel electrophoresis (Chapter 14)) are present per exponentially growing cell. In multicopy states, however, and in contrast to the situation in diploid eukaryotes, the daughter chromosomes in excess of two formed at the onset of a cellular division will segregate randomly in the dividing cells. No rule will control their partitioning, and this will lead to complex segregation patterns of the overall genetic content among the descendent population when different alleles are present on different copies of the chromosomes.

Prokaryotes are thus essentially haploid, with no stable diploid phase in their reproduction cycle. Exceptions to this rule can sometimes be found in certain species after some particular forms of genetic exchange, e.g. in certain cases of conjugation and transduction, the transfer may lead to a partial, though stable, diploidy (Chapters 11 and 12). Similarly, cell fusions, possible in a few species (Chapter 12), create diploid states which can be maintained under appropriate conditions.

1.3 Circularity

Both genetic and physical arguments have led to the idea that the bacterial chromosome is circular. The construction of the 'genetic map' of the *E. coli* chromosome (1964) (Chapter 11) demonstrated that all the genes could be located on a single molecular unit but could also be linked to each other through circular permutation.

This hypothetical circular structure received convincing support when Cairns (1963) succeeded in making an electron microscopic autoradiograph of a complete *E. coli* chromosome (Fig. 2.2). A suspension of *E. coli* cells was incubated in a medium containing tritiated thymidine, so that, upon replication, the new DNA molecules would

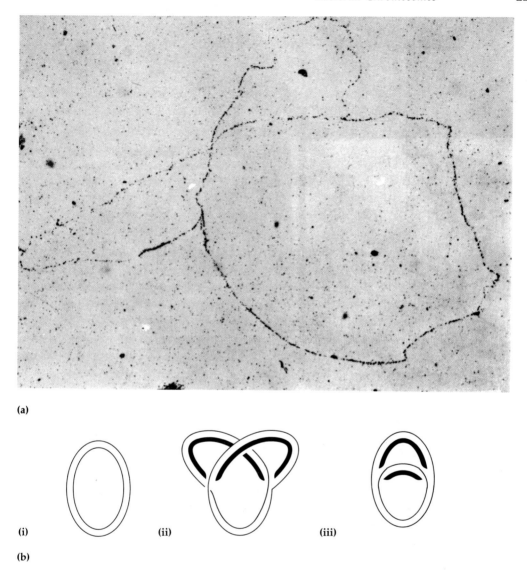

(a)

(i)   **(ii)**   **(iii)**

(b)

Fig. 2.2 Circularity of the *E. coli* chromosome. See text for experimental principle. (a) The autoradiograph shows the so-called θ configuration. Approximately one half of the circular chromosome has already been replicated (Cairns, 1963). (b) Diagrammatic interpretation: (i) resting chromosome; (ii) replicating chromosome; (iii) the same, but represented as a θ letter.

incorporate this radioactive precursor, which could thus be used as a marker to detect the DNA. The chromosomes were then very carefully extracted from the bacterial cells, layered onto a solid support, and exposed to an appropriate film. Due to the radioactive emission of the ^{3}H, a 'picture' (autoradiograph) of the molecule was obtained and photographed through an electron microscope. Although most autoradiographs showed linear structures resulting from the breakage of the molecule, a few circular forms could be found. Figure 2.2(a) shows one such picture, in which part of the molecule (approximately half here) has already been replicated and forms two daughter strands joined to the rest of the parental molecule. A diagrammatic representation of this structure (Fig. 2.2biii), although not very accurate as to the relative sizes of the different parts of the molecule, has since been widely accepted as the symbol of this process and is known as the θ configuration because of its shape.

The circularity of the bacterial chromosome has since been established in a few other strains through genetic arguments similar to those mentioned above for *E. coli*. Thus it is now well demonstrated in *Bacillus subtilis, Streptomyces coelicolor, Salmonella typhimurium* and *Streptococcus pneumoniae*. A similar circular structure has been suggested in a number of other cases, although often with much less well-documented data (a *Pseudomonas* sp., *Caulobacter, Erwinia, Vibrio cholerae* (two strains), *Rhodobacter capsulatus*). It was therefore assumed that all prokaryotic chromosomes consist of covalently closed circles of DNA. A recently developed method to obtain physical (restriction) maps of chromosomes has rapidly enlarged the number of strains where circularity could be demonstrated (Chapter 14). It is interesting to note that circularity has also been established in the case of the chromosomes of chloroplasts (several plants and algae) and mito-chondria (yeast), since these organelles show a number of homologies with bacteria.

Linear chromosomes have, however, been described in some spirochaetes and in two *Borrelia* strains. For instance, the largest DNA molecule present in *Borrelia burgdorferi* cells, considered to be its chromosome, appears as a 10±3 kb-long unit when observed by electrophoretic techniques which facilitate the observation of very large molecules (Chapter 14). The molecule is double-stranded. Nothing is known at present as to the structure of its extremities, a crucial problem with regard to its protection against nucleolytic enzymes. All organisms code for a variety of such enzymes, in particular exonucleases, which attack unprotected DNA extremities (Section 2; Chapter 8). If this is true for *Borrelia*, then a device must be present that prevents access of the nucleases to the extremities of the molecule. This strain possesses five other DNA molecules, currently classified as plasmids (Chapter 3), three of them also being linear double-stranded DNA (Chapter 14). It has been shown for one of them that covalent bonds link the extremities of the two antiparallel strands at each end of the molecule, a structure also encountered in the vaccinia virus but which is in contrast to the systems used in the linear genomes of phages, in which a protein ensures the stability of the structure (Chapter 4), and of other eukaryotic viruses. The same situation may exist for the chromosome. This linear structure also opens to question the classical mechanism of replication demonstrated for circular chromosomes (Section 2).

Linear DNA molecules, which are common in bacteriophage (Chapter 4) and viral genomes, are also encountered in some bacterial and eukaryotic plasmids (Chapter 3). In contrast to bacteriophage genomes, for which the information was often established through genetic data, the linearity of plasmids, for which no genetic information was available, was demonstrated only recently, when new methods allowed the isolation and analysis of unbroken large molecules (Chapter 14). The sizes of the molecules, although sometimes large (e.g. 70–520 kb in two known *Streptomyces* plasmids), are always smaller than that of the *B. burgdorferi* chromosome.

1.4 Size

The size of a DNA molecule can be expressed in different units. The molecular mass (in daltons) is meaningful only for molecules that are sufficiently free of associated proteins, that is, those that consist (mostly) of DNA, which is the case for bacterial and viral chromosomes, plasmids and organelle chromosomes. The development of molecular technologies has promoted the nucleotide (or base, or base pair in case of double-stranded molecules) as the unit. Taking into account the average molecular weight of a nucleotide pair (1 purine + 1 pyrimidine + 2 phosphates + 2 deoxyriboses), the following relationship can be made between these two units:

1 base pair (1 bp) = 0.65×10^3 dalton (Da) = 0.65 kD.

Thus the *E. coli* chromosome measures 3.1×10^9 Da (3.1×10^6 kDa), or 4.7×10^6 bp (4.7×10^3 kbp). (The mean number of nucleotides necessary for the definition of a protein of medium size is usually taken as about 1000.)

Small molecules, such as some plasmids and viral chromosomes, can be directly measured from their contour length on electron micrographs. A molecule 1 µm long corresponds to about 2×10^6 Da, or 3×10^3 bp (3 kbp) of DNA.

More than a 10-fold variation separates the extreme sizes of known prokaryotic chromosomes (Fig. 2.3). Examination of Figure 2.3 points to a rough correlation between the apparent genetic complexity of a species, a notion representing its expected total genetic content deduced from its known physiological (and morphological) capacities, and the size of its genome. Mycoplasmas have genomes of around 0.5×10^9 Da. The largest genomes (up to 9×10^9 Da) are often found in large cells showing rather complex morphologies or physiological

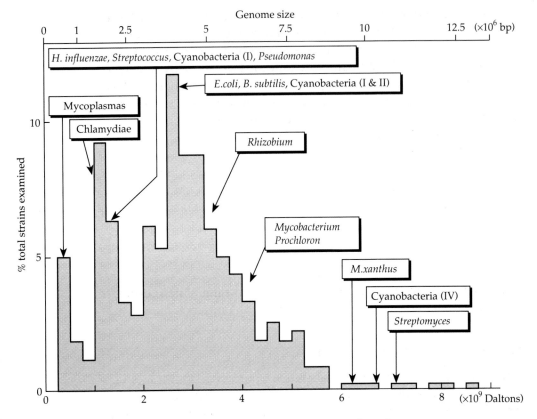

Fig. 2.3 Distribution of genome sizes among some eubacteria.

capacities such as formation of filaments or differentiation of specialized cells (filamentous cyanobacteria, streptomycetes, myxobacteria).

A survey of the genome sizes of approximately 128 strains of cyanobacteria, performed in 1979, has led to an interesting observation: the various sizes, ranging from 1.6 to 9×10^9 Da, can be organized into discrete groups, which are rough multiples of a small unit size (half of the smallest determined genome). A fairly striking parallelism between increases in genome sizes and in genetic capacities has suggested that acquisition of genetic complexity may have been realized through duplications, followed by modifications, of minimal genome units. Similar hypotheses are currently considered to be plausible processes of evolution.

Most registered prokaryote genome sizes range from 2 to 4×10^9 Da. *E. coli*, with a genome size of 3.1×10^9 Da, is considered to possess 3000–5000 genes, of which approximately one-third have been functionally identified. Even though more complex prokaryotes, capable of particular traits such as cell differentiation, might be expected to require more genetic information, this supplementary information has been estimated as smaller than the excess sizes found in the larger genomes (above 6×10^9 Da). It may thus be speculated that not all this DNA is necessary in these cells. Whether this supposed excess DNA is merely blank, or includes additional copies of necessary functions, or has other roles, is still unresolved. Duplications, known to exist for certain categories of genetic information (e.g. tRNA and rRNA genes, penicillin resistance genes in *Pneumococcus*, certain photosynthesis genes in cyanobacteria) may help the cells to cope with risks of deleterious mutations in non-dispensable genes. *Streptomyces* strains, known for the instability of their genome, frequently show tandemly amplified regions, usually 3 to 35 kb long, present in up to a few hundred copies when they are maintained in laboratory conditions. The presence of such structures is often associated with increased levels of resistance to antibiotics or toxic metals. High numbers of duplications of particular DNA sequences (or redundancies), which are very abundant in eukaryotes and supposed to be involved in the control of gene expression, do not seem to have their equivalent in prokaryotes.

1.5 Organization inside the cell: the nucleoid

It has been calculated that the *E. coli* chromosome, if completely spread, would look like a thread approximately 2 nm wide (the diameter of the DNA double helix, not including possible proteins) and 1 mm long. These extremely disproportionate dimensions explain the difficulties encountered in obtaining whole molecules upon extraction. Thus Cairns' observation of whole circles was possible only because he managed not to have to manipulate the molecule outside the cells, and even then the yield of unbroken pictures was very low.

Even more striking is the comparison of the length of this DNA molecule with the size of the cell that harbours it. *E. coli* is rod-shaped, with dimensions 0.5 µm in diameter and 1–2 µm in length. Thus the DNA molecule is approximately 1000 times as long as the cell in which it is found. The necessity of tightly folding the DNA is thus obvious.

The first image of such an organization was obtained for *E. coli* (1972). After gentle treatment of the wall with lysozyme (an enzyme that destroys the integrity of the peptidoglycan layer), the cell burst open, due to the slight overpressure of the cytoplasmic compartment. The DNA-containing fraction, separated through a sucrose gradient, was then examined under an electron microscope. A similar organization is shown for the *Bacillus subtilis* chromosome in Figs 2.4 and 2.5, on which regions being transcribed into RNA molecules (Chapter 15) are also visible (Fig. 2.5b). Dense regions of tightly twisted filaments can be seen, showing supercoiled structures. The filaments represent

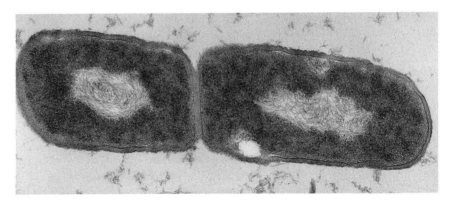

Fig. 2.4 Electron micrograph of the *B. subtilis* chromosome showing the nucleoid organization. Ultra-thin sections of glutaraldehyde fixed and OsO₄ postfixed cells. Magnification ×22680 (Courtesy of F. Le Hegarat & R. Charret, Laboratories of Structure et Expression du Génome Bactérien and Biologie Cellulaire, CNRS and Université Paris XI, Orsay).

broken segments of the whole molecule. They always appear associated with a large protein unit, most probably a part of the cytoplasmic membrane. Linkage of the DNA to a specific site of the membrane is thought to play an important role in the partitioning of the daughter chromosomes at cell division (Section 3.2). Careful analysis of this DNA–protein complex has revealed the presence of about 50 super-helical domains maintained as such by both protein and RNA molecules. The whole structure has received the name 'nucleoid'. Whether the RNA moiety is specific for the cohesion of the nucleoid or reflects a chance association of RNA molecules with other physiological roles is not known.

Such nucleoids have also been visualized in other bacterial species as well as in a number of bacterial viruses during their replication stage. The generality of this type of structure among prokaryotes can thus be accepted.

In eukaryotic cells, in which a similar problem of DNA folding must be coped with, an elaborate structure has been achieved. Specific basic DNA-binding proteins, the histones, ensure supercoiling of the DNA into a succession of discrete nucleosomes, themselves organized in a second level of coiling, making the dense filaments visualized as chromosomes. The discovery of histone-like proteins has prompted more careful examination of the structure of the chromosomes of bacteria and bacterial viruses. Such investigations on gently lysed cells have allowed observation of nucleoprotein structures comprising supercoiled DNA and histone-like proteins, reminiscent of the eukaryotic nucleosomes, and denoted nucleosome-like particles. It is supposed that the histone-like proteins wrap the DNA, thus con-

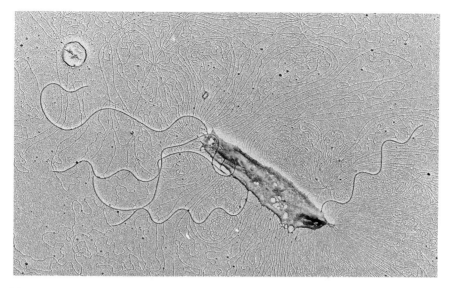

(a)

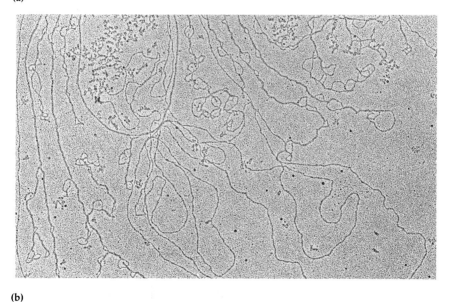

(b)

Fig. 2.5 The nucleoid of *B. subtilis*. (a) Cells treated by RNase and pronase. The ghost of the cell is visible at the centre. Magnification ×10 230. (b) Detail of the outskirt of the nucleoid as in (a), but no RNase treatment was applied. The heavily dotted regions are zones of RNA synthesis. Supercoiled stretches are visible as successive loops. Magnification ×40 000 (Courtesy of F. Le Hegarat and R. Charret, Laboratories of Structure et Expression du Génome Bactérien and Biologie Cellulaire, CNRS and Université Paris XI, Orsay).

tributing to its packaging. However, these nucleosome-like particles are very labile upon extraction, and do not seem to involve more than 15–25% of the whole DNA molecules. It is thus conjectured that the

histone-like proteins are responsible for the packaging of DNA, but –
perhaps more – for the recognition of specific sequences involved in
transcription (Chapter 15) or in particular processes (Chapter 8).

2 DNA synthesis: enzymic mechanisms

This short review of the enzymic steps involved in DNA synthesis,
and of the semiconservative mode of synthesis, applies to all known
double-stranded DNA molecules, bacterial, eukaryotic or viral, but
specific deviations from this general model do exist for some bacterial
viruses and plasmids, and for bacterial conjugational replication. These
exceptions will be described when relevant (Chapters 3, 4 and 11).

2.1 The semiconservative mode

Maintenance and transmission of the integrity of the genetic information contained in a double-stranded DNA molecule are elegantly
achieved, thanks to the complementary base composition of the two
strands. The semiconservative replication involves the partitioning of
the two parental strands, one to each daughter molecule. The synthesis
of two new strands, complementary to each parental one, regenerates
in duplicate both the double-stranded structure and its original chemical
and genetic integrity. This model was first demonstrated in *E. coli* by
Meselson and Stahl (1958). Although no specific demonstration has
been performed in other prokaryotes, the assumption of the generality
of the model has never been doubted. Indeed, in the case of bacterial
viruses with single-stranded DNA genomes, the first step in their
reproduction consists in the making up of a double-stranded form
which is then used as a template (Chapter 4).

Meselson and Stahl's rationale was the following: depending on
whether the two strands of the double helix separated or not during
replication, the two daughter molecules should consist of either half-
new and half-old material each, or of one fully new and one fully old
molecule. By differentially labelling new and old DNA with a density
marker, Meselson and Stahl could demonstrate the separation of the
two initial strands and thus the semiconservative model (Fig. 2.6). It is
interesting to note that the non-ambiguity of the results reflected not
only the validity of the model, as predicted by its authors, but also
another characteristic of the process of replication which was foreseen and demonstrated only later, namely the existence of a control
mechanism regulating precisely the initiation schedule and the start
sites of the successive cycles of replication (Sections 2.6 and 3). This
condition was indeed necessary to allow the separate observation of

individual rounds of synthesis. In spite of this restriction, and taking into account the limited technical means and conceptual knowledge available at that time, Meselson and Stahl's contribution constitutes a remarkable piece of work for its logical and elegant demonstration.

2.2 The process of DNA synthesis: the enzymic machinery involved

Here again the process of DNA synthesis is best known in *E. coli* and in some of its viruses. In particular, both the number and the roles of most of the proteins involved have been completely identified in this bacterium, either after purification and *in vitro* or semi-*in vivo* studies, or from the analysis of mutants. While overall similar processes have been established for other species, at most only a limited number of enzymes have been identified, or of mutants examined, in these organisms. It is thus at present impossible to infer some generality to the detailed genetic and enzymic organization as known in *E. coli* although similarities are extremely likely.

2.2.1 *Formation of the replication fork*

Separation of the two parental strands, that is the formation of a loop by melting of the H-bonds between the complementary bases, constitutes the first step. In a previously resting chromosome, this always takes place at the same location on the genetic map, the origin of replication (called *oriC* in *E. coli*) (Fig. 2.7). It is achieved by proteins known as topoisomerases, or helicases (four are known in *E. coli*: Res, and helicases I, II, III), acting inside a complex structure including, besides the origin region of the DNA molecule, several other proteins. The helicase is responsible for the local melting of the duplex molecule, by forming a loop, kept open by basic proteins (single-strand binding, or Ssb, proteins) which coat the emerging single strands (Fig. 2.7c). In *E. coli* the function of most of the other proteins involved (at least seven, DnaA, B, C, I, P, X, Z) is not completely elucidated, although anchorage to the cell membrane might be controlled by proteins DnaP and DnaA. As we shall see, proteins DnaP, X and Z are also considered as part of the PolIII enzyme.

  All these constituents are associated in an intricate, three-dimensional structure, the initiation complex or primosome, which facilitates the access of the polymerizing enzymes to their sites of action.

2.2.2 *Synthesis of the leading strand, complementary to the 3' → 5' chain*

The DNA molecule is now ready for the second step, the initiation of synthesis *per se*. The synthesis starts at a precise location of the loop,

(a)

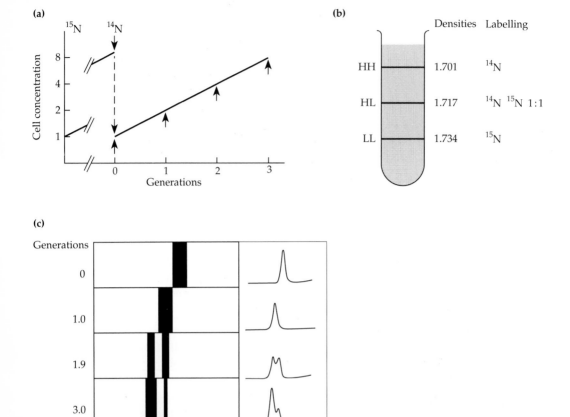

(b)

(c)

Fig. 2.6 (*Above and opposite.*) Meselson and Stahl's demonstration of the semi-conservative mode of replication of DNA. *E. coli* cells were grown in ^{15}N-containing medium, so that both strands of their DNA had a 'heavy' density (HH). They were then transferred to ^{14}N-medium (Time 0). (a) Samples were taken at the indicated times during the following generations. (b) The density and the relative proportions of the molecules present in the cells (old and newly synthesized) were determined after their separation, on a density basis, in a CsCl equilibrium density gradient. The DNA extracts were mixed with a CsCl solution at a concentration such that its density equals that of a Cs-DNA salt formed in the mixture. Centrifugation under appropriate conditions resulted in the establishment of a gradient of CsCl concentrations (and thus of densities). HL, hybrid DNA; LL, light DNA. (c) The Cs-DNA salts banded at levels corresponding to their own densities. The bands were recovered and their DNA content measured by absorption at 260 nm. Normal, light ^{14}N- (LL) and fully labelled, heavy ^{15}N- (HH) DNA samples were used as density markers. (*continued on facing page*)

(d)

Generations

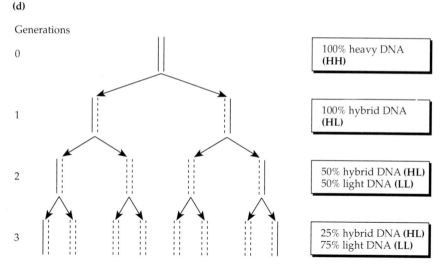

0	100% heavy DNA **(HH)**
1	100% hybrid DNA **(HL)**
2	50% hybrid DNA **(HL)** 50% light DNA **(LL)**
3	25% hybrid DNA **(HL)** 75% light DNA **(LL)**

Fig. 2.6 (*Continued.*) (d) Interpretation of the results is schematically represented here. Confirmation of the nature of the hybrid molecules was obtained by denaturing them by heating and running them on a CsCl gradient along with single-strand molecules prepared from both light and heavy samples (single-strands do not show the same density-banding pattern as double-strands) (Adapted from Meselson & Stahl, 1958.)

the origin of replication, characterized by a specific structure (Section 2.6).

Considering one side of the DNA duplex from this point (Fig. 2.8), it is obvious that the two strands have opposite orientations. This anti-parallel structure introduces further complications for the synthesizing process. All known polymerizing enzymes, be they RNA- or DNA-specific, always synthesize in the 5′ to 3′ direction exclusively (Fig. 2.9a). Only one strand of DNA, that orientated 3′ to 5′, can thus be copied directly from the starting-point and is thus synthesized in a continuous manner. In the absence of any 3′ to 5′ polymerizing enzyme, a more complex strategy has to be devised to copy the other strand, which is synthesized in a discontinuous manner (Section 2.2.3). The first strand is called the 'leading strand', while the other one is the 'lagging strand'.

As the new strand is laid down, the new nucleoside-triphosphate to be added is bound, through its 5′-phosphate residue, to the free 3′-OH radical of the last inserted unit:

$$5'\text{-P-dN}_n\text{-3'-OH} + \text{PPP-5'-dN-3'-OH} \rightarrow 5'\text{-P-dN}_{n+1}\text{-3'-OH} + \text{PP}.$$

The enzyme that performs replication in *E. coli*, that is the con-tinuous synthesis of a whole chromosome, is the DNA polymerase III

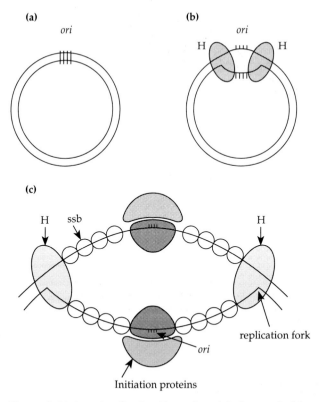

(a) *ori*

(b) *ori*

H   H

(c)

H   ssb   H

ori

replication fork

Initiation proteins

Fig. 2.7 Initiation of replication. Formation of the loop and of the replication fork of a circular chromosome. (a) Resting chromosome. (b) Opening of the loop, by helicases, at the origin of replication, *ori*. (c) Constituents and structure of the stabilized loop. H, helicases; ssb, single-strand DNA binding proteins.

(PolIII), or replicase, a polymer of five to seven subunits, according to how it is defined (three for the core, and two to four for the holoenzyme), for which only six genes have been identified. However, this enzyme is unable to start a new chain and needs a primer from which to elongate the DNA chain. The same is true of the other DNA polymerases, PolI and PolII, known in *E. coli* and of all known DNA polymerases.

Another common characteristic of these three *E. coli* enzymes is their exonuclease activity: they can hydrolyse one strand of a duplex molecule from its 5'-phosphate extremity, leaving 3'-OH ends (Fig. 2.9b). PolI and PolIII also show a 3'-OH → 5'-P exonuclease activity (Fig. 2.9c). Curiously, PolI and PolII are formed of only one polypeptide each, which is sufficient to catalyse the same reactions as does the complex PolIII enzyme. The rate of DNA elongation, however, is 15 times slower for PolI (17 nucleotides added/sec) than that of PolIII (250 bases/sec). Although PolI could formally synthesize long stretches of DNA, it does not replace PolIII, since PolIII-deficient mutants are not

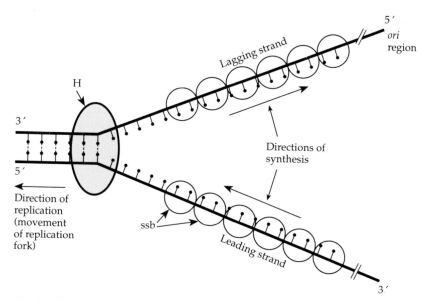

Fig. 2.8 The replication fork. Melting of the H bonds by a helicase and local unwinding of the double helix liberate single-strand regions, stabilized by ssb proteins. Directions of synthesis on each strand as compared to direction of replication (i.e. actual movement of the fork) are shown. ssb, single-strand DNA binding proteins.

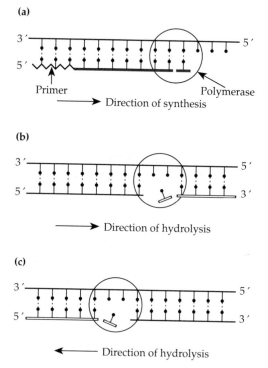

Fig. 2.9 Activities of DNA polymerases. (a) Polymerization: in the 5′ → 3′ direction, along a template, and from a primer. (b) 5′ → 3′ exonuclease: hydrolysis of phosphate-deoxyribose bonds, on the front side of movement, from a free 5′ extremity. (c) 3′ → 5′ exonuclease: hydrolysis of phosphate-deoxyribose bonds, rear of movement, from a free 3′ extremity.

Table 2.2 DNA polymerases of several bacterial species.

Species	Number of known polymerases	Exonuclease activities	
		5′–3′	3′–5′
E. coli	3 PolI	+	+
	PolII	+	?
	PolIII	+	+
B. subtilis	3 PolI	+	
	PolII	+	–
	PolIII	+	
Mycoplasma orale	1	–	–
M. hyorhinis	1		
Spiroplasma citri	3	+/–?	–?
Spiroplasma sp. BC3	3	+/–?	–?
S. floricola	3	+/–?	–?

viable, indicating that it probably does not play a true replicating role *in vivo*. The role of PolII is not known (it displays a very slow rate of synthesis: 2.5 bases/sec), although it may be implicated in local synthesis associated with DNA repair (Chapter 8).

Not all prokaryotes have a complete set of three DNA-polymerizing enzymes (Table 2.2). Bacterial viruses, when they code for their own replicating system, have only one enzyme (Chapter 4).

The primer necessary for DNA polymerization to start is a small segment of RNA. It is probably synthesized by a specific RNA polymerase, a primase. The classical RNA polymerase, responsible for mRNA transcription, may play an indirect role in this initiation, under certain conditions. Both enzymes are capable of initiating a new strand from a single-stranded DNA template in the 5′ → 3′ direction (Fig. 2.10a). After approximately 60 ribonucleotides have been linked, the polymerizing enzyme replaces the primase, and elongates the RNA primer into a DNA chain, by moving along the 3′ → 5′ strand continuously, while the replicating fork complex maintains the template in a proper structure (Fig. 2.10b).

The opening of the loop and its movement along the replicating strands implies the displacement of a local unwinding of the helix, compensated for by extra coiling in the adjacent region of the duplex. This soon forces the normally negatively supercoiled double helix to wind into positive supercoils, which hinder further processing of the replicating complex. A gyrase is then necessary, which releases the extra constraint by opening one or both of the duplex chains, unwinding it around the other, and resealing the two chains (Fig. 2.11).

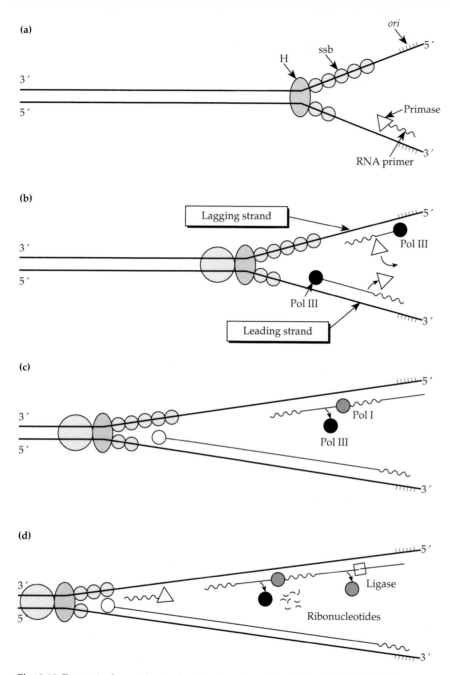

Fig. 2.10 Events in the synthesis of DNA, from the origin. (a) Synthesis of the RNA primer along the 3′ → 5′ strand. (b) PolIII has replaced the primase and moved forward along the 3′ → 5′ strand. A first primer polymerized on the lagging strand is being extended by a PolIII molecule showing the effect of delay in the start of synthesis on the lagging strand. (c) Continuous synthesis proceeds on the leading strand. A second fragment is being finished on the lagging strand. A molecule of PolI has displaced PolIII, when the previous primer was reached. (d) PolI hydrolyses an RNA primer (5′ → 3′ exonuclease activity), while extending the 3′-OH extremity of the third fragment (concomitant processes, called nick translation). A ligase joins the first two fragments, on the lagging strand.

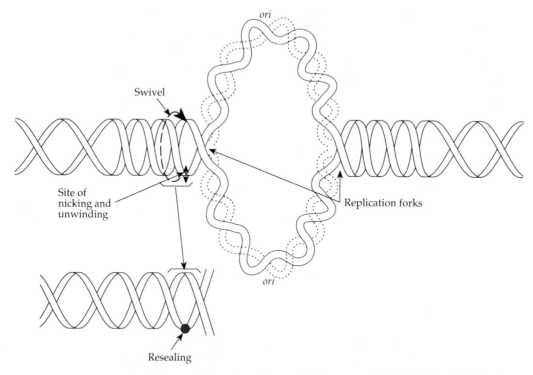

Fig. 2.11 Diagram of the action of gyrase. Displacement of the replicating fork introduces positive winding on the adjacent regions of the DNA molecule, usually involved in a negative supercoiled structure. A gyrase is shown here to cut one of the double-stranded chains, rotates the extremities around the other one, and reseals them. It is very important that the enzyme prevents the DNA molecules from dispersing during this process.

The process ends at a 'terminator' region (Section 2.7).

The initial RNA primer at the origin region will eventually be eliminated. A molecule of PolI will recognize the region because of its non-sealed structure, degrade the RNA primer through its 5' → 3' exonuclease activity and simultaneously replace it by a DNA one. The final step is the ligation of the extremities by the ligase, which covalently joins the 3'-OH radical of the terminal deoxyribose moiety to the 5'-P of the adjacent one.

2.2.3 *Synthesis of the lagging strand, complementary to the 5' → 3' chain*

Synthesis of this chain is performed by the same enzymatic machinery as forms the complementary other one (Fig. 2.10). Due to the specificity of function of the enzymes, synthesis must take place in a backward direction, compared with the movement of the replicating fork. It can start only after a delay, during which the fork has moved forward so as to generate a length of single-stranded DNA. In *E. coli*, when approximately 1000 nucleotides have been uncovered, synthesis of an RNA primer by the primase takes place. It is followed by elongation into DNA by PolIII, as described above, until it reaches the origin region (Fig. 2.10b). In between, the fork has progressed, liberating another length of approximately 1000 bases of the adjacent region of the 5′ → 3′ chain. A second set of RNA priming and DNA elongation fills this space up to the edge of the previous RNA primer (Fig. 2.10c). The whole process is repeated continuously as the replicating fork and replication of the other strand proceed.

When the PolIII enzyme approaches or reaches the preceding RNA primer, it is displaced by a molecule of PolI (Fig. 2.10c). This enzyme, while continuing to elongate the 3′ extremity of the new chain if needed, simultaneously degrades the RNA primer, when it reaches it, in the same process as for the leading strand. The RNA segment is thus replaced by a DNA one, and no gap is left in the continuity of the base sequence.

This strand is therefore synthesized discontinuously.

The final step of the synthetic process is the linkage of the successive segments to each other. This is performed by the DNA ligase (Fig. 2.10d). A continuous DNA chain is obtained. Similarly, PolI and ligase terminate the replication of the strand synthesized continuously.

The discontinuous process of synthesis is probably less rapid than the continuous one on the other strand, due to its higher complexity and the participation of PolI, which is slower than PolIII. This delay generates a stretch of single-stranded DNA downstream from the fork, on which several initiations of discontinuous synthesis can take place more or less simultaneously, utilizing the requisite number of enzymic units, which are constantly recycled.

The transient, 1000-nucleotide-long segments of DNA synthesized backwards were first described in *E. coli* by Okasaki & Okasaki (1969), and were named 'Okasaki fragments'. Although the generality of this discontinuous replication in other organisms has been little studied, no available information challenges it, since all known polymerizing enzymes progress in only one direction.

It has been questioned whether the leading strand might also be synthesized discontinuously. The answer is that normally it is not, although breaks in the leading strand can be detected. These are due to chance integration of uracil instead of thymine (an anomaly which is not detected by the polymerases since it does not introduce base-pairing impairments – see Section 2.4); another correcting mechanism eliminates the uracil nucleotides, thus creating temporary fragmentation. The size of these occasional fragments is very variable, in contrast to that of the Okasaki fragments.

2.3 Bidirectionality of replication

The loop formed at the origin of replication by the initiating complex has a symmetrical structure on both strands (Fig. 2.7), and replication can proceed in opposite directions from the origin. This is indeed what happens not only in *E. coli*, but in all bacterial and eukaryotic chromosomes known (in this latter case, the situation is more complex due to the huge length of the molecule), and in most viral and plasmid genomes. Direct demonstration of bidirectional replication was first obtained for bacterial viruses, since the smaller size of their genomes facilitates their purification as whole, unbroken molecules. Autoradiography of replicating molecules, in which low-intensity followed by high-intensity radioactive labels had been successively introduced at initiation and a few minutes later, respectively, showed two regions of dense labelling, symmetrically located, at each extremity of the loop (Fig. 2.12). This method was also used to show bidirectional replication in *B. subtilis*.

In fact, bidirectional replication was first proposed because of genetic evidence in *B. subtilis*. This organism can be genetically transformed when its own DNA is added to suitable suspensions (Chapter 10). Transformation frequencies can be quantified for different genetic markers. This property thus provided a means of assessing gene dosage in transforming DNA isolated from replicating cultures and in DNA from synchronously replicating cells: the frequency of transformation for a given marker was expected to double as the corresponding gene (the region of the DNA molecule) was replicated. Synchronously replicating DNA is easily obtained in this bacterium by extracting the DNA from spores that have germinated synchronously. So it was possible to show bidirectional replication and to localize the origin of replication on the genetic map of this bacterium (1963). (Obviously, this required the previous establishment of a genetic map).

A different approach was used in *E. coli*. The cells were synchronized for their replication, and pulse-treated (1–2 min) with a mutagen at various times during the course of a replication cycle. Counts were made of the mutants appearing for a large set of characters, chosen because their localizations covered the whole chromosome. The expectation was that any genetic character situated in a newly replicated region, present in two copies per cell, would itself be present in two copies, and thus have twice as many chances to be the target of the mutagenic agent at any given time, as compared with the unreplicated part of the chromosome or with a resting chromosome. The sequential increase of mutation frequencies reflected the advance of the replicating fork along the chromosome, as a function of time. The surprise was the observation that correlation with the genetic map could be met only if bidirectional replication was postulated (Fig. 2.13).

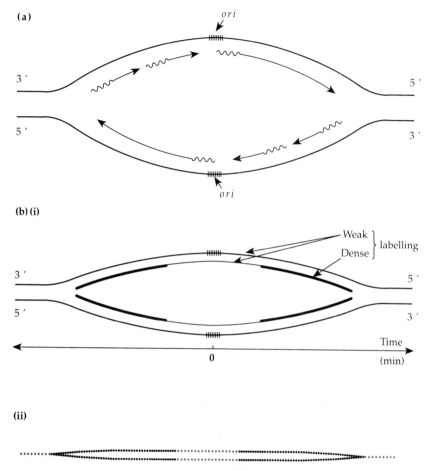

Fig. 2.12 Bidirectional replication of DNA, pictured at the origin of replication. (a) Each parental strand is replicated continuously along its 3′ → 5′ direction, and discontinuously on its 5′ → 3′ direction, from the two sides of the origin region. (b) Pulse-labelling demonstration. Cells pre-grown in the presence of a low specific activity of ³H-thymidine (so that all the DNA are labelled) are set to replicate synchronously in the presence of the same concentration of ³H-thymidine for a few minutes; a higher concentration of ³H-thymidine is then added. Autoradiographs of the origin region differentiate the two densities of labelling. (i) Diagram of experiment. (ii) Electron microscopy autoradiograph (adapted from Prescott & Kuempel, 1973).

Also, using a density-shift technique similar to that of Meselson and Stahl on synchronized cultures of *Streptococcus pneumoniae*, followed by genetic transformation using DNA samples taken at various times after the round of replication had begun, an order of replication of the markers was obtained which could only be reconciled on a chromosome map with gene linkage data by implicating bidirectional replication (Butler & Nicholas, 1973).

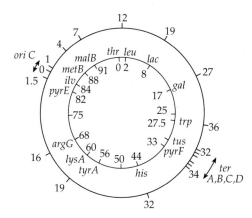

Fig. 2.13 Genetic demonstration of bidirectional replication in *E. coli*. The diagram represents the alignment of the replication map (outer circle) with the genetic map (inner circle). Figures indicate the distances on the chromosome (genetic map) and the times at which mutation frequencies doubled (replication map) in a synchronized population pulse-mutagenized (see text for details).

2.4 The proof-reading role of PolIII and PolI

The maintenance of the integrity of the genetic information of a cell, a necessary condition for its survival and that of the species, requires a precision close to 100% in the complementary copying of the template DNA at each round of replication. However, the replicating enzyme, PolIII, is known to introduce errors, that is mispairing of bases, with an average frequency of 10^{-6}. At least part of these errors may be due to the faculty of the bases to switch, occasionally, to unstable tautomeric forms which modify their hydrogen-bonding capacities and thus their pairing specificities. This mispairing, which is revealed by the subsequent failure of the newly incorporated base to form hydrogen bonds with its complement, constitutes a substrate for the $3' \to 5'$ exonuclease activity of PolIII. The wrong base is thus eliminated and replaced by the correct one. A similar editing process can be repeated, if needed, by PolI. Certain polymerase mutants are unable to perform DNA editing. Such Mut clones show a high rate of spontaneous mutations (Chapters 7 and 8).

As shown in Table 2.2, the polymerizing enzymes of some strains do not possess exonuclease activity. The editing function, most probably a necessary one, must then be fulfilled through another process. Whether enzymic systems, such as those involved in DNA repair (Chapter 8) or mismatch repair following recombination (Chapters 8 and 9), may play this role is not yet known.

2.5 Genetic approach to the understanding of DNA replication

As for many other systems, the genetic approach has been a powerful tool for the investigation of the mechanism of DNA synthesis and chromosome replication. This approach implies the study of mutants deficient for the various steps of the process. Indeed, the enzymes involved have been identified through comparisons between such mutants and wild-type cells. However, since all (or most of) these enzymes are absolutely necessary for replication, and thus for re-production of the cell, the corresponding mutants can be obtained and maintained only if belonging to a certain category, called conditional mutants. That is, their deficiency is expressed (enzymic activity absent) under certain conditions (non-permissive), while the deficiency is not expressed (enzymic activity normal or only slightly modified) under other conditions (permissive). A classical type of such mutants is called temperature-conditional. Thus most DNA-synthesis-deficient mutants (Dna⁻) are temperature-sensitive (ts). In *E. coli*, such mutants typically show normal activity at 30 °C and complete blockage at 40 °C. Only *polA* mutants, deficient for PolI activity, could be isolated as deficient at all temperatures, either because other processes can replace the enzyme, or due to a residual activity in the mutants that have been isolated.

Most of the genes involved in the replication of the *E. coli* chromosome that have been identified (about 40) have been mapped. It is striking to note that they are scattered over the whole genome (Fig. 2.14). Some of them, however, seem to be organized in operons, e.g. units showing coregulated expression (Chapter 15).

2.6 The origin

Several lines of evidence, including biochemical, genetic and molecular approaches, have ascertained the existence of a defined, fixed, origin, from which replication starts, on the chromosome of several bacterial species and in all sufficiently studied bacterial, plasmid and viral genomes. The strategy used to identify and characterize this region has consisted in inserting it into another DNA molecule and analysing the new mode of replication of this second molecule. This has allowed the determination of the minimal size and of the structure (base sequence) of this minimal region in the chromosomes of a few species.

In the *E. coli* origin, *oriC* (Fig. 2.15), a minimal size of 235–245 bases is sufficient to control initiation of monodirectional replication. A somewhat longer sequence on one side of this region ensures replication in the second direction. A set of four or five small repeated sequences, known as *dnaA* boxes, included in this region probably participate in the setting up of the initiation complex (Section 3).

Sequences for the binding of the primase (or the RNA polymerase) and for the initiation site of this specific type of transcription have been particularly looked for. In the *oriC* region of *E. coli*, at least one RNA polymerase promoter-like structure (binding site) (Chapter 15) has been found. However, since the role of RNA polymerase in replication

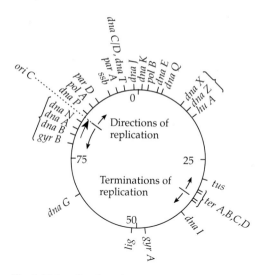

Fig. 2.14 Localization of genes implicated in DNA synthesis, regulation of replication and partitioning of the chromosomes in *E. coli* K12. Symbols: *dnaA*, initiation regulation; *dnaB*, initiation complex; *dnaC(D)*, initiation complex; *dnaE*, *(polC)*, sub-unit α of PolIII; *dnaG* primase; *dnaI*, elongation; *dnaJ*, initiation; *dnaK*, initiation; *dnaN*, PolIII sub-unit; *dnaP*, PolIII sub-unit, linkage to the membrane; *dnaQ*, PolIII sub-unit; *dnaT* termination; *dnaX*, sub-unit of PolIII; *dnaY*, unspecified; *dnaZ*, sub-unit of PolIII; *gyrA*, sub-unit α of gyrase; *gyrB*, sub-unit β of gyrase; *huB*, histone-like protein B; *lig*, ligase; *oriC*, origin in *E. coli*; *parA*, partitioning of chromosomes; *parD*, partitioning of chromosomes; *polA*, polI; *polB*, PolII; *ssb*, single-strand binding protein; *terA, B, C, D*, termination sites in *E. coli*; *tus*, termination protein in *E. coli*. Genes *dnaN*, *dnaA*, *dnaB* and *gyrB* probably form a coordinated unit of expression, overlapping *oriC*.

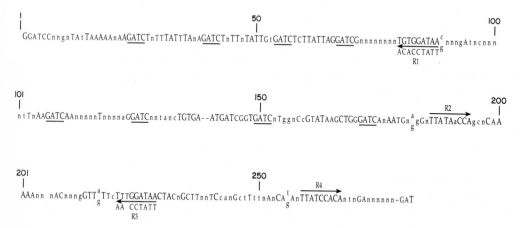

Fig. 2.15 Base sequences of the consensus origin of replication of six Enterobacteria. The sequences of the following six organisms are compared: *E. coli, Salmonella typhimurium, Enterobacter aerogenes, Klebsiella pneumoniae, Erwinia carotovora, Vibrio harveyi*. The consensus sequence that can be derived is shown. Degrees of conservation of the bases among these strains are symbolized as follows: large capital letters, 6/6; small capital letters 5/6; lower case letters, 3–4/6; the n letter, less than 3/6. The four 9 bp repeated regions, R1, R2, R3 and R4 are the *dnaA* boxes (Sections 2.6 and 3.2).

initiation is not certain, and since no recognition sequence for the primase has been postulated, the exact role of such a possible structure is not confirmed. The observed heterogeneity of the sequences at which Okasaki fragments start suggests that such binding sites may indeed not exist.

Comparison of the nucleotide sequences of other *ori* regions in Gram-negative bacteria indicates a high degree of conservation (Fig. 2.15). Moreover, there is enough functional homology to allow the *Vibrio harveyi* origin to operate in an *E. coli* cell.

2.7 The terminus

Termination of replication of the *E. coli* chromosome takes place when the two replicating forks meet. This happens in a fixed region of the chromosome in which few genetic markers have been identified, close to the *trp* operon and exactly diametrically opposite to *oriC*.

The location of the termination region in *B. subtilis* divides the chromosome into two unequal arms from the origin, one being approximately one-third and the other two-thirds of the whole genome. It is not known whether this correlates to different rates of progression of the two replicating forks, or whether one waits for the other to reach this region.

Is termination the mere consequence of the collision of the two opposite replicating forks, or is there a signal or a specific structure? Both suggested models have been verified, at least in the few cases so far elucidated – *E. coli*, *B. subtilis* and two *E. coli* plasmids. The *E. coli* *ter* region contains four terminus sites, *terA, B, C, D*, as defined from the presence of four repeated conserved sequences approximately 20 bp long, spread over about 1000 bp. The *ter* sites are organized into two couples, with opposite orientations, each responsible for inhibition of one direction of DNA synthesis. They function as replication fork traps, blocking the progress of the replicating forks and preventing them from leaving the terminus. A basic protein, coded by a gene, *tus*, located in the same region, participates in the process by its specific binding to the *ter* sites and by its antihelicase activity interfering with further DNA unwinding.

The three other known terminus regions may be less elaborate, since only two *ter* sites have been recognized.

Some viral or plasmid molecules do not terminate their replication at fixed points. It is not known whether this may be true also for some bacteria.

3 Coordination of replication cycles with cell division

It is clear that, in order for an organism to maintain the equilibrium one chromosome/one cell under any growth conditions, a coordinating mechanism must function. Table 2.3 shows that variations of a factor of 4 can be forced in the division times of E. coli when fed with different carbon sources at the same temperature. The completion of a round of replication requires 40 min at 37 °C under all these conditions, and the mass ratio of DNA/cell remains constant. When the cell mass becomes unbalanced, DNA synthesis is initiated. The coordination process does not involve a modification of the rate of synthesis of DNA.

On the other hand, modifying this rate results in a parallel modification of the generation time. Mutants of E. coli deficient in the synthesis of thymidine will depend on an exogenous supply of this base for the synthesis of their DNA. By decreasing the external concentration of the precursor, the availability of thymidine becomes the rate-limiting step of the replication process and can eventually slow down the polymerizing rate of PolIII. The duration of the division cycle increases in parallel, even if other conditions, such as the nature of the carbon source, would allow faster growth rates. The generation times then correspond exactly to the time necessary for the completion of a doubling of the amount of DNA.

Table 2.3 Variation of the generation time of E. coli at 37 °C in a mineral medium with different carbon sources (A) and in a very rich organic (complex) medium (B).

		Generation time (min)	Replication time (min)
A	Carbon source		
	Glucose	40	40
	Glycerol	70	40
B	Broth	25	40

3.1 Control at the initiation of the replication cycles

The experiments described above suggested that the coordination between cell division cycles and replication cycles was controlled at the initiation step. This was demonstrated by studying the kinetics of DNA synthesis and the state of the chromosomes in the E. coli cells mentioned above, in which growth rate was controlled through the addition of different carbon sources (Fig. 2.16).

Studies of E. coli cultures under conditions when DNA synthesis was not the limiting factor and occurred at a constant rate gave rise

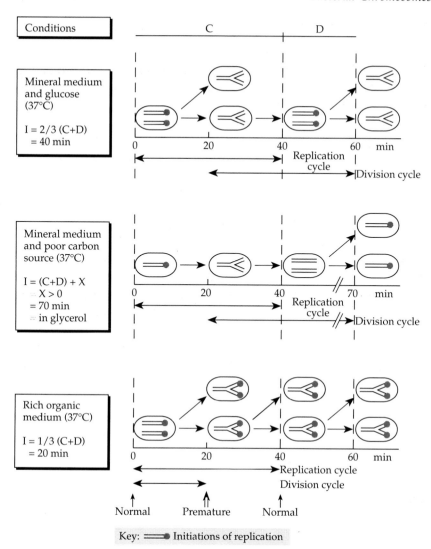

Fig. 2.16 Regulation of initiation of rounds of replication, in correlation with generation times in *E. coli*. For clarity of the diagram, the chromosome is represented as a linear molecule, and only one direction of replication is shown from the origin. The other direction is symmetric from *oriC*, and the molecule is obviously circular.

to a hypothesis which allowed the understanding of the relationship regarding the number and state of replication of the chromosomes at the time of cell division. It was known as the I+C+D hypothesis, where I = the time of accumulation of the protein initiation complex; in exponential growth this equals the division time; C = the time of chromosomal DNA replication = 40 min at 37 °C; D = the time after completion of a round of replication which was found necessary to trigger division of the cell = 20 min at 37 °C.

Thus, when I, the division time, equals (C+D) = 60 min, then, if initiation occurs at time zero, replication ends at 40 min, which then acts as a trigger for cell division to occur after a gap of 20 min. Then cell division coincides with the initiation of the new round of replication. At the time of cell division, therefore, each daughter cell contains one chromosome, which immediately starts to replicate.

During growth on glucose at 37 °C, both chromosomal replication and cell division require 40 min. Since the time for replication equals the division time, initiation takes place approximately 20 min before cell division occurs (Fig. 2.16a): I = 2/3(C+D) = 40 min.

Under conditions of growth resulting in generation times shorter than 40 min, e.g. in rich organic medium, the division time could be as short as 20 min (i.e. I = 1/3(C+D)). In these cases, initiations of new rounds of replication, known as 'premature initiations', occur before the completion of the existing rounds, giving rise to multiple forks. At division of the cell, the chromosome will already be half replicated, and at the 20-min division time the daughter cells will contain two half-replicated chromosomes. These cycles are otherwise normal, in that they start at the origin, and proceed symmetrically, on the two un-finished replicating regions. At each division, each daughter cell con-stantly inherits the same amount of DNA; only this amount is larger than one chromosome equivalent, and the actual molecule is a structure undergoing replication (Fig. 2.16c). The frequency of these premature initiations is regulated so as to result in an average doubling per generation of the amount of DNA per cell. This mode of replication, with premature initiations always starting from the origin, has been established from the examination of chromosome-replicating figures obtained after double-labelling experiments. The origin region is labelled by a short pulse of synthesis in the presence of a radioactive precursor. The application of pulses of a second type of label, at different times afterwards, leads to an overlap of the two labels, indicating that the origin has been replicated a second time in the short time allowed for the second labelling, to the exclusion of other regions.

When I is longer than (C+D), then there is a longer gap than 20 min before division of the cell and also before initiation of a new round of DNA synthesis. Under growth on glycerol for instance, the cell cycle is divided into two periods, one lasting 40 min, during which replication takes place starting at the origin and going to completion, and a second one (30 min in this example (Fig. 2.16b)) appearing as a lag before the onset of the next cycle of replication. This situation is formally equivalent to the S and G phases of mitosis in a eukaryotic cell.

3.2 Mechanism of control

The molecular process of this control, which can be expected to be complex, since it probably involves a reciprocal coordination between replication and division, is still poorly understood. Little is known about the coordination of synthesis of the replicating enzymes in connection with the cell cycle. Several proteins control either the initiation of replication (DnaA, GyrB, DnaG in *E. coli*) or the partitioning of the chromosomes at cell division (ParA and D in *E. coli*).

The generation time is obviously correlated to the overall rate of protein synthesis, and thus of cell mass increase between two divisions. Current hypotheses for the control of initiation imply the positive effect of (at least) DnaA proteins; their binding to the *dnaA* boxes (Section 2.6) present in the *oriC* region would be necessary to ensure a proper configuration of the initiation complex and thus the fixation of the primase (or RNA polymerase?). The level of Ade methylation of the seven GATC sequences of *oriC* (Fig. 2.15) may play a role, though this has not been clearly established (Chapter 6).

The crucial point of this model (and of all models) is the regulation of synthesis of the protein(s) themselves since initiation is triggered by the attainment of a critical mass/DNA ratio: their concentration must be tightly controlled so as to be maintained below a critical value (i.e. to prevent early initiations), and increased only immediately before needed. Several (non-exclusive) procedures seem to be encountered: a correlation of the rate of synthesis of the protein(s) with the overall cellular rate of protein synthesis; a negative mechanism of autoregulation; and a localization of the gene(s) close to the origin, the structural modification of this region during/after initiation acting as a signal for its (their) expression.

Inverse models suggesting negative control have been advanced: the regulatory protein(s) would act as repressors of initiation, their 'dilution' due to cell mass increase releasing the inhibitory effect just prior to cell division.

Correct partitioning of the two chromosomes in the two daughter cells is another crucial aspect of the coordination. A positive control has been proposed. It has been mentioned that the chromosome appears linked to the membrane. The sites of fixation would be, respectively, the origin of replication and a specific region of the membrane correlated to the septation site. Several proteins (ParA and D) have been recognized as being directly involved in this partitioning, although the mechanism is still highly speculative and vague.

Whatever the trigger(s) of the control of initiation, the molecules or the structures involved show a high degree of specificity over the DNA molecule they control. This will be apparent when regulation of plasmid replication is considered (Chapter 3). But it can also be inferred from the characteristics of the phenomenon of integrative suppression described in the following section.

3.3 Definition of a replicon

The presence on a DNA molecule of an origin of replication and the information necessary to control the frequency of initiations of new replication cycles constitute the definitive requirements of a replicon, that is an autonomously regulated unit of replication (Jacob, Brenner and Cuzin, 1963).

The reality of such a unit was demonstrated from the analysis of temperature-sensitive initiation mutants of *E. coli*, now called *dnaA* mutants (Section 2.5). When transferred to a non-permissive temperature, these cells carry on a residual synthesis of DNA, which has been shown to correspond to the completion of the cycles of replication underway at the time of transfer. No further synthesis occurs, due to lack of initiation of new cycles.

Other DNA molecules present in the cell are not influenced by the mutation, and thus constitute independent replicons. One of these, the F plasmid (Chapters 3 and 11), can integrate into the chromosome. This results in suppression of the DnaA temperature-sensitive phenotype: replication of the chromosome can be achieved at high temperature. However, the new synthesis starts at the origin of the plasmid, and not of the chromosome, and continues through the whole molecule in a way characteristic of that of the plasmid. Since synthesis of the plasmid, once initiated, is performed by the cell enzymic machinery previously described, coded by chromosomal genes, the only function the plasmid can furnish to overcome the deficiency of the chromosome is its capacity to initiate and regulate replication. The specificity of the initiation-controlling system is clearly demonstrated by the fact that the F system functions at its own origin, and can overcome the chromosomal mutation only when it forms a continuous molecule with it. This phenomenon has been called 'integrative suppression'.

A membrane initiation complex has not been isolated or even precisely characterized. That such a structure exists is suggested by the capacity of *in vitro* reconstituted complexes, formed of the *E. coli oriC* sequence associated with topoisomerases, DnaA, primase or RNA polymerase and the DNA-synthesizing enzymes, to initiate synthesis. On the other hand, DnaA has been shown to attach to the membrane and to the origin.

All the molecular, biochemical and genetic information accumulated since the replicon model was first proposed have verified it. The next chapters will show that indeed any autonomous DNA molecule must have the characteristics of a replicon in order to persist as such during successive cellular divisions.

References

Butler L. O. & Nicholas G. (1973) Mapping of the *Pneumococcus* chromosome. Linkage between the genes conferring resistances to erythromycin and tetracycline and its implication to the replication of the chromosome. *Journal of General Microbiology,* 79, 31–44.

Cairns J. (1963) *Cold Spring Harbor Symposium on Quantitative Biology,* 28, 43.

Jacob F., Brenner S. & Cuzin F. (1963) On the regulation of DNA replication in bacteria. *Cold Spring Harbor Syposium of Quantitative Biology,* 28, 329.

Meselson M. & Stahl F.W. (1958) *Proceedings of the National Academy of Science, USA,* 44, 671–82.

Okasaki T. & Okasaki R. (1969) Mechanism of DNA chain growth. IV – Direction of synthesis of T4 short DNA chains as revealed by exonucleolytic degradation. *Proceedings of the National Academy of Science, USA,* 64, 1242–8.

Prescott D.M. & Kuempel P.L. (1973) *Methods in Cell Biology,* 7, 147.

3: Other Autonomously Replicating Genetic Entities: the Plasmids

In addition to the chromosome, other autonomously replicating, stably maintained DNA units may also be found in prokaryotic cells: these are the plasmids. A large fraction of them are considered as facultative elements, since plasmid-free cells of most species either occur frequently under natural conditions or can be kept viable under laboratory conditions. In spite of their dispensable nature, and often because of it, plasmids are of considerable medical, ecological and practical interest and have been thoroughly studied ever since they were discovered.

There is a huge number of already known naturally occurring plasmids, and even more that have been engineered. It would be impossible to present an exhaustive account here. Anyway, such a review would be tedious and of limited interest, since the list of both newly discovered and engineered plasmids increases constantly. In this chapter we shall describe the general properties of these elements and review a few relevant examples, so as to illustrate all their typical characteristics, and also pick out some particular cases.

As we shall see, specific plasmids are usually maintained in only one or a few host bacteria, usually of the same or related species. This is not true, however, for the so-called promiscuous, or broad-host-range, plasmids, to which special reference will be made. The ability of the latter to replicate in numerous species confers on them a great advantage for applied purposes. In this respect, Gram-negative bacteria appear as a quite homogeneous group, since a number of promiscuous plasmids can replicate in most of them, with a few exceptions such as *Bacteroides* spp., *Campylobacter* and myxobacteria, whereas Gram-positive bacteria seem more heterogeneous in this respect.

The first- and best-studied plasmids are, as usual, those found in *Escherichia coli*, whilst those from Gram-positive bacteria have been studied for a much shorter period of time.

Most plasmids are composed of circular supercoiled double-stranded DNA, the so-called CCC (covalently closed circular) molecules, but linear plasmids have been described, for instance in some *Streptomyces* species and in *Borrelia burgdorferi* (Chapter 2), and stable single-stranded DNAs present in some species such as *Myxococcus xanthus* may also be considered as plasmids. Open-circle (OC) plasmids, in which one of the DNA chains has been nicked, may also be seen but these are probably formed as a result of the preparative and analytical procedures used.

The size of plasmids is extremely variable, ranging from hundreds of bp to several hundreds of kilobase pairs. Their base composition (% G+C content) may vary widely, and is often different from that of their usual host, indicating a variety of sources from which plasmids (or parts of them) have been derived.

1 Plasmid replication

1.1 Review of the problems

By definition, a plasmid is a replicon. This means that it contains a specific origin of replication susceptible to regulation, and the functions necessary for this regulation. Coordination of replication must be realized both with the cell cycle and in equilibrium with the actual number of chromosomes present in the cell (Chapter 2). Study of this replication entails analysis of the nature of the origin of replication and its functioning in terms of the proteins required, be they plasmid- or host-encoded, and elucidation of the regulatory process.

Regulation is monitored by two different criteria:

1   The copy number of a plasmid, or number of copies of the plasmid simultaneously and stably present per cell. This number, characteristic of the plasmid, is fairly constant in a given host under constant physiological conditions. Plasmids may be divided into high-copy-number plasmids, which are generally small, and of which up to 100 copies (even 1000 in *Streptomyces*) may be found in a single bacterium; low-copy-number ones (these are most often large plasmids and there may be as few as one plasmid per chromosome); and oligocopy (5–20 copies per cell) plasmids (Table 3.1).

2   Plasmid incompatibility, defined as the inability of two particular plasmids to be maintained stably in the same cell line. All incompatible plasmids form an incompatibility group, e.g. IncP, IncFII, etc. (Table 3.1). Members of a group generally share a high degree of homology, at least in the regions responsible for the control of replication. All known low-copy-number plasmids (and one group of high-copy-number plasmids, of which a member is RSF1010) of *E. coli* can be classified into 30 incompatibility groups. Incompatibility may result from the inability of the maintenance systems of two incompatible plasmids to distinguish between them. Random selection of plasmid DNA molecules used as templates for replication introduces an imbalance into the proportion of the two plasmids, and eventually the loss of one of them from a cell line.

Table 3.1 Some of the best-known plasmids.

Plasmid designation (host)	Size in kbp	Copy number	Number of different OriV	Uni- or bidirectional replication	Host proteins required	Positively acting, plasmid-encoded proteins	Regulation
ColE1 (CloDF13, P1517, RSF1050, pMB1 . . .) (*E. coli*)	6.4	High	1	Unidir.	PolI PolIII, RNA polymerase DnaE, B, C, Z RNase H	None	Negative regulation of primer RNA synthesis by a protein and an RNA complementary to the preprimer RNA
R6K (*E. coli*)	38	13–40	3 (α, β, γ) direct repeats	Bidir. and asymmetric (terminus) (direct repeats)	RNA polymerase	Pi protein (*pir* gene) binds to *oriV*	Pi protein binds to the direct repeats of *oriV* and is an autoregulated protein. At low concentration it regulates replication positively; at high concentration it exerts a negative regulation
IncFI: F (*E. coli*)	94.5	1	2 direct repeats	Bidir. (direct repeats)	DnaB, C, E and A	Protein E binds to *oriV*	Negative autoregulation of protein E synthesis. Protein E might be titrated down by the direct repeats near or in *oriV*, which are responsible for both copy number and incompatibility (*copB* or *incC*)
IncP: RK2, RP4 RP1, R68, R18 (Gram-negative, promiscuous)	60	4–7 in *E. coli* 3 in *P. aeruginosa*	1 direct repeat	Unidir.	DnaA	TrfA protein binds to *oriV*	Complex negative regulation of the TrfA protein by several genes also involved in cell killing or rescue (*kilA, B, C, D* or *kilABCD*)

	Size (kb)	Copy no.	Iterons	Direction	Host factors	Rep proteins	Regulation / Comments
IncQ: RSF1010 (Gram-negative, promiscuous)	8.7	10–12	1 direct repeat	Bidir.		RepA, RepB, RepC. RepC binds to *oriV*	Negative regulation of *repC*. *repC* has two promoters, negatively regulated by the product of a small gene located downstream of each of them
IncFII: R1 (R100, R6-5) (*E. coli*)	100	2	1	Unidir.	DnaB, C, E, F and G	RepA1 (*rif* RNA polymerase)	Negative regulation of *repA1* expression at two levels. The protein product of *copB* represses transcription of *repA1*, the RNA product of *copA* (or *repA2*) prevents translation of *repA1* mRNA and negatively regulates *copB*
PT181 (*Staphylococcus aureus*)	4.4	20–25	1 (*repC*)	Unidir.	Unknown	RepC (313 aa) binds to a region within the origin	Negative regulation of *repC* by attenuation at the transcriptional level, and by two RNAs at the translational level. These RNAs are complementary to the leader region of the *repC* mRNA
pIJ101 (*Streptomyces*)	8.9	High	1		Unknown		Comments: broad host-range, no incompatibility system, conjugative, no active partition

All plasmids are, to various degrees, dependent on their host for their replication. Besides availability of precursors and a number of enzymes involved in DNA synthesis, they probably require structures equivalent to the replicating initiation complex known for bacterial chromosomes (Chapter 2). Such multimeric structures, at least partly coded by the host, function through recognition processes between specific plasmid DNA regions and their constituent proteins. This capacity of recognition (the authorization for replication) by a given host is referred to as the host-range of the plasmid. Whether host-range involves proteins implicated in other steps of replication is not clear.

Thus complete study of the process of plasmid replication involves the analysis of the origin of replication, *oriV*, of the plasmid-encoded proteins required for replication (products of genes often called *rep*), of the genes involved in copy number control (*cop*) and in incompatibility (*inc*) (as we shall see, there may be overlap between these different functions), and of the host proteins required.

The first step generally used to study these functions is to construct a mini-replicon out of the original plasmid. Mini-replicons include only the replication origin and possible positively acting genes, and may lack other characteristics of the parental plasmid implicated in replication, which must then be analysed by other means. Complementation assays among mutant plasmids, *in vitro* replication assays and use of host *dna* mutants are the other tools usually utilized for this type of research.

1.2 The different mechanisms of regulation of plasmid replication

Table 3.1 summarizes the replication mechanisms known so far among plasmids. As may be seen, the copy number is roughly an inverse function of the plasmid size. It also depends on the host and the culture conditions. The figures given in Table 3.1 are those found in *E. coli*, unless specified.

The nature of the host proteins required for replication of the different plasmids tested varies from one to the other. ColE1, a very narrow-host-range *E. coli* plasmid, requires the *E. coli*-specific polymerase I (coded by gene *polA*). However, other narrow-host-range plasmids do not. On the other hand, several broad-host-range, i.e. promiscuous, plasmids differ in their host protein requirement.

All the plasmids listed in Table 3.1 replicate following the classical semiconservative model described in the preceding chapter. Replication is either monodirectional or bidirectional, and several plasmids possess more than one origin of replication. In the latter cases, *in vivo* and *in vitro* studies have shown that these are closely linked. Their sequences, which are often known, have revealed the presence in several non-related plasmids of one or several sets of short tandem repeats at the origin site and sometimes also close to it. These regions are involved in the regulation of replication.

All plasmids listed, except those belonging to the ColE1 group, code for at least one protein required for replication. In all cases this protein – or one of them – is the target of the regulating mechanism. The ColE1 group is an interesting exception at two levels. It does not code for a protein necessary (that is, with a positive action) for its replication. Its synthesis is performed by the PolI, and not the usual PolIII, host enzyme. Its regulation mechanism acts at the level of the formation of the RNA primer. A preprimer is synthesized from a region close to *oriV* (RNA II) and is cleaved by the host RNase H to yield the primer from which replication proceeds (Fig. 3.1a). This cleavage, however, is prevented if another RNA molecule, RNA I, complementary to the 5′ end of RNA II, binds to RNA II in a complex involving a protein, Rop, produced from a distant gene, *rop*. RNA I is constitutively transcribed but is unstable and thus acts as the regulator of the frequency of initiation of replication, and thus of copy number. The absence of requirement for a plasmid-encoded protein with positive action for its replication has a very useful application: protein synthesis inhibitors, e.g. chloramphenicol, by preventing the synthesis of Rop, favour an increase of its copy number, the large cellular pool of PolI being sufficient to ensure many rounds of replication. The copy number can reach a thousand, even in host cells which do not replicate or divide. This amplification phenomenon is widely used for preparation of plasmids belonging to this group and has been a major issue that has led to the creation of an important family of engineered vectors derived from ColE1 (Chapter 16).

The variety of devices involved in plasmid regulation in all the other plasmids is quite amazing. For example, in plasmid R6K, only one gene, *pir*, coding for the C protein, is required, but the regulation is nevertheless quite sophisticated. Not only is *pir* an autoregulated gene, but C regulates replication positively at low concentration and negatively at high concentration, by binding to the tandem repeats close to the origin. Current models suggest that the binding titrates down the number of C proteins.

The same overall model accounts for the very-low-copy-number plasmid, F, whereas R6K is moderately high-copy-numbered. This suffices to show that all is not yet understood in the regulation of these plasmids.

In the *Staphylococcus aureus* plasmid pT181, the synthesis of the necessary *repC* product is regulated both at the transcriptional level, by an attenuation-like mechanism, and at the translational level, by

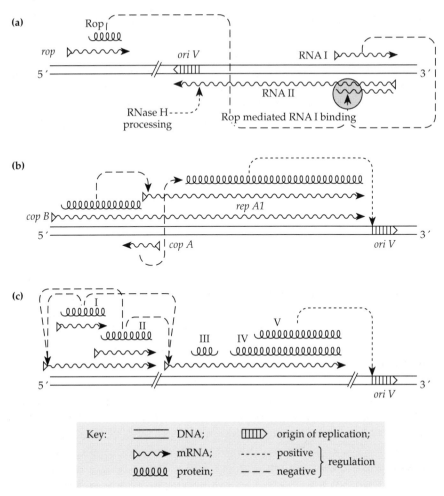

Key:

═══════	DNA;	▥▥▥▷	origin of replication;
◟◠◠◠➤	mRNA;	- - - - -	positive ⎫
◟◟◟◟◟◟	protein;	– – –	negative ⎬ regulation

Fig. 3.1 Three examples of replication regulatory processes in plasmids. See text for explanations. (a) Plasmid colE1. Negative regulation: (i) synthesis of protein Rop and of RNAI; (ii) Rop and RNAI inhibit RNAII cleavage. Positive regulation: (i) synthesis of RNAII; (ii) processing of RNAII by RNase H allows initiation of DNA synthesis. (b) Plasmid R1. Positive regulation: synthesis of RepA1 allows initiation of replication. Negative regulation: the CopB protein inhibits transcription of *repA1* RNA, and thus synthesis of RepA1 and the complementary *copA* RNA inhibits *repA1* mRNA translation. (c) Plasmid RK2. Positive regulation: protein V allows initiation of replication. Negative regulation: proteins I and II inhibit synthesis of protein V.

two 'antisense' RNAs complementary to the leader region of the *repC* mRNA. (Antisense RNAs are RNA molecules synthesized from the DNA strand complementary to that coding for the protein of interest, which can prevent expression of the gene either at the transcription level, by forming a double-stranded structure with the coding DNA strand, or at the translation level, by forming a double-stranded RNA molecule with the normal messenger RNA. Both double-stranded structures block further transcription or translation processes. Gene expression, e.g. transcription and translation leading to the synthesis of protein, is reviewed in Chapter 15.)

Such a double level of regulation is also described for the plasmids belonging to the IncFII group, such as R1. Expression of gene *repA* is negatively controlled at the transcriptional level by the protein product of gene *copA*, and at the translational level by a complementary RNA, product of gene *copB* (Fig. 3.1b).

As for the regulation of replication of IncP plasmids such as RK2, the target is the protein product of gene *trfA*. Several other genes are involved. Figure 3.1c gives a brief idea of the complexity of this regulatory pattern.

Although RSF1010, a representative of the IncQ incompatibility group, is a very small molecule, it encodes three Rep proteins. RepA is a helicase, RepB is a rifampicin-resistant RNA polymerase and RepC is another protein that binds to the *oriV* sequence. Regulation of replication is realized through modulation of the intracellular concentration of protein RepC. However, there might well be a coordinated production of all three proteins. In this case, as for the RK2 plasmid, a tandem repeat adjacent to the origin of replication seems to control copy number.

An additional complexity stems from the fact that replication of broad-host-range plasmids may depend on the host. For instance, the *trfA* gene in plasmid RK2 can be translated into two different polypeptides. The smaller one (285 amino acids) is sufficent in *E. coli* and *Pseudomonas putida*, since mutations in the *trfA* gene outside this region still allow RK2 replication in these species. However, the large polypeptide (382 aa) is necessary for plasmid replication in *Pseudomonas aeruginosa*. This difference is paralleled by a different definition (size) of *oriV* in the corresponding strains. But this property certainly does not explain away the nature of plasmid promiscuity, which so far can only be deduced empirically and has resisted rational analysis of the replicative mechanism.

Besides, however complex and diverse the systems listed in Table 3.1 may be, they cannot be supposed to encompass the overall diversity of regulatory mechanisms. Further studies will no doubt increase the range of variety already recorded.

Even the overall mode of replication is not unique. Thus plasmid pC194, present in Gram-positive bacteria, has been shown to require a single-strand nick at its *oriV* as an early step of replication. This is reminiscent of the rolling circle mechanism described for some phages (Chapter 4), and also attested during the conjugative replication of autotransferable plasmids (Section 2.1). As for linear plasmids, they have proteins associated with their 5' ends which are believed to serve as a primer for the DNA polymerase, as is the case for some DNA phages with linear replicative forms (ϕ29, T4 and T7) and for the eukaryotic adenoviruses.

1.3 Other genes involved in stable maintenance

In addition to the need for accurate regulation of replication and control of copy number, stable plasmid maintenance requires that, when cells divide, each daughter cell receives at least one plasmid copy. This is apparently achieved by random distribution when the copy number is high enough and the multimer resolution system helps to keep this copy number. Low-copy-number plasmids, however, must be partitioned actively.

1.3.1 *Multimer resolution systems*

It has been shown, using certain plasmid derivatives, that plasmid multimerization may occur in recombinant-proficient (Rec$^+$) cells (Chapter 8) by recombination events involving an odd number of crossovers between homologous molecules. This causes biases in plasmid transmission by reducing the number of segregating units. Determinants responsible for maintenance of stable copy number have been found in several plasmids which convert multimeric forms into monomers. In ColE1 (and also CloDF13 and ColK) the determinant consists of a small sequence (*Scer*) devoid of any open reading frame. When two such sequences are in direct repeat on the same replicon, as will result from the formation of a dimer, site-specific recombination actively takes place at these sites and yields a high ratio of monomers (Fig. 3.2). A plasmid encoded function (a resolvase, coded by gene *Pcer*) is responsible for this specific recombination, but a host-encoded function (Pxer) can replace it. The low-copy-number plasmid P1 (belonging to the IncY group, and which is actually the prophage state of the P1 bacteriophage (Chapter 4)), also possesses a multimer resolution system. The region necessary for site-specific recombination, positioned in *cis*, is the target of a resolvase coded by the plasmid itself.

1.3.2 *Plasmid partitioning*

As stated above, active partitioning of the plasmids between daughter cells is an absolute requisite for the stable maintenance of low-copy-number plasmids. Sequences involved in partitioning (the *par* sites) have been discovered in several plasmids, such as R1, pSC101, F and P1. Lack of the partition region results in random distribution of the

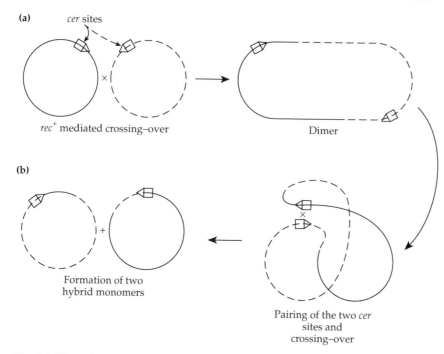

Fig. 3.2 Plasmid multimer resolution. (a) Rec-mediated dimerization between two identical plasmids. (b) Resolvase-mediated monomerization.

plasmids. As expected, the rate of loss is dependent on the actual copy number (from 0.2% per generation for pSC101 to 15% for P1).

Although not homologous among independent plasmids, these regions are analogous and can be active when transferred onto any plasmid deprived of its own partitioning system. In each case, a small region (e.g. 270 bp in pSC101) is required in *cis*.

In F and P1 plasmids the *par* sites are also involved in incompatibility determination (*incB* and *incD* genes, respectively). The properties of *incB* and *incD* strongly suggest that they consist of sites recognized by the cellular partitioning machinery, in a way that may be analogous to the eukaryote centromeres.

Partition seems to be accomplished by linkage of the plasmid to a cellular structure (e.g. the cell envelope) through the binding to one or several proteins. Thus proteins A and B of the F plasmid are known to be involved in the *par* function of this plasmid.

1.3.3 *The* hok–sok *stability system*

In addition to the above-mentioned systems, some plasmids possess devices that kill the cells that have lost it. The R1 plasmid (IncFII), in addition to a low-copy-number partitioning system (*parA*), possesses another region (*parB*) involved in stability in a completely different manner. This region, in fact, covers two genes. The product of gene *hok* is a highly toxic protein whose overexpression causes rapid killing of the host cell. Gene *sok* encodes a labile protein which counteracts the *hok* gene-mediated killing. If R1 fails to be correctly partitioned in a cell, the plasmid-free daughter cell will nevertheless inherit copies of the stable Hok protein. These will express their toxicity, because of the rapid disappearance of Sok proteins, the synthesis of which cannot be renewed in the absence of template genome. The cell is then killed.

The generality of such a device is not yet documented, but sequences homologous to the R1 *parB* region occur in plasmids belonging to different incompatibility groups: R100 (IncFII), F (IncFI), RP1 (IncP) and R6K. However, none has been observed in pSC101 and RSF1010 (IncQ).

1.3.4 *Curing and instability*

Thus natural plasmids are quite stable. As a matter of fact, it is often difficult to 'cure' a strain of its plasmids, i.e. to permanently remove the plasmid from all the host cells in the culture. Different devices have been established for this purpose, which are intended to interfere with the replication mechanism. For instance, ColE1-type plasmids are lost in *polA* mutants. F plasmids are unable to replicate in the presence of the intercalating dye acridine orange, and ethidium bromide has been used for some plasmids resistant to acridine orange. Sensitivities to phages adsorbing specifically on plasmid-encoded pili (Section 2; Chapters 4 and 11) may also be used to select host cells cured from certain conjugative plasmids (the IncP group, for instance).

On the other hand, engineered or recombinant plasmids may have lost their natural stabilizing devices and become unstable to various degrees. They may also confer a severe counterselective pressure on their host. This has led to the frequent necessity of finding new devices to stabilize them in industrial-scale cultures.

2 Plasmid-borne functions

In addition to autonomous regulation of replication, many plasmids display other functions. Some of them obviously confer selective advantages on their hosts, while others favour plasmid dissemination. However, in many cases, especially among plasmids from Gram-positive species, no known accessory functions have so far been identified, either because they do not possess or display any, or because the appropriate tests have not yet been applied. These plasmids are said to be cryptic. Their direct utilization, or even their study, is difficult, unless a marker gene is inserted into them.

2.1 Transfer functions: conjugative plasmids

Some plasmids are able to promote their autonomous transfer into bacteria belonging either to the same species or to a different one and which are devoid of a copy of the same plasmid, without anticipation of whether the plasmid will or will not be able to replicate in the recipient cell. Plasmids harbouring these functions are said to be autonomously transferable, or conjugative, and the whole process is called conjugation. Chapter 11 will describe in detail the characteristics and consequences such transfer systems may confer at the cellular level.

Briefly, conjugational transfer implies the actual contact between the donor, plasmid-bearing cell, and the recipient, plasmid-free cell. A specific structure, the conjugative bridge, joins the two cells, and serves as the channel for transfer of the genetic material. When the first conjugative plasmid was discovered, the nature of the phenomenon led to the description of the conjugating cells as male and female, but this terminology is now only used in the context of male-specific bacteriophages.

Self-transfer was actually the first plasmid function to be discovered. Its presence in the *E. coli* F plasmid also leads to the cotransfer of the host DNA and hence to quasi-sexual genetic exchange. This explains why the plasmid was called F (for fertility) factor. Numerous conjugative plasmids have now been studied in both *E. coli* and other species, including Gram-positive bacteria. The F plasmid remains by far the most studied and can so far be considered the prototype of conjugative plasmids in Gram-negative bacteria.

2.1.1 *The conjugative system of F*

All the genes involved in the conjugational transfer of F, denoted *tra* and *trb*, are located on a 33.3-kb region, representing approximately one-third of the plasmid (Fig. 3.3). Isolation of mutants unable to transfer (Tra⁻) and complementation analyses have led to the identification of 20 genetic units (cistrons). The existence of two additional essential genes was postulated from assays for activities associated with transfer. An exhaustive study of F-encoded polypeptides has up to now allowed the identification of seven other genes, but their possible role in F transfer is still not ascertained (Fig. 3.3 and Table 3.2).

Most of the *tra* gene products have been purified, generally after cloning of the genes in an accessory plasmid vector and allowing

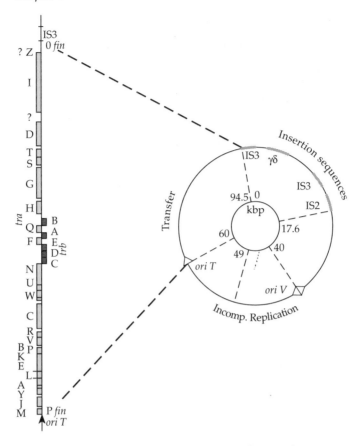

Fig. 3.3 Map of the F plasmid, showing the main functional regions.

controlled amplification. Several genes have been sequenced. All the known functions fall into four classes: pili synthesis, surface exclusion, pairing stability and DNA transfer, described below.

In addition, two regulatory genes, denoted *fin* for fertility inhibition, were discovered when F-like plasmids were studied. *finP* codes for a negative regulator of transfer that is not expressed in the F plasmid, since its expression requires the product of *finO*, which, in F, is inactivated by a particular DNA structure, an IS3 insertion element (Chapter 5). However, F-like plasmids have an intact *finO* gene that can complement *finO* in F in *trans*, thus resulting in a decrease of the transfer frequency. F is thus a derepressed conjugative mutant plasmid. This most probably explains the early discovery of its conjugative functions and the subsequent success of its utilization. This property of fertility inhibition (fi) was once used as a classification system for plasmids, dividing them into fi$^+$ or fi$^-$ groups, and was largely associated with the production by the host of 20-μm pili (e.g. F

Table 3.2 Products of the transfer region of the F plasmid, as ordered on the DNA molecule.

Product	MW (KD)	Function
TraM	14.491	DNA transfer
FinP		Fertility inhibition; TraI repressor if FinO is present
FinO		Fertility inhibition; mutated by IS3 insertion in F; present in F-like plasmids
TraJ	27.031	Positive regulator for the transcription of all other *tra* and *trb* genes
TraY	13.846	DNA transfer, *OriT* nicking exonuclease
TraA	13.200	F pili
TraL	10.350	F pili
TraE	21.200	F pili
TraK	(24)	F pili
TraB	(55–64)	F pili
TraP	(21.5–23.5)	Unknown
TraV	(21)	F pili
TraR	(9)	Unknown
TraC	(48–85)	F pili
TraW	(23)	F pili
TraU	(20)	F pili
TraN	(66)	Stabilization of mating pairs
TrbC	(21.5)	Unknown
TrbD	(23.5)	Unknown
TrbE	(10)	Unknown
TraF	(25–26)	F pili
TrbA	12.947	Unknown
TraQ	10.867	F pili (pilin maturation)
TrbB	(18.4)	Unknown
TraH	(39–45)	F pili
TraG	(100–116)	F pili + stabilization of mating pairs
TraS	16.861	Surface exclusion protein
TraT	26.017	Surface exclusion protein
TraD	(77–90)	DNA transfer
TraI		Helicase
TraZ		*OriT* exonuclease

Molecular weights were either deduced from DNA sequencing data or determined as the apparent MW from SDS-PAGE analysis (SDS-PAGE figures in parentheses).

pili) by those possessing fi$^+$ plasmids or with 2-µm pili (ColI pili) by those possessing fi$^-$ plasmids.

2.1.1.1 F pili Bacteria containing an F plasmid possess external structures known as F pili. These are present if culture conditions are adequate. F pili are thin, flexible, hollow filaments, 8 nm in diameter, which extend from the cell surface.

The study of F pili has been helped by the discovery of various F-pilus-specific bacteriophages. One class, the single-stranded RNA-containing bacteriophages (f2, R17, Qβ (Chapter 4)), attach along the pilus length. Another class, the filamentous single-stranded DNA phages (f1, M13, fd (Chapter 4)) are equally F-pilus-specific, but attach to the tip.

F pili are formed by the polymerization of a 7 kDa polypeptide unit, the pilin, possessing an acetylated amino terminus. Very rapid polymerization–depolymerization can take place, in a manner reminiscent of the eukaryotic tubulin. Thirteen genes are involved in pilus synthesis (Table 3.2). The *traA* gene codes for a large non-acetylated precursor of the pilin. The extra 50-aa-long polypeptide of the precursor possesses a hydrophobic region and a typical signal peptidase cleaving site. Processing (cleaving of the signal peptide (Chapter 16)) and exportation through the cytoplasmic membrane require the presence of the product of gene *traQ*. The products of the 11 other genes are required to add the acetyl residue to the N terminus and to correctly assemble the mature pilin into an F-specific complex spanning from the membrane.

2.1.1.2 The first step in mating: donor–recipient surface interactions
In a pure culture of F-containing bacteria, the *traS* and *traT* gene products prevent cellular interactions. *traT* codes for an abundant outer-membrane protein, which is highly conserved between all F-like plasmids and is responsible for preventing stable cellular associations and hence mating between cells harbouring any of these plasmids. The phenomenon has been called surface exclusion. It is completely different from, and genetically unrelated to, incompatibility reactions. The TraS protein is an inner-membrane protein that prevents the entry of DNA in the occasionally occurring donor–donor pairs.

2.1.1.3 Pairing stability When F-free (F$^-$) *E. coli* cells are mixed with F-containing (F$^+$) ones, conjugal contact is initiated from the F-pilus tips and favours the formation of aggregates. Initially fragile, these associations are stabilized by the *traN* and *traG* gene products. This contact allows both transmission of a 'mating signal' to the donor cell and formation of an intermembrane channel between the F$^+$ and F$^-$ cells. It is not clear whether the F pilus participates in the formation of this channel (Fig. 3.4).

This step is dependent upon the integrity of the recipient outer membrane. The LPS (lipopolysaccharide) layer is the target for F-pilus recognition. The major outer-membrane protein, OmpA, is also required in matings occurring in liquid suspensions, but can be dispensed with if mating is performed on solid medium.

In the absence of TraN and of the C-terminal part of TraG, piliated pairs remain unstable and DNA transfer cannot take place, but the 'mating signal' is transmitted to the donor cell.

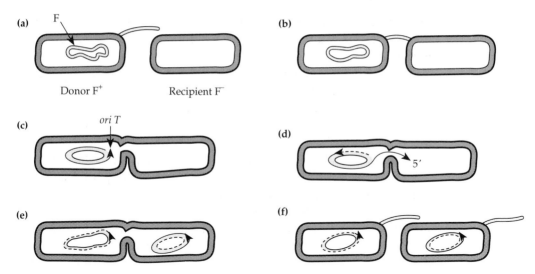

Fig. 3.4 Model of F transfer in *E. coli.* (a) Mixing of donor and recipient cells. For clarity, the chromosomes have not been represented. (b) The pilus of the donor makes contact with the recipient. Proteins TraN and TraG stabilize the pairs. (c) Mobilization of the plasmid: nicking of one strand at *oriT*. (d) Transfer of the nicked strand to the F⁻; DNA replacement synthesis on the remaining strand in the F⁺. (e) Complementary strand synthesis in the recipient cell. (f) Separation of the cells. Synthesis of pili in the ex-recipient; reformation of pili in the donor.

2.1.1.4 DNA conjugal transfer: rolling-circle replication The 'mating signal' promotes a specific single-strand nicking at a particular site, the 'origin of transfer', *oriT*, located at one end of the *tra* region (Fig. 3.3). The nicked strand is then unwound in the 5' to 3' direction (away from the *tra* genes) and transferred in its single-stranded form through the membrane channel into the recipient cell. The process of single-strand nicking and preparation of the transfer has been termed mobilization. Replacement of this strand in the donor occurs by synthesis, according to a modified rolling-circle mode of replication (Chapter 4). No energizing of DNA transport is necessary. Complementary strand synthesis initiated by the recipient cell primosomes and recircularization of the F-plasmid strand transferred take place in the recipient cell (Fig. 3.4). When mating is performed at 37°C, under usual growth incubation conditions, the F plasmid received by the recipient strain is ready to be transferred again after 50 min. The transfer itself requires only a few minutes, in contrast to the case when F is integrated into the chromosome (Section 2.1.5).

Nicking of the plasmid DNA is *traY*- and *traZ*-dependent. The products of these two genes may form a specific endonuclease, although *traZ* is not yet fully characterized.

If *oriT* is defined as the minimal sequence that confers high frequency of mobilization by F upon an otherwise non-mobilizable plasmid (Section 2.1.3), it consists of a 373-bp fragment. This means that it is the only region required in *cis* for plasmid mobilization. Its specific sequence is characteristic of particular classes of mobilizable plasmids. As we shall see below, the presence of an *oriT* region on a non-conjugative plasmid may also occur naturally.

traM and *traI* are *oriT*-specific. They are required for the replacement synthesis in the donor. TraI is a helicase, an unwinding enzyme. It aggregates into a complex of 70–80 fibrous protein molecules that moves in the 5' to 3' direction along the unwound transported DNA strand. The role of TraM is unclear (it may also transmit the mating signal).

Exit of DNA through the donor cell envelope seems to be promoted by the inner-membrane protein coded by *traD*.

2.1.2 *Other conjugative plasmids from Gram-negative bacteria*

Numerous conjugative plasmids of Gram-negative species are now known and it can be expected that even more will be discovered when new species become genetically studied. The best-documented group is that of the F-like plasmids, in which the *tra* region shares a good homology with that of F. The F-like plasmids appear otherwise unrelated: they belong to five different incompatibility groups (IncFI to IncFV) and we have seen previously that their mode of regulation of vegetative replication may vary widely (see IncFI and IncFII, for instance).

These plasmids show narrow host-ranges, but they seem to conjugate into some Gram-negative species in which they cannot be stably replicated. The IncP group is the most promiscuous of all and the next best studied. Their overall conjugation systems appear quite similar to that of the F plasmid. The pili are specific for each group. IncP pili are rigid and fragile, and good efficiencies of mating are obtained if crosses are performed on solid surfaces such as agar or even on filter pads (filter mating). IncI plasmids display both flexible and rigid pili. Each type of pilus is recognized by specific phages (phage GU5 for IncP pili, for instance). But the role of the pilus termini in the formation of aggregates, as well as the involvement of numerous genes in their synthesis, is very similar to what is known about F.

Each conjugative plasmid possesses an *oriT* with a specific nick site. DNA transfer appears quite similar to that in F. Conjugation is negatively regulated in most of these plasmids (including F-like plasmids, as seen above). Thus the frequency of transfer may be quite low in some instances.

Thus, to increase transfer frequencies, 'three-partner matings' are

Table 3.3 Promiscuous plasmids from Gram-negative bacteria.

Incompatibility group	Main representatives	Size (kbp)	Resistances	Remarks
IncI	ColIb-P9			Forms two types of conjugative pili
IncN	N3	41		kik = 'killing of *Klebsiella*'
	pCU1	39		capacity present on all IncN
	pKM101	35.4		plasmids
	RP6	51.7		
IncP	RK2 (RP4, R68, RP1)	60	Km, Tc, Ap	The broadest-host-range plasmids known
IncQ	R300B RSF1010	8.7	Su, Sm	Mobilizable
IncW	R388	33	Tm, Sm	Rigid conjugative pili; lack of
	pSa	39	Tm, Sm, Sp, Cm, Km, Su	restriction sites on the replicative and transfer regions

often utilized in heterologous crosses. After transfer into an intermediate host that acquires the plasmid with rather high efficiency (an *E. coli* strain for instance), the plasmids remain unrepressed for long enough to amplify transfer into cells of a less efficient recipient species.

It has also to be mentioned that the recipient strain may influence the overall efficiency of conjugation. In particular, in most species, they may possess specific devices for recognition–protection (Chapter 6) of their DNA, devices which may limit the maintenance of the newly transferred DNA. Although this DNA is transferred as single strand, it is not completely modified (protected) during the process of complementary strand synthesis, and may remain sensitive to the degradative (restriction) enzymes of the recipient strain.

Several classes of promiscuous conjugative plasmids have received great attention lately, due to their utilization to promote genetic studies of other Gram-negative species (Section 2.1.4 and Chapter 16) (Table 3.3).

Specific narrow-host-range conjugative plasmids have also been found in Gram-negative species other than *E. coli*.

2.1.3 Mobilizable plasmids

Some plasmids – sometimes of very small size – may possess an *oriT* region but without the complete corresponding *tra* region: this lack may be complemented in *trans* by conjugative plasmids from one or several groups. These plasmids are classified as non-conjugative but mobilizable. ColE1 for instance is mobilizable both by F and by the

IncP plasmids. RSF1010 is a promiscuous plasmid mobilizable by IncP plasmids and has thus often been utilized for genetic studies of non-*E. coli* species. In addition, numerous plasmids containing the *oriT* of known conjugative plasmids (mainly that of IncP plasmids) have been constructed lately.

2.1.4 *Conjugative plasmids of Gram-positive bacteria*

These plasmids have been known for a much shorter time and are still under extensive study.

Plasmids from *Streptococcus* strains fall into two groups: those that transfer at high frequency in liquid cultures (10^{-1}) and confer a mating response to the sex pheromones excreted by the possible recipient strains, and those that transfer only on solid surfaces (filter mating) and with low frequencies (10^{-4}–10^{-6}). The latter type often exhibits a broad host-range among *Streptococcus* species and some may transfer to numerous other Gram-positive bacteria (though not including actinomycetes). The former are specific to *Streptococcus faecalis*.

Streptococcus faecalis recipient cells (devoid of the corresponding transferable plasmids) excrete five or more peptide molecules, known as the sex pheromones. Each can be specifically recognized by a donor harbouring a given plasmid. When two compatible strains are mixed, the donor strain responds to the pheromone by the synthesis of a proteic, aggregative substance, the adhesin, which uniformly coats the surface of the cells and allows the formation of mating aggregates upon collision with recipients. The aggregative substance is believed to bind to a substance present on both cell surfaces (lipotechoic acid). Once the plasmid has been acquired by the recipient cell, the corresponding pheromone, but not the others, stops being synthesized. If other plasmids are present in the donor strain, they can be transferred during the process, together with the pheromone-sensitive one. The mechanism of transfer itself is still under study, and no general scheme can be presented yet.

Conjugative plasmids are also known in streptomycetes, where they have been used for genetic purposes (Chapter 11), but their mechanism of transfer is still poorly understood.

2.1.5 *Interactions between conjugative plasmids of Gram-negative bacteria and the host chromosome*

The property of conjugative plasmids that has been most extensively utilized is that, in most cases, DNA transfer takes place irrespective of the size of the replicon that contains the *oriT* sequence. Thus, if a conjugative plasmid is integrated into the host chromosome, DNA unwinding and transfer start normally at the plasmid *oriT* nick, in the 5' to 3' direction, but may continue until the whole chromosome and the terminal extremity of the plasmid have been transferred into the recipient cell. The whole transfer requires a much longer time than the mere transfer of a free plasmid. Thus, when F is inserted into the *E. coli* donor chromosome, 100 min, at 37 °C, are necessary to complete the conjugation. However, transfer of the whole chromosome is seldom complete, because breakage often interrupts it, thus resulting predominantly in the transfer of the genes located near the origin region of the inserted plasmid.

This process is called polarized chromosome mobilization, and the donor called an Hfr (for high frequency of recombination). The complete description of this process is given in Chapter 11. It is the operational basis for genetic mapping using conjugation, and has for a long time been the most powerful genetic tool that could be used for chromosome mapping. Another phenomenon also displayed by Hfr strains has been extremely useful for genetic analysis: the formation of F-prime (F') plasmids, arising when the plasmid is improperly excised from the chromosome and carries away a piece of neighbouring chromosomal DNA with it. The mechanism leading to the formation of such structures is described in Chapter 11. Many other conjugative plasmids may promote polarized chromosome mobilization and/or prime-plasmid formation.

In many instances chromosome mobilization is known to be due to plasmid integration. However, in some cases, such as the *Pseudomonas aeruginosa* P2 plasmid, such an integration has not been demonstrated and transfer may result from the complementation of an *oriT* chromosomal sequence.

Plasmid integration into the chromosome was originally believed to be restricted to a special class of plasmids termed episomes. This belief resulted from the comparison between the F factor capacity for integration, and the absence of such a capacity in a few other plasmids then described. Indeed, F integrates with rather high frequencies into as many as 22 different sites in the *E. coli* chromosome. This integration is dependent on the *recA* recombination system (Chapter 9) and takes place between one of the three different insertion sequences present on the plasmid (IS2, IS3 and γδ) (Chapter 5) and one of the numerous copies of these ISs present on the *E. coli* chromosome. In Rec⁻ strains, F integration (Hfr formation) can still happen – but with 100- to 1000-fold lower frequencies – and results from cointegrate formation mediated by the ISs themselves (Chapter 5).

The presence of copies of the ISs on the *E. coli* chromosome is also responsible for the formation of the F' plasmids.

Now that the properties of transposable elements are known, the distinctive classification between episomes and plasmids appears useless. Any plasmid can be inserted into any chromosome, provided a suitable transposable element is present on either of the DNA molecules. However, the integration frequency may be very low. Several strategies have been developed to promote efficient insertion, and sometimes plasmid-prime formation, with the most useful broad-host-range plasmids coming from the IncP group (Chapter 11).

2.2 Resistances to antibiotics and toxic ions

'Resistance plasmids' were discovered in Japan during the 1950s in clinical isolates of *Shigella* strains which had gained simultaneous resistances to several different antibiotics. Certain combinations of resistances tended to recur. It was then shown that these resistances were generally unlinked to the chromosome and often infectiously transmissible. The plasmids were originally called R factors, then R plasmids, and were later found in clinical isolates of both Gram-negative and Gram-positive organisms. Resistances to heavy metals were also discovered to be linked to plasmids of strains isolated from industrially polluted areas. Toxic-ion resistance genes are still under active scrutiny.

Most of the known plasmid-borne resistances are now well understood with regard to their mechanism of action, the proteins involved and their regulation. In some cases, the determinants have been sequenced. Table 3.4 summarizes some of the results obtained for the antibiotics that are used most frequently. With the exception of tetracycline resistance, most of the others consist of enzymes that hydrolyse or modify the antibiotic. Some toxic-ion resistances also correspond to a neutralization (probably by binding to a protein) of the toxic effect of the ion. Except for the fact that in most cases the resistance determinants found in Gram-negative and Gram-positive bacteria are not closely related, there is no obvious classification of the R plasmids according to the nature of their resistance. Different resistances may be borne by related plasmids, and homologous resistance determinants can be found on plasmids from quite different incompatibility groups. Many resistance determinants are found on conjugative plasmids of all kinds. This indicates recombination and genetic exchange, which are partly explained by the fact that many of these resistances are borne by mobile DNA structures, transposons (Chapter 5). The pattern of resistances carried by multiresistant plasmids may change with the spontaneous loss of a resistance gene or the acquisition of another transposon.

The intensive usage of antibiotics during the last 50 years has led to the spread of the corresponding resistances, at least among human commensal species, which may become a dangerous reservoir for transfer to pathogenic organisms. However, although the indiscriminate use of antibiotics, especially in animal foodstuffs, appears to have contributed to the spread of R factors, the resistance determinants must have existed earlier, since R factors mediating resistance to streptomycin and tetracycline together, through enzymic action, have been identified in strains of Enterobacteriaceae isolated from natives of the Solomon Islands who had never been subjected to prescribed antibiotics. Furthermore, an *E. coli* freeze-dried before the introduction of antibiotics was found to possess an R factor which also mediated resistance to both streptomycin and tetracycline. Some resistance plasmids can be found in soil bacteria. A current hypothesis states that most of them may originate from the resistance determinants present in the antibiotic-producing strains. Similar resistance mechanisms have indeed been found in some instances, but the resistance determinants of most of the antibiotic-producing *Streptomyces* species, for instance, seem to be chromosomally located. However, most plasmid-borne resistance genes are not very closely similar to the genes described so far from antibiotic-producers. In only one case (fosfomycin) is the resemblance very close.

2.3 Bacteriocins and toxins production

Bacteriocins are antibacterial proteins elaborated by a given bacterium and active against bacteria of the same species. Their name stems from the species that produces them (i.e. colicins for *E. coli*, subtilisin for *B. subtilis*, etc.). They are coded for by plasmids that also encode a function that prevents the producing strain from being destroyed by its own bacteriocin (immunity function). In the best known instances in Gram-negative bacteria (mainly colicinogenic plasmids), a function allowing the secretion of the bacteriocin has been described on the same plasmid. A recent review lists 35 different colicinogenic plasmids belonging to different incompatibility groups, some of which are conjugative.

Interest in bacteriocins extends beyond the mere knowledge of the plasmids bearing them, since their modes of action, of release from the cells and of immunity are challenging.

Bacteriocins belong to the more general class of toxins, i.e. proteins produced by a bacterium and active against specific target cells, several of which are also encoded by plasmids. In the best-studied cases (for instance, the *E. coli* haemolysin), the plasmid has been shown to code also for protein(s) specific for secretion of the toxin. It can be noted that some haemolysin determinants, homologous to the plasmid ones, may be found on the bacterial chromosome and may be presumed to be carried by a transposon.

Table 3.4 Some examples of plasmid-borne resistances to antibiotics.

Antibiotic	Mechanism of resistance	Classification	Observations
β-lactams (penicillins and cephalosporins)	*b-lactamases* hydrolyse the *b*-lactam ring at a specific site. Their classification depends on their spectrum of action on the different β-lactams	Several constitutive enzymes located on plasmids of Gram-negative bacteria[a]: TEM (2), SHV (1), HMS (1), ROB (1), OXA (14), CARB (3), CEP (2). Inducible enzymes coded on plasmids of Gram-positive bacteria: A, B, C, D of *S. aureus*	Gram-negative bacteria are often intrinsically resistant to many β-lactam drugs due to their outer membrane barrier and partly to low levels of chromosomally encoded β-lactamase expression
Chloramphenicol	*Cam-acetyltransferases* (CAT). Their mechanism of action and their regulation have been extensively studied	Three types, expressed constitutively but repressed by catabolic repression, are encoded by Gram-negative plasmids. Five inducible closely related CAT types are encoded by *Staphylococcus* plasmids	Type I also confers resistance to fusidic acid to Gram-negative strains which have lost their natural resistance to this drug
Tetracycline	Decrease of drug accumulation by the creation of a drug efflux mechanism. In addition a ribosome protecting mechanism may be involved	Five inducible, plasmid-specified Tc determinants (*tetA* to *tetE*) have been distinguished in enteric bacteria, but they have significant sequence homologies. Tc plasmids have been found in *Staphylococcus*, in several soil Bacilli, and in the streptococci. No hybridization occurs between these determinants	Gram-negative bacteria are generally more sensitive to tetracycline, which penetrates through the OmpF porin.

Aminoglycoside aminocyclitol (kanamycin, neomycin, streptomycin...)	*Acetyltransferases (AAC)*, using acetyl-CoA as cofactor. The acetylation takes place at a specific site indicated in the denomination of the enzyme, e.g. AAC(6′) acetylates the 6′-amino group on aminohexose I of susceptible drugs	Nine enzymes differing by their spectrum of action and the acetylated site. Eight are coded only by plasmids of Gram-negative bacteria	Chromosomally located resistance to these antibiotics reduces drug binding to the ribosome or impairs transport across the cytoplasmic membrane
	Nucleotidyltransferases (AAD). Use ATP as a cofactor	Seven enzymes of different spectrum and site of action. Three are specfic to Gram-positive strains, four to Gram-negative strains	
	Phosphotransferases (APH). Use ATP as a cofactor	Eight types of enzymes differing by their spectrum and site of action. Two are found on plasmids of Gram-positive and Gram-negative bacteria, but it is not known whether they are the same enzymes	
MLS group. Structurally diverse group. Bind to the 30S ribosome subunit (erythromycin (macrolide), lincomycin (lincosamide))	Specific N6,N6-dimethylation of adenine residues in 23S RNA, which prevents the binding of the drugs	Numerous conjugative plasmids of Gram-positive bacteria (*S. aureus*, *Streptococcus*) and of the Gram-negative *Bacteroides* spp. Transposition of transposon Tn917 Em is inducible by erythromycin. Three distinct groups of MLS resistance determinants are known in Gram-positive species	Many Gram-negative bacteria are intrinsically resistant to MLS drugs because of their impermeable outer membrane and possibly the level of methylation of the rRNA

a, Names and numbers refer to different types of enzymes.

2.4 Plasmids involved in host metabolism

All the functions listed above, and hence the plasmids, are truly facultative, and the presence or absence of the plasmids encoding them has no taxonomic significance for the strains. Some plasmids, however, encode functions that confer substrate specificities to the strains harbouring them, or functions that confer the most important of their metabolic properties.

2.4.1 *Degradative plasmids of* Pseudomonas *strains*

Pseudomonas species are nutritionally very versatile bacteria, able to feed upon extremely diverse and unusual substrates, such as organic solvents and many kinds of phenolic derivatives. These properties often depend on the presence of specific plasmids, called degradative plasmids, which encode all the enzymes needed for the catabolism of a family of those compounds, generally organized in a few operons.

The biochemistry and genetics of these degradative operons have been extensively studied. The best known are the TOL plasmids from *Pseudomonas putida*. The genes coding for the enzymes that degrade toluene are organized in two operons. Several plasmids that otherwise differ from each other have been found to bear the same set of degradative genes, which may again be borne by a transposon. They all belong to the same *Pseudomonas putida* species.

2.4.2 *Plasmids of the lactic bacteria*

Lactic bacteria of the group *Streptococcus lactis* harbour a diversity of plasmids (from 2 to 11 per bacterial strain), some of which code for functions useful for the dairy industry, such as lactose metabolism, galactose metabolism, proteolytic activities and citrate utilization. These activities are not always plasmid-encoded, depending on the strains, but many of the plasmids involved are conjugative.

2.4.3 *Symbiotic plasmids of* Rhizobium

Rhizobium spp. are legume symbiotic bacteria responsible for the formation of N_2-fixing root nodules. The genus comprises numerous species, defined so far by their specificity towards a given legume species. Most of the functions involved in nodulation and N_2 fixation, including host specificity, are coded on giant plasmids (200 MD), the Sym plasmids.

The operon for the N_2-fixation enzymic complex is highly conserved in all nitrogen-fixing prokaryotes (be they eubacteria or archaebacteria) (Chapter 16). The corresponding genes are chromosomally located in most of the free-living N$_2$-fixing prokaryotes, but also in some symbiotic species. Thus, the evolutionary pathway that led to the formation of the Sym plasmids must have been quite complex. One might speculate whether these molecules are true plasmids or rather 'minichromosomes'.

2.4.4 *Tumour induction plasmid of* Agrobacterium

Crown-gall is a disease that affects most dicotyledonous plants. The inducing agent is a soil bacterium, *Agrobacterium tumefaciens*, a close relative of *Rhizobium*. Most of the functions related to crown-gall induction are borne on a giant conjugative plasmid, the Ti plasmid. Four types of Ti plasmids have been demonstrated, which confer either different types of tumour, or slight metabolic variations in the parasitic relation of the bacteria with the plant. All the plasmids, however, share large regions of homology. Figure 3.5 gives the map of two typical examples.

The Ti plasmid can conjugate into *Agrobacterium* or *Rhizobium* species, but, in addition to its replication functions and transfer region, the plasmid encodes numerous functions related to oncogenesis. The

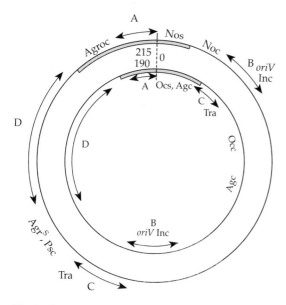

Fig. 3.5 Maps of two Ti plasmids. Homologous regions A–D (arrows) between plasmids pTiAch5 (octopine, inner circle) and pTiC58 (nopaline, outer circle): Tra, transfer function; Inc, incompatibility; *oriV*, origin of replication. Specific regions involved in opine metabolism are indicated: nopaline metabolism, Nos, Noc; agrocinopine metabolism, Agroc, Agr^s; octopine metabolism, Ocs, Occ; agropine catabolism, Agc. The T regions are indicated by a shaded line. The figures indicate the sizes of the plasmids (in kilo base-pairs, kp).

vir (or *onc*) regions are involved in plant infection. The T region is absolutely unique in that it is responsible for tumour formation by inserting into the plant chromosomes. It codes for the synthesis of an auxin, which is responsible for the induction of the tumour, and of a specific metabolite, an opine. Depending on the Ti plasmid, different sorts of opines have been described, such as nopaline (Nd) (N-α-(1,3-dicarboxypropyl)-L-arginine) and octopine (N-α-(D-1-carboxyethyl)-L-arginine).

Agrobacterium strains that induce the synthesis of one of the opines can use it (but not the other) as their sole carbon and nitrogen source, thanks to the presence of genes located outside the T region coding for catabolism of the opine on the corresponding Ti plasmid.

3 Nomenclature of plasmids and of plasmid-encoded functions

Originally plasmids were named from what was considered their most important function: F for fertility, R for resistance, Col for colicin production, Sym for symbiosis, etc. Additional letters and numbers

Table 3.5 Nomenclature of antibiotic resistances (phenotypes and genes) of chromosomal and plasmid locations.

Antibiotic (abbreviation)	Phenotype conferred by a chromosomal mutation	Phenotype conferred by a plasmid	Plasmid-borne genes
β-lactams			
Penicillin (Pen)	Pen^r		
Ampicillin (Amp)	Amp^r	Ap	*bla*
Carbenicillin (Carb)		Cb	
Cephalosporin (Cep)			
Chloramphenicol (Cam)	Cam^r	Cm	*cat*
Aminoglycoside aminocyclitol			
Kanamycin (Kan)	Kan^r	Km	*aphX*
Neomycin (Neo)		Ne	*aphX*
Streptomycin (Str)	Str^r	Sm	*aadX*
Spectinomycin (Spc)	Spc^r	Sp	
Tetracycline (Tet)	Tet^r	Tc	*tetX*
MLS group			
Erythromycin (Ery)	Ery^r	Em	*emrX*
Lincosamin			*emrX*
Sulphonamide		Su	
Trimethoprim		Tp	
Toxic ions			
Hg$^+$		Hg	*mer*

soon gave a very confusing pattern (RP4, RK2, RSF1010 . . .) and the present difficulty over assigning any given function to most classes of plasmids led to an international agreement: new plasmids, be they natural or engineered, are named as pXY followed by order figures, e.g. pSC101, pIP400.

The international convention has also fixed the rules for the designation of known functions such as resistances. The genes coding for these functions are designated, like the chromosomal ones, by three italic lower-case letters followed by a capital one: *tetA*, *aphA*, etc. The resistance phenotypes conferred by the plasmids to the strains are designated by a capital letter followed by a lower case one, e.g. Tc, Km (Table 3.5). The resistance character can be specified (e.g. Tc^r, Km^r) if ambiguity is possible, for instance if a sensitivity state (Tc^s, Km^s) has to be compared with it.

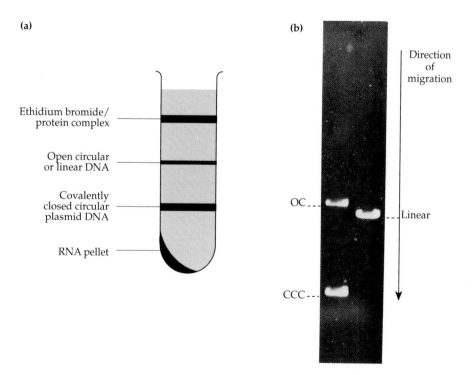

Fig. 3.6 Separation of covalently closed circles (CCC), open circles (OC) and linear forms of DNAs. The various forms of plasmid DNAs, CCCs, OCs and linear molecules can be separated either on a CsCl/EthBr gradient (a) or by electrophoresis on an agarose gel (b). The intercalating dye ethidium bromide (EthBr) fluoresces on exposure to UV light, thus allowing visualization of the DNA bands.

4 Preparation and analysis of plasmids

4.1 Plasmid preparation

Plasmid DNA may be separated from chromosomal DNA in suitably lysed cell preparations, after extraction of the proteins, by using differential centrifugation – density-gradient centrifugation in CsCl (Chapter 2) in the presence of the dye ethidium bromide – as for other DNA molecules (Chapter 14). Most of the chromosomal DNA can be previously eliminated by sedimentation at 20 000 rev/min, yielding a 'cleared lysate'. The dye intercalates differentially in superhelical (circular covalently closed (CCC)) molecules and relaxed (linear and circular with a single-strand nick, i.e. open circles (OC)) ones, thus changing their apparent density. The various forms will migrate separately (Fig. 3.6a). The band for the linear form may also contain some residual chromosomal DNA. Since molecular biology technologies have developed, miniaturized adaptations of most classical techniques have become available, and this is particularly true for plasmid preparations. Very rapid methods, known as 'minipreps', which necessitate small volumes (e.g. with *E. coli* only 1.5 ml) of cultures, are now widely used.

The cleared lysate or the CsCl/EthBr fractions may then be run on an agarose-gel electrophoresis system. Smaller molecules migrate more rapidly than larger ones (Fig. 3.6b). The DNA in these bands may, if needed, be extracted from the agarose gel by diverse techniques such as electroelution.

4.2 Physical analysis of plasmids

Plasmids may be characterized by the number of base pairs making up the molecule and by their pattern of cleavage (digestion) by site-specific restriction endonucleases (Chapter 6), yielding a 'restriction', or physical, map. Such a map is characteristic of the plasmid.

4: Further Autonomous Genetic Elements: the Bacteriophages

1 Definition

Bacteriophages, often shortened to phages, are bacterial viruses, their name meaning 'eaters of bacteria' since they reproduce at the expense of the bacterial host, which usually does not survive the process. The genome of a bacteriophage is, according to the definition given earlier, a replicon, in the same way that a plasmid is one. Like plasmids, these genomes are dependent on another, more elaborate, organism for their reproduction, in this case a bacterium. They differ from plasmids, however, in the sense that they are coated by a protein envelope, which allows them a larger degree of freedom from their host than plasmids, since they can both survive outside the host and reassociate with it. The free state corresponds to a metabolically inert phase, but is important for the propagation of the bacteriophage. We shall see, however, that this clear-cut distinction between phages and plasmids may sometimes be less straightforward.

Several types of relationships exist between bacteriophages and bacteria, some of which may lead to modifications of the bacterial genetic content. Bacteriophages are thus one of the factors determining the genetics of the bacterium. Besides their intrinsic interest, their study is also important in this respect. A minimum knowledge of the structure, physiology and genetics of bacteriophages is necessary to understand the mechanisms of these relationships and of their role in bacterial genetics, and it is this that will be considered in this chapter. No description of the genetics of bacteriophages *per se* will be given here.

Their existence in a free, though metabolically inert, form makes phages, and viruses of eukaryotes, the smallest and simplest self-reproducing entities known. However, their inert state as free units has led to long controversies as to whether they should be considered as living organisms or as inanimate autocatalytic agents. Due to the high similarity of their biochemical and metabolic capacities when in a host cell with those of all living beings, they have been given the status of living entities, though of an extreme parasitic type. Viruses infecting eukaryotic cells share many properties with bacteriophages, and their widespread importance and the extent of the present development of research on these 'organisms' emphasize the importance of bacteriophages as paradigms for these studies.

The first mention of their existence was made simultaneously, around 1915, by Twort and by d'Hérelle, who classified them as viruses. It was only about 25 years later that interest in these 'nucleoprotein complexes' was raised, mostly by the pioneering work of Delbrück. Since then, bacteriophages have been found to be very numerous and widespread. One or several have been isolated for almost all known bacterial groups (if not species or strains) in which they have been looked for, including archaebacteria.

A huge gap exists between the level of knowledge accumulated on the *E. coli* bacteriophages and those of most other species, some phages of *B. subtilis* excepted. So, once again, the information presented in this chapter will mostly concern phages of the former species. Even among these, however, a large span of differences are to be observed. Thus, although a priori limited, a description of this group of phages can most probably be considered as a representative sample of the various types to be expected among still unknown ones.

2 Structural characteristics

2.1 Specific nature

The chemical composition of viruses, and hence of bacteriophages, allows their unambiguous identification among all other living cells: (i) they are comprised of only two types of molecules, proteins and nucleic acids (a few specific exceptions will be mentioned); (ii) each viral particle possesses only one type of nucleic acid.

2.2 Morphology and size

The common morphological feature of all bacteriophages is the coating of their nucleic acid by a protein structure, called the capsid or coat, to form a structure which determines the shape of the particle. Two main morphologies are encountered: polyhedral and filamentous (Table 4.1 and Fig. 4.1).

The capsid can form a polyhedral (icosahedral, octahedral or oblong) head, the size of which varies from 25 to 130 nm in diameter, in rough correlation with genome sizes. It is found either bare (MS2, Qβ) or surrounded by a number of spikes (φX174, G4, S13) or attached to a tail. The tailed phages can be organized into several groups according to the morphology and anatomy of this appendage: it can form a simple tubular filament, shorter (T7) up to two to three times longer (λ,

T1, T5, T2, T4, P1) than the head diameter, and is either rigid (T2, T4) or flexible (T5, λ). A number of phages have a complex tail structure, with elaborate extremities, as exemplified by P1 and the T-even phages (T2, T4 and T6 are related, and named collectively as T-even): in T4, the tail *per se* comprises two coaxial cylinders, the core and a sheath, topped by a collar, and terminated by a basal plate holding six thin fibres and six spikes. The whole head + tail structure necessitates more than 50 different proteins, each present in different amounts depending on their function.

Filamentous capsids are found exclusively among single-stranded DNA phages. They are composed of a helical arrangement of 2000–3000 copies of a single protein and are usually very long compared with the size of more complex bacteriophages and with the size of their host (Table 4.1, pp. 82–3; Figs 4.1, pp. 84–6; 4.11b, p. 114). A few copies of other phage-coded proteins, attached to each of the extremities of the capsid, are important for infection.

One case of structure reminiscent of viruses of eukaryotes is φ6, a phage of a *Pseudomonas* strain, in which the capsid is surrounded by an envelope, made of a fragment of its host cell membrane.

Some phages contain 'internal' proteins, located inside the capsid. These molecules do not participate in the structure of the capsid.

2.3 Genomes

2.3.1 *Size*

Greater variation exists in the dimensions of the nucleic acid molecule forming the genome of these viruses than in the size of the particle. The only common trait is that the length of the molecule is much greater than the internal dimensions of the head of the particle (filamentous phages being an exception). Thus the T4 genome is about 10^3 times longer than the diameter of the phage capsid (60 µm as against 95 nm). A compacting process is thus expected to control the coiling and packaging of the nucleic acid molecule inside the capsid. While the smallest genomes known are only long enough to code for four or five proteins (for instance phages Qβ and MS2, specific for F-bearing strains), several reach a size up to 10% of that of their host's chromosome (Table 4.1). These phages belong to the class showing the more complex external morphologies (PSB1, T-even), and they are also found to be less dependent on bacterial functions for their reproduction.

All intermediate sizes can be found. Large and medium-size genomes correspond to double-stranded DNA, small ones to single-stranded DNA or RNA. A single case of a double-stranded RNA genome has been described in the case of phage φ6 of *Pseudomonas*.

Table 4.1 General characteristics of some phages.

Phage particle morphology and size (nm)		Genome structure, size, approx. coding capacity	Type	Examples		Observations
Head	Tail			Phage	Host	
Polyhedral oblong 120 × 80	Double cylinder Base plate Contractile 110 × 20	ds DNA linear Term. redund. Circ. permut. 100 MD, ≅130 genes	V	T2, **T4**, T6	*E. coli*	T4 has two introns
Icosahedral 93 120	Contractile 220 240	ds DNA linear Circ. permut. Term. redund. 60 MD, 60–80 genes 200 MD	T	**P1** P7 **PBS1** PBS2	*E. coli* *B. subtilis*	Prophage as free plasmid
Icosahedral 46–60	Contractile	Unique sequence ss extremities 22 MD, ≅20 genes	T	**P2** **P4**	*E. coli*, *Shigella*, *Serratia*	Prophage structure similar to that of λ prophage
Icosahedral 54 × 61	Contractile 100 × 18	ds DNA linear Unique sequence Host DNA extremities 25 MD, 37 genes	T	**Mu** D108	*E. coli* K12	Colinear transposon Can also infect *Serratia* and *Citrobacter*

Morphology	Tail	Nucleic acid	Type	Phages	Host	Remarks
Icosahedral 55–60	Flexible Non-contractile 152 × 17	ds DNA linear ss extremities 31.6 MD, ≈ 60 genes	T	**λ**, φ80, φ82 φ434, φ424	*E. coli* K12	Prophage and phage maps permuted
Icosahedral 32 × 42	Non-contractile 32 × 6	ds DNA linear Unique sequence 11 MD	V	**φ29** X15, GA1 M2Y	*B. subtilis*	
Octahedral 50–100	Non-contractile (tail fibres) 150 × 10 15 × 15 200 × 20	ds DNA linear Unique term. redund. 26.3 MD, >25 genes	V	T1, T5, BF23 **T7**, T3 SP01	*E. coli* *E. coli* *B. subtilis*	Infect other enterobacteria
Icosahedral 60	Short, with spikes	ds DNA linear Circ. permut. Term. redund. 26 MD, ≈20 genes	T	**P22**	*Salmonella* *E. coli*	Prophage structure and map similar to that of λ prophage
Icosahedral 25	Spikes	ss DNA circular 1.7 MD, 10 genes	V	**φX174**, S13, φR, G4	*E. coli*	
Icosahedral 25	–	ss RNA linear Unique sequence 1 MD, 4 genes	V	R17, f2, Qβ, **MS2**, M12, GA, SP, F1	*E. coli*	Infect other male enterobacteria
Filamentous (2000–875) × 5	–	ss DNA linear 1.9 MD, 10 genes	V	**M13**, Ff, If groups	*E. coli*	Infect other male enterobacteria also

ds, double-stranded; ss, single-stranded; term., terminal; redund., redundancies; circ. permut., circularly permuted; V, virulent; T, temperate. **Bold characters**, model for the group.

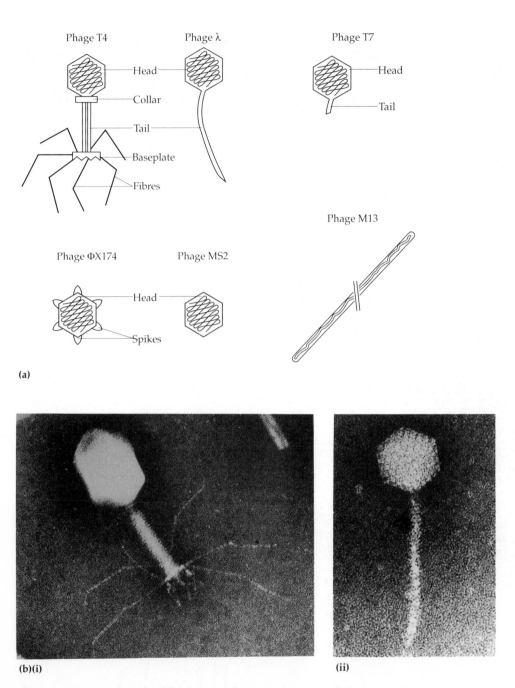

(a)

(b)(i)   **(ii)**

Fig. 4.1 (*Above, opposite and on following page.*) Common phages morphologies. (a) Schematic representation of the main morphologies encountered. (b) Electron micrographs. (i) T4 (1 cm = 50 nm); (ii) λ (1 cm = 40 nm); (iii) (*see facing page*) T7 (1 cm = 120 nm); (iv) φ × 174 (1 cm = 60 nm); (v) MS2 (1 cm = 50 nm); (vi) M13 (1 cm = 30 nm) (Friedfelder, 1983). (c) (*see* p. 86) Detailed diagram of phage T4 structure based on electron micrography to a resolution of about 2–3 nm. Figures and symbols refer to encoding gene nomenclature. (Eiserling, 1983.)

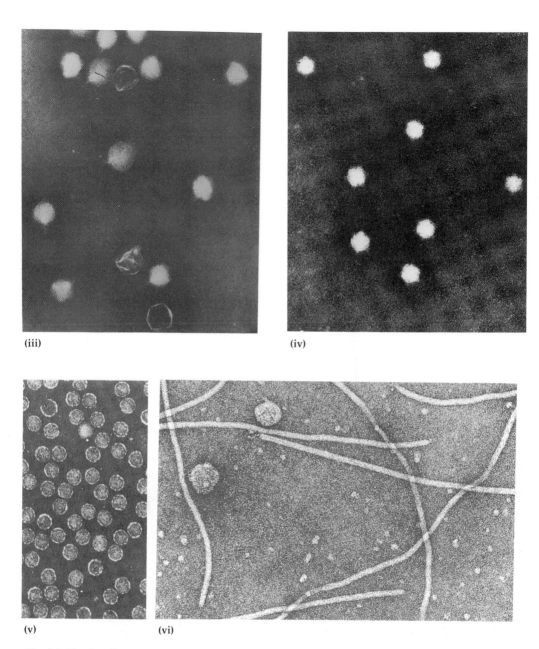

(iii)

(iv)

(v)

(vi)

Fig. 4.1 *(Continued.)*

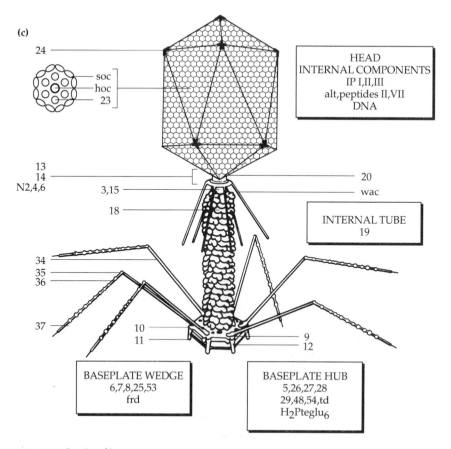

(c)

24

soc
hoc
23

HEAD
INTERNAL COMPONENTS
IP I,II,III
alt,peptides II,VII
DNA

13
14
N2,4,6

3,15

20
wac

18

INTERNAL TUBE
19

34
35
36

37   10
11

9
12

BASEPLATE WEDGE
6,7,8,25,53
frd

BASEPLATE HUB
5,26,27,28
29,48,54,td
$H_2Pteglu_6$

Fig. 4.1 (*Continued.*)

2.3.2 *Structure*

Most genomes are present as linear molecules inside the capsid. The only exceptions are single-stranded DNA phages, which have a circular molecule. Similarly to the bacterial chromosome, however, replication is in many cases dependent on a circular structure, and a number of strategies are found to solve the problem of the passage from linear to circular forms.

Most double-stranded DNA phages terminate their DNA by varied structures such as cohesive extremities or terminal redundancies which facilitate their conversion into non-covalently closed circles. Base-pairing between the complementary terminal sequences of the two antiparallel chains allows the formation of a circle, maintained by hydrogen bonding between these bases. Covalent closing of these 'open circles' is realized by either the cellular ligase or a phage-coded one, after it has been synthesized. In several phages, such as λ or P2, the recognition region is constituted of complementary single-stranded sequences, 10–15 bases long, on each chain of DNA, called cohesive (or sticky) ends (Fig. 4.10a, p. 112). Other phages possess double-stranded terminally redundant extremities. Circularity may then be achieved through homologous pairing and elimination of one copy of the redundant region. This redundant region is either a unique one (T1, T7, Fig. 4.9a, p. 110) or a circularly permuted portion of the genome (T-even, P22, P1). In the latter case, the phages of a population will have differently circularly permuted linear maps (T4, Fig. 4.8d, p. 109). The duplicated lengths range from 100 to 10^4 base pairs, representing up to 2% (T4) and 9% (T5) of the genome.

Phage T4 (Fig. 4.8, p. 108) and the other T-even phages T2 and T6 represent a particularly complex situation: the circularly permuted redundant extremities of their DNAs allow the formation of circular structures which are present in the cell together with linear monomeric or multimeric forms during their whole reproductive cycle. While the linear forms are used as transcriptional and replicational templates, the circular ones might be only replicative intermediates.

An exception to this process is the *E. coli* T7 phage (and the closely related T3); although it has large enough terminal redundancies (260 bp), it does not form circles, and is both transcribed and replicated as a linear molecule (Fig. 4.9, p. 110).

The process leading to the circularization of linear unique sequences had, for a long time, remained an enigma. Phage φ29 of *B. subtilis* has been found to circularize its genome by means of a covalent binding of a protein to the two extremities of the DNA molecule.

The double-stranded DNA genome of T5 includes five single-strand interruptions, located at precise, fixed sites, but their role, if any, is still a mystery.

As mentioned, single-stranded DNA genomes are always found as circular structures. In contrast, single-stranded RNA ones form unique linear sequences.

The only double-stranded RNA phage known, φ6 of *Pseudomonas*, has two peculiarities which recall features of some eukaryotic viruses: (i) its capsid is coated by an envelope originating from the membrane of the cell in which it developed; and (ii) its genome is fragmented into three molecules (segmented genome).

2.3.3 *Unusual bases in DNA*

Most phages have a classical base composition; a few, though, show modified bases. In the T-even series, a large fraction of the cytosines is replaced by hydroxymethylcytosines (HMC) (Fig. 4.2), synthesized by a specific phage-encoded pathway. The modified base makes normal H

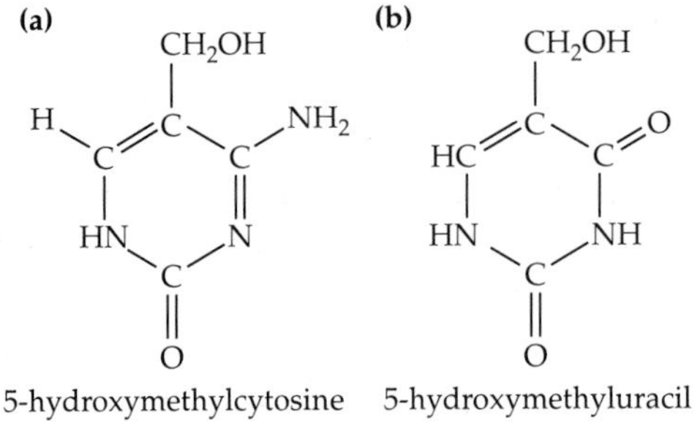

(a)

CH$_2$OH

(b)

CH$_2$OH

5-hydroxymethylcytosine   5-hydroxymethyluracil

Fig. 4.2 Modified bases found in certain phage DNAs.

bonds with guanine, and thus does not result in an abnormal structure of the DNA helix. The hydroxymethylcytosines may be totally (T4) or partially (T2, T6) glycosylated on the added –CH$_2$OH radical. Other special nucleotides are hydroxymethyluracil (a thymine with an –OH radical on its 5-methyl group (Fig. 4.2)) and uracil in the *B. subtilis* phages SPO1 and PBS1, respectively. Like the T-even ones, they are large phages whose genomes comprise 100–200 genes, a number of which are responsible for the establishment of these unusual features. These modifications protect the DNA against the nucleolytic action of certain endonucleases that may be present in the host cell (Chapter 6).

2.4 Classification

The characteristics described above have been used to build classes, either among the phages of a family or group of bacteria or even among the phages of different bacterial groups. Further physiological and genetic analyses have often supported the formulation of such groups. These studies have led to the determination of the relationships of the phages with their bacterial hosts: their degree of dependence on bacterial metabolism, their strategy of propagation and the possible consequences for the fate of the bacterial genome or of the cell itself. In fact, these considerations appear more fruitful than those based on strict morphology.

If modes of propagation are taken as the main criteria of classification, two groups of phages are defined. Virulent phages display only one mode of reproduction, namely the lytic, or virulent, cycle: production of free viral particles is the only issue of an infection of the host bacterium, which in most cases does not survive the infection. On the other hand, the second group, temperate phages, have, in addition to this lytic reproductive capacity, functions which allow them to propagate only their genome without destruction of the host. Their physiological capacities are thus richer than those of virulent phages. The stable phage–host state is named lysogeny.

In spite of its simplicity, this classification has proved convenient and rich in predictive capacity.

3 Experimental methodologies

A key advantage in studying phages is the possibility to rapidly isolate
and produce genetically pure clones and to quantify phage suspensions, so that one can follow an infectious cycle and observe the effect
of any treatment or conditions which might interfere with it. In order
to understand the setting up of these techniques, a brief outline of the
process we have referred to as a virulent cycle is necessary. Detailed
descriptions of the physiological and molecular events taking place
during the cycle will be developed in Section 4.

About a parasite, a phage needs to be associated with a bacterial host in
order to develop. This association starts with the adsorption of the viral
particle on to a susceptible bacterium (Fig. 4.3). After completion of
the infectious cycle, the resulting infected cell, called an infectious
centre, leads to the production of mature infectious particles, which are
liberated into the medium, usually due to concomitant lysis of the cell.
If other bacteria are available, successive similar rounds of infection can
take place.

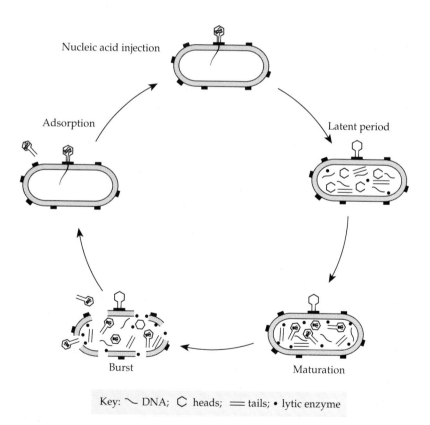

Key: ⌒ DNA; ☾ heads; ═ tails; • lytic enzyme

Fig. 4.3 The reproductive cycle of a double-stranded DNA phage.

3.1 Phage titration

The usual method for phage titration takes advantage of the possibility of achieving successive rounds of infection on solid medium. Samples containing 50–500 particles from appropriate dilutions of the phage suspension to be titrated are mixed with a large number (10^7–10^8) of susceptible bacteria, called indicator bacteria, in a small volume (0.2–0.5 ml) so as to facilitate phage–bacterial collisions. After a few minutes at optimal temperature to allow adsorption, during which the addition of specific cofactors may be useful or necessary, the suspension is spread, usually together with soft agar, onto solid nutrient medium and incubated under appropriate conditions. Each infected bacterium will lead to the *in situ* liberation of progeny phages, which will in turn infect and lyse a new set of neighbouring cells. Up to 10^5 such cycles can take place. The phages produced, although very numerous (10^7–10^9), cannot be detected by the naked eye, but the resulting lysis *in situ* of a large number of bacteria results in a hole in the otherwise fully grown lawn of surrounding cells. These holes, called plaques, reach sizes of 0.5–5 mm in diameter, and can easily be counted (Fig. 4.4a). Each plaque represents one input plaque-forming unit (pfu), originating from either a phage or an infected bacterium. Appropriate calculations yield the concentration of the initial suspension. An identical procedure is used to numerate infectious centres.

3.2 Production of phages

The production of phages is achieved by allowing a sufficient number of successive rounds of infection to take place, usually in a liquid suspension of metabolically active bacteria inoculated with an initial small aliquot of a previous phage preparation and incubated under appropriate conditions. The yield of each cycle depends on the phage and the conditions, but an average production is 100–200 particles/infected cell. Using a host suspension in exponential phase of growth (10^7–10^8 cells/ml) even a small phage inoculum (10^3–10^5/ml) will yield 10^9–10^{11} phages/ml after only a few cycles, all bacteria having been infected and lysed. A variation of this method is to allow the successive cycles of infection to take place in a soft agar layer containing the phage and bacteria, spread on a solidified nutrient medium. Ideally, the number of phages should be such as to give almost, but not quite, complete, or confluent, lysis, with only small, irregular areas of bacterial growth being visible (Fig. 4.4a). The whole soft agar layer is harvested.

In either method, the phage is purified from the bacterial debris, non-lysed cells and soft agar by centrifugation, usually in the presence

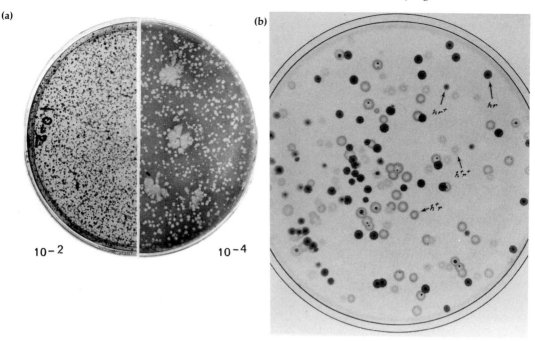

Fig. 4.4 Phage titration and plaque morphologies. (a) Titration. A suspension of λ phage was diluted as indicated, and equal volumes spread on a lawn of susceptible bacteria. Set up of 2 plates with different dilutions. (b) Plaque morphologies. The different plaques of T2 shown correspond to host-range mutants plated as a mixture on two hosts, also both spread on the plate. The four types of plaques correspond to the possible associations of two alleles of each of two phage characters. The h^+/h^- alleles control host-range: each allele allows infection of only one of the two hosts present, thus leading to apparently turbid plaques. The r^+/r^- (for rapid lysis) alleles control time of lysis of the host, and thus size of the plaques.

of chloroform to complete cell lysis. Further purification can be obtained by various methods. One consists of centrifugation on a CsCl density gradient, which separates the viral particles from soluble proteins (lighter) and nucleic acid molecules (heavier). The phage suspension can be kept with little loss either frozen (with the addition of an osmo-regulator) or at low temperature, as long as a bacterial inhibitor prevents contamination.

3.3 Phage cloning – plaque morphology mutants

Under strictly constant conditions the morphology of a plaque is characteristic of a phage. Mutants of functions controlling the size, shape or turbidity of plaques can readily be recognized and cloned, i.e. isolated as a genetically pure line. The characteristics of the plaque can be used as genetic markers (Fig. 4.4b).

Plaques can be treated by techniques similar to those used for colonies. Phage cloning is achieved by picking the phages from a plaque, with a toothpick or platinum wire, and then propagating them by inoculation to a new cell suspension, as described above.

Replica-plating methods, originally used in bacterial genetic studies (Chapter 7), are also feasible with plaques. They allow rapid screening of large numbers of phage clones for genetic traits such as growth capacities on different bacterial hosts or at different temperatures, etc.: the plate containing the plaques to be tested is printed (replica-plated) on to plates spread with the suitable indicator cells. Direct comparison of the appearance of plaques after incubation of the new plates allows both definition of the genetic character and isolation of the clones of interest.

3.4 Phage crosses

Upon mixed infection of a bacterium with two different phages (i.e. two mutants of the same phage species), exchanges of genetic material, leading to the formation of recombinant clones, can take place during the reproduction cycle. The progeny can be readily isolated, if recognizable under appropriate conditions. This technique has been widely used for the understanding of phage genetics.

3.5 Physiological and molecular analyses

Since a phage reproduces inside a bacterium by using most of the host synthesizing capacities, usual methods (inhibitors, molecular labelling, etc.) developed for bacterial studies apply to phage studies. All techniques available for the analysis of proteins and nucleic acids, including fractionation, are obviously usable for phages. The phage constituents can be prepared either from mature particles or from infected bacteria lysed during the reproduction cycle.

4 The lytic cycle of infection

Lytic cycles, which lead to the production of free progeny particles, can be performed by all phages, be they virulent or temperate. The lytic cycles of temperate phages are similar to those of virulent ones, and thus are included in this description, but their temperate propagation will be presented later (Section 5).

4.1 The one-step cycle

The first overall description of the events occurring during the course of an infectious cycle was presented by Ellis and Delbrück in 1939 and completed in 1952 by Doerman, both groups working with the *E. coli* bacteriophage T4.

A suspension of *E. coli* (about 10^8 cells/ml) was synchronously infected by T4 phages at a multiplicity of infection (moi) (i.e. the ratio of phage particles/bacterium) close to 10, which is sufficient to allow all cells to be infected*. The suspension was incubated for a few minutes, so that the probability for each bacterium to have been in contact with a phage approached 1. Synchrony was obtained by blocking cellular metabolism, letting only phage adsorption go to completion. Upon release of the inhibition of metabolism, the mixture was massively diluted, in order to avoid any late adsorptions, and incubated under appropriate conditions (37 °C, aeration). Aliquots taken at various times were assayed for pfu concentrations. Direct plating of the samples gave information as to the evolution of the total concentration of pfus, formed from either free phages or infected bacteria. Plating after elimination of the bacteria (infected or not) by differential centrifugation yielded the concentration of free phages (initially present and liberated). Doerman's contribution to this experiment consisted in performing a premature lysis of the bacteria, so as to time the initial appearance of mature phage particles inside the cells.

A schematic representation of the results is shown in Fig. 4.5. The whole T4 cycle lasts 26–30 min at 37 °C. It can be divided into several phases: during the first 12 min, the bacterial host genome is frag-

---

* *The Poisson distribution* The Poisson distribution allows an approximate determination of the proportions of a population in which 1, 2, 3, *n* events have occurred. It defines the probability of an occurrence of this event.

In the case of the infection of a bacterial population by a phage suspension, it allows the determination of the number of cells actually infected by 0, 1, 2, *n* phages, according to the formula:

$$P_n = \frac{m^n \cdot e^{-m}}{n}$$

where m = average moi, n = number of phages infecting a cell, P_n = probability of a cell to be infected by n phages and e = the basis for Naeperian logarithms = 2.7182828.

The following table gives the distribution of cells, P_n, expressed as a proportion (100% = 1), infected by 0, 1 or >1 phages in the case of a few mois.

moi	n		
	0	1	>1
0.1	0.905	0.090	0.005
0.3	0.741	0.222	0.037
0.5	0.607	0.303	0.090
1	0.364	0.364	0.272
2	0.135	0.270	0.595
5	0.007	0.035	0.958

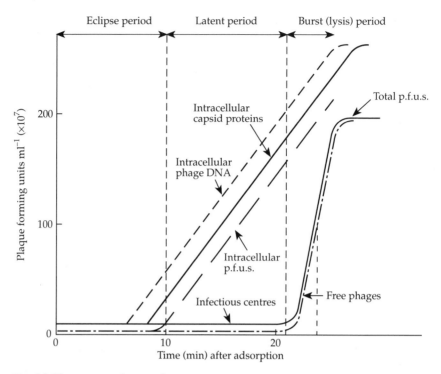

Fig. 4.5 The one-step phage cycle.

mented, and no free phage is found either in the medium or intra-cellularly. The total input phage can be titrated as infectious centres. No modification of their number is visible. This is called the eclipse period, on account of the disappearance of all free particles. Between minutes 12 and 22, mature particles start being detected but only intracellularly. Although their number increases, direct numerations from the suspensions still yield a constant number of pfus, each infected cell being counted as one infectious centre. This period, from time 0 to minute 22, the latent period, ends when the first extracellular progeny phage begin to appear. This leads to the lysis period, or burst, a very rapid (almost synchronous) lysis ending with complete disappearance of the cells and liberation of all mature particles. The increase in phage concentration is a measure of the average yield, or burst size – approximately 200 phages/infected centre in Fig. 4.5.

   Phage DNA synthesis, indicated on the figure, obviously starts earlier than the appearance of intracellular particles but continues simultaneously with maturation of progeny phage until the host cell is lysed. This is true also of the synthesis of the capsid proteins, which start to form into heads as soon as available. Cell lysis usually occurs while a large proportion of DNA and coat-protein molecules have still

Table 4.2 General characteristics of the infectious cycles of some *E. coli* phages.

Type of phage	Receptor	Duration of cycle (min, 37 °C)	Cell lysis	Mean yield per cell
T4 (T6, T2)	LPS	26	+	200
T5 (T1)	Siderophore on cell wall	12	+	200
P1	LPS		+	100–200
λ	LamB porin in outer membrane	65	+	100–150
φX174	LPS + membrane		+	
MS2 (Qβ) (ss RNA)	Side of F pili	Continuous	+	≈ 1000
M13 (F1) (ss DNA)	Tip of F pili	Continuous	−	≈ 1000

not been organized into mature particles, and these phage components are lost upon lysis.

The infection described was done under conditions such that all cells had adsorbed at least one phage. The whole population was thus lysed after one cycle, hence the name given to the experiment. If any non-infected cells had remained, a second cycle of infection could have started as soon as free progeny phage were liberated, from 22 minutes on. On the other hand, infection under conditions of larger mois changes neither the timing nor the final yield. Only very high mois (50–100 or more for *E. coli* infected by T4) disturb this process by provoking almost immediate lysis before any progeny phage can be synthesized. This 'lysis from without' probably results from degradation of the cell wall.

In its general outline, the format of this cycle is true for all phage infections resulting in a lytic event. Table 4.2 shows examples of the variations of this basic theme in different phages. Temperature has been indicated since the time schedule of the cycle is influenced by the overall metabolic rate of the bacteria. A more precise description of the successive steps of the infectious cycle is presented below, emphasizing the specificities of each phage type.

4.2 Adsorption

Random collision of a phage and a bacterium will result in efficient adsorption only if a number of conditions are fulfilled.

4.2.1 *Specific receptors*

Molecular structures present on both the bacteriophage and the bacterium ensure the stabilization of the pair. The phage possesses a
recognition structure: several proteins situated at the tip or on the basal
plate of the tail for tailed phages (Fig. 4.8a), one particular protein, the
maturase, for tailless icosahedral phages, or a protein situated at one of
their extremities for filamentous phages. These interact with particular
structures of the cell envelopes, the nature of which varies with the
phages. It can be a protein or a complex from the LPS (in Gram-
negative bacteria), or a constituent of the cell wall or of the cytoplasmic
membrane, or even a more elaborate complex involving elements
belonging to two or three of these layers. These structures, referred to
as phage receptors, are in fact cellular components which, at least in a
number of cases, have known functions in cellular metabolism (Table
4.2). It is probable that selection and evolution have led to the adaptation of structures on the phage that allowed their efficient, and thus
almost compulsorily specific, fixation to a host so that each phage can
infect only one or a few related bacteria. T4 infects *E. coli*, λ is specific
to the K12 strains of the same bacteria, Mu can adsorb on to several
enterobacteria, P1 on numerous Gram-negative bacteria, and the Ff
group and single-stranded RNA phages need the pili coded by the F
plasmid of *E. coli* or *Salmonella* (they are 'male-specific' (Fig. 4.11b)).

As for any molecular interactions in which the exact structures of
the components ensure the success of the process, modifications of
either the phage recognition complex or the receptor decrease the
efficiency of adsorption. Mutants resistant to a phage, resulting in lack
of fixation of the phage on a modified receptor, can usually be isolated
from a bacterium sensitive to this phage. These mutants have been
important for the analysis of the adsorption process and of the nature
of the receptors. Obviously, appropriate phage mutants can, in turn,
achieve efficient adsorption on the modified bacterial receptors. Such
modified phages, known as host-range mutants, are easy to find, since
they will selectively develop on bacteria resistant to the wild-type
phage population (Fig. 4.4b).

Since in most cases different phages use different receptors, a bacterial mutation leading to resistance to a given phage, implicating only
a specific receptor, will maintain the previous sensitivity of the cell to
any other phages.

4.2.2 *Mechanism, energetics and cofactors*

No energy is required for the fixation of the phage on the receptor. The
interaction involves the formation of electrostatic bonds. Some phages,

like the T-even series of *E. coli*, carry a few molecules of ATP at the extremity of their tail, but this is probably utilized for the step following adsorption.

Cofactors may be necessary for successful adsorption. The most frequent requirement is a divalent cation, Ca^{2+} or Mg^{2+}, but other products are known to be important, e.g. in a mineral medium T4 is successfully attached only if tryptophan is added. While cation participation in modifications of the bacterial external charges is easily explained, the role of a molecule like tryptophan is totally unknown.

Specific conditions may be required. Thus, the receptor of λ is the maltodextrin porin of *E. coli*, a protein synthesized upon induction by maltose of the expression of the corresponding gene. Although not a cofactor (it does not participate directly in the adsorption process), maltose increases the probability of fixation when added to the suspension long enough prior to the infection to allow the porin to be synthesized.

4.2.3 *Kinetics*

The probability that a collision between a phage and a bacterium takes place is directly dependent on the total concentration of both particles (phages + bacteria). Efficiencies approaching 100% are obtained under usual conditions – that is, when working with suspensions containing $10^7–10^9$ cells/ml – whatever the phage concentration.

The adsorption itself is extremely rapid, as compared with the average time required to ensure a high probability of collision between the appropriate structures of the two organisms. Mean adsorption times vary with the phages, probably as a function of the number of receptors present on the bacterial surface and on the recognition process.

As was indicated in the one-step experiment, adsorption is usually favoured by incubating the phage-plus-bacteria mixture for a few minutes under optimal conditions (high concentrations, physiological temperature). This duration must be increased if lower absolute concentrations of both phage and bacteria have to be used.

4.3 Nucleic acid injection

After the phage particle has attached to a receptor, the cycle *per se* is considered to start, the next step being the injection of the phage genome inside the cell (Fig. 4.3). The constant characteristic of this step is the exclusive injection of the genome, the capsid remaining outside, where it was adsorbed. It is interesting to note that a totally different situation exists in the case of eukaryotic viruses: the capsid, with its nucleic acid content, enters the cell either via a phagocytosis-like process or after fusion of the protein–lipid envelope of the virus with the cell membrane.

The different fates of the two molecular species (i.e. proteins and nucleic acid) of a phage during the injection process were elegantly demonstrated by Hershey and Chase in 1952, using T4 and *E. coli*. Two phage suspensions were obtained after development of infected *E. coli* cells incubated in each of two media containing either radioactive phosphorus or radioactive sulphur. These molecules, when incorporated in the newly made phage particles, were selectively distributed in the nucleic acid and protein fractions, respectively, and could be used as specific markers of each type of macromolecule. After adsorption to new host cells, possible residual components of the phages were separated from the surface of the cells by vigorous shaking with a blender and differential centrifugation. This did not disturb the development of the infection. The phosphorus radioactive label was totally recovered inside the cells whilst the sulphur labelling was found in the cell-free fraction. This experiment, besides elucidating an important aspect of phage infection, was the first demonstration to be readily accepted by the whole scientific community that nucleic acids (DNA in this case) are sufficient to allow total and faithful reproduction of an organism, a statement which implies that these molecules contain all the genetic information, supporting the finding shown by the transformation experiments of Avery, McCarty and McLeod in 1944 that DNA is the genetic material (Chapter 10).

In fact, more precise determinations performed later indicated that one or a few proteins are injected together with the genome of some phages (RNA phages, single-stranded DNA phages, λ, T-even phages). In most cases, these proteins are thought to be directly involved in the injection process. Other T4 internal proteins play a role in the initial metabolism of DNA. Some may serve to protect the phage DNA upon entry into the cell.

The mechanisms and the energetics of the injection process are poorly understood. It is clear, however, that different methods have been devised by different phages. The first problem to solve is the creation of a pore in the cellular outer layers. T4 is thought to carry lysozyme, an enzyme that hydrolyses the peptidoglycan of the cell wall. Proteins involved in the recognition structure of some phages have been shown to possess an endoglycosidase activity, responsible for local hydrolysis of the LPS (T-even, φ29 and λ of *E. coli*). Some phages may use the natural, non-specific, paths of the porins.

The injection *per se* is supposed to start spontaneously, using the energy associated with the highly condensed state of the nucleic acid molecule in the phage head. Recent information suggests that the DNA is pulled from the cell cytoplasm. It is known, at least for a few phages (and assumed to be a general rule), that the linear non-permuted DNA genomes are always positioned with the same orientation in the phage head and thus always injected from the same extremity. How this is controlled is not always clear. In the case of the T-even phages of *E. coli* and of a few others (Table 4.1 and Fig. 4.1), the tail comprises two tubular layers. Initiation of injection inside the cell is due to the contraction of the outer sheath, accompanied by the hydrolysis of the ATP molecules present in the tail (Fig. 4.8a). The tip of the inner cylinder then comes into contact with the cell membrane, and directs the DNA into the cell by an unknown mechanism.

Some peculiar situations are known. Injection of the DNA of T5 is achieved in two steps: the distal part of the molecule can enter the host only after a specific protein coded for by the proximal part has been synthesized. The single-stranded interruptions present on the phage genome seem to have no role in the injection, which is supported by the fact that T7 is also dependent on transcription of the proximal extremity of its genome to complete injection, despite the fact that it has a continuous double-stranded DNA with no single-stranded interruptions.

It is not known, either, how the circular single-stranded DNA genome of icosahedral phages (φX174) is transferred. In the case of the male-specific filamentous phages (the Ff and If groups) and of the single-stranded RNA polyhedral phages, which adsorb, respectively, to the tip and the side of the F plasmid-mediated pili (Fig. 4.11b), it has been proposed that the nucleic acid molecule is transferred through a central hole of the pilus, thanks to the action of a special protein present in the capsid, the maturase. This protein, sometimes called the 'pilot protein', is located in φX174 in spikes jutting out of the 12 corners of the head and in M13 at one end of the filament. It is responsible for several functions (Figs 4.6 & 4.7): (i) it causes adsorption of the phages to the host cell receptors; (ii) it carries the phage DNA into the cell after the protein has undergone certain conformational changes; (iii) it initiates replication of the phage DNA, probably by linking it to the host replicative machinery at the cell membrane; and (iv) it plays a part in morphogenesis. The versatility of the protein emphasizes the economy necessary in such small particles (the φX174 genome is 5.4 kb, carrying nine genes, whilst that of M13 is 6.4 kb, carrying eight genes). As will be seen, this economy is also shown in the arrangement of the genes, since several of them overlap.

The filamentous phages constitute exceptions to the rule that the capsid remains outside the cells: the proteins forming the capsid of the infecting particle are transferred, probably through the pilus, to the cell membrane, together with the maturase protein, the gene III pro-

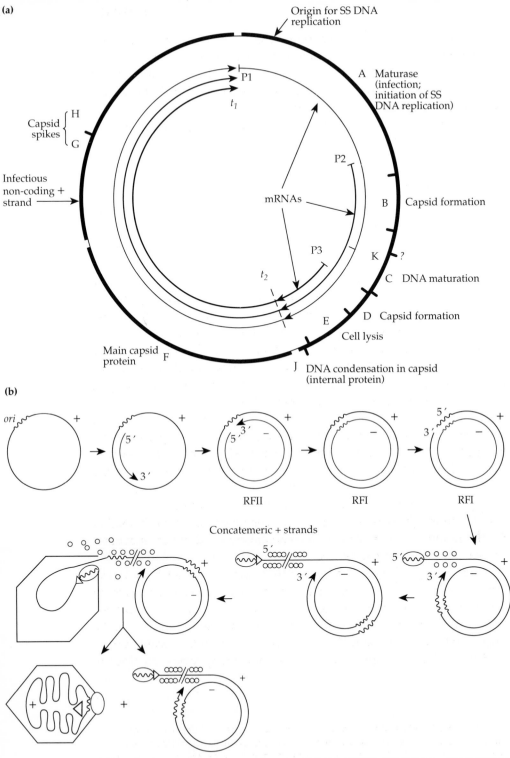

(a)

Origin for SS DNA replication

P1

t_1

A Maturase (infection; initiation of SS DNA replication)

Capsid spikes { H / G

Infectious non-coding + strand

mRNAs

P2

B Capsid formation

P3

K ?

t_2

C DNA maturation

E

D Capsid formation

Cell lysis

Main capsid protein F

J DNA condensation in capsid (internal protein)

(b)

ori

RFII

RFI

RFI

Concatemeric + strands

Key: ⬭ Maturase (gene A); ○○○ single strand binding (ssb) proteins (gene C); △ J protein duct. After the DNA has been replicated in the host cell, the newly synthesized '+' strand is protected by the gene V protein and taken to the cell membrane where it loses the gene V protein but picks up the coat proteins, both those left behind and the newly made ones, and also the gene III protein at the tip (Fig. 4.7b).

In most cases, the injection process is very rapid, with exceptions such as T5 and SP01.

4.4 The latent period

In spite of its name, the latent period, which starts immediately after the entrance of the phage genome in the host cell, is a very active phase of the cycle, metabolically speaking. Reproduction of a phage implies the synthesis of new genome copies and of large numbers of all the capsid proteins. The functions involved in replication of the nucleic acid molecule, when coded for by the phage genome, are performed by proteins not present in the mature particles (except in a few cases). All the structural proteins of the capsid are coded by phage genes.

All reproduction cycles imply the synthesis of phage proteins, and thus synthesis of mRNAs, a process which requires an RNA polmerase. Since no such protein is present in mature particles of the majority of phages, transcription is dependent on the host RNA polymerase. Some large phages, such as the *E. coli* T-even ones (Table 4.3), however, are capable of more or less releasing this dependency by coding for protein(s) which replace (e.g. phage T7) or modify (e.g. T-even phages) the host RNA polymerase. Again, these proteins are not present in the phage particle and will be made only during the course of the lytic cycle. No phages are known to code for their own protein-synthesizing system, although some (e.g. the *E. coli* T-even ones) have information for the modification of some of the constituents and can thus express some specificity at this level also.

Large phages also possess genetic information for other functions considered as non-participating *per se* in their reproduction, such as protection of their DNA against possible degradation by host restriction systems (Chapter 6) and lysis of the cell wall. These will be mentioned when relevant. Functions involved with the establishment or release of the lysogenic state, since they are specific to temperate

Fig. 4.6 (*Opposite.*) Phage φX174 reproduction characteristics. (a) Genetic map and transcription pattern. Three mRNAs are synthesized from the newly-made strand, starting at promoters P1, P2 and P3, and terminating at the same strong (t1) or secondary (t2) termination sites. Note that genes B, K and E overlap genes A, C and D, and are translated on different reading frames. (b) Replication and maturation.

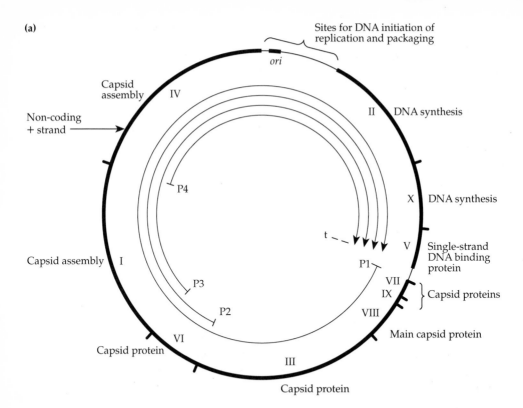

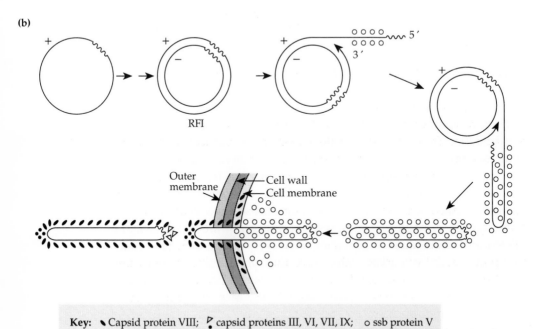

Key: ⟋ Capsid protein VIII; ⁊ capsid proteins III, VI, VII, IX; ○ ssb protein V

Fig. 4.7 Phage M13 reproduction characteristics. (a) Genetic map and transcription pattern. (b) Replication and maturation. P1–P4 are four transcriptional promoters; t, the single termination site.

bacteriophages, will be dealt with when lysogenic processes are described (Section 5).

4.4.1 *Transcription*

4.4.1.1 Expression of the genomes of single-stranded DNA phages
The DNA strand present in the phage head, or + strand, is the non-coding one, and synthesis of the − strand must thus be performed early. Also, single-stranded DNA cannot be used as a template by RNA polymerase. The first event taking place as these circular DNAs enter the cell is their conversion to double-stranded forms (Section 4.5.1).

Transcription is performed by the host RNA polymerase. Several promoter sites have been identified in all cases, yielding a number of different polycistronic mRNAs (Figs 4.6 & 4.7), which are translated by the bacterial system. No elaborate regulatory process seems to exist to either quantitate or introduce a timing in the expression of the various genes. But the presence of several promoters and terminators (Chapter 15) leads to a higher level of expression of the genes for capsid proteins than of the others.

Among the 10 genes identified in both ϕX174 and the Ff phages, seven to eight are mostly implicated in the elaboration of the capsid and in cell lysis and two or three code for proteins involved in DNA replication.

4.4.1.2 Sequential transcription of the double-stranded DNA genomes of virulent phages The strategies of transcription are similar enough among these phages to allow the construction of a general model, the main characteristics of which are the capacity to gain some degree of independence from the host system and the existence of elaborate regulatory processes controlling transcription of classes of genes.

As in the case for single-stranded phages, transcription is usually initiated by the host RNA polymerase. This is, however, restricted to a few genes, called 'immediate early'. The discrimination is made thanks to the specificity of their promoters, recognized by the host σ factor (the subunit which directs specificity of transcription by the core of the RNA polymerase by binding to the appropriate promoters, i.e. increases the affinity of the enzyme for these sites), while those of all other genes are not. The few functions thus expressed vary among phages (Table 4.3) and may involve: a phage-specific σ factor; a new, monomeric, RNA polymerase (T7); a ligase (T-even phages); endo- and exonucleases (which hydrolyse host DNA); enzymes responsible for the synthesis of modified bases (T-even phages); inhibitors of host cell restriction processes, etc.

Table 4.3 Principal functions and their temporal expression during the lytic cycle of some *E. coli* phages.

Phage	Immediate early	Early	Late	Coding strand
T4	Modif. of σ of RNA polymerase Endo- and exonucleases Ligase Cytosine-modifying enzmes Prevention of host restriction Phosphorylation of DNA precursors	DNA replicating enzymes Modification of RNA polymerase α, β, β' SUs Replacement of σ Modif. of ribosomes Specific tRNAs	Capsid structural proteins Assembly Cell lysis	*l* for early and some late genes *r* for the other late genes
T7	Nucleases Ligase Protein kinase RNA polymerase Modif. of host RNA polymerase Prevention of host restriction	DNA replicating enzymes	Capsid structural proteins Assembly Cell lysis	
T5	DNA injection (pre-early) RNA polymerase Modif. of host RNA polymerase core Nucleases	DNA replicating enzymes	Capsid structural proteins Assembly Cell lysis	*l* for early genes *h* for immediate early, some early and late genes

λ	Antiterminator N cI repressor	Recombination system Prevention of host restriction DNA replication control O and P Antiterminator Q	Capsid structural proteins Cell lysis	*l* for immediate early *r* for delayed early and late genes
P1	σ factor Ligase Cytosine-modifying enzymes Inhibition of host restriction	Some DNA-replicating enzymes	Capsid structural proteins Assembly Cell lysis	
φX174		DNA replicating enzymes Capsid structural proteins Assembly Cell lysis		minus
M13		DNA-replicating proteins Capsid structural proteins Assembly (maturase)		minus
MS2		Replicase Capsid structural protein Maturase Cell lysis		plus

l and *r* or *h*; the two strands of the double-stranded DNA genomes; SU, subunit.

The consequence of the first modifications of the host RNA polymerase by T-even phages is the sequential onset of transcription of the phage genes involved in replication. These form the 'early' genes. Their number varies widely (Table 4.3). Replication of the DNA starts as soon as these enzymes are synthesized and continues until lysis of the host cells occurs. Taking T4 as an example, the host RNA polymerase (RNAP) starts the transcription process, attaching via its σ factor to the relatively limited 'immediate early' promoter sites. Changes are then induced in the RNAP as development proceeds. Firstly, an 'altered RNAP' appears, in which ADP-ribosylation of its α subunits occurs. Then further modifications occur on the β and β' subunits, followed by the attachment at two different stages of various phage proteins. In this way, transcription is controlled by the phage (Fig. 4.8c).

The processes involved in replication are described in Section 4.5.1.

Most of the double-stranded DNA-containing virulent phages degrade the chromosome of their host cell. They usually do so early during the cycle, using one (or several) nucleases transcribed as immediate early genes (Table 4.3). This process, of variable efficiency and rapidity, provides a large fraction of the precursors for the synthesis of their own genetic material and constitutes a drastic means of preventing any further synthesis of host mRNAs. The cell machinery for the metabolism of macromolecules can thus be utilized exclusively for phage biosynthetic needs, as soon as cellular mRNAs are degraded.

In some phages another set of early genes brings about further modifications of the host transcriptional or translational systems, resulting in different specificities of recognition of transcription or translation signals (Table 4.3). A consequence of the expression of these functions is the onset of transcription of the last class of genes, the late ones, mostly responsible for the synthesis of the structural and assembly proteins making up the capsid of the mature particles, and also for the cell-lysing system.

In some phages (T3, T7) one of the immediate early proteins is a particular RNA polymerase consisting of a single protein responsible for transcription of the early and late genes (Table 4.3, Fig. 4.9). No further modification of the transcriptional system seems to exist. It is not clear what controls the timing of expression of these functions, but promoters showing differential strengths allow for different levels of transcription, rather than for strict timing.

Two interesting features seem to be common among most of the phages of this category:

1 The genes belonging to a given class of transcription, and thus involved in the functions necessary at a given time of the cycle, are

usually grouped on the genomic map of the phage (Figs 4.8 & 4.9), even if they are not cotranscribed.

2   The immediate early genes are frequently transcribed on one strand of DNA, while all others are transcribed on the other strand (Table 4.3). T7 and the related phages (T3, ϕXII) are exceptions in that only one strand is used as template.

The grouping of genes required at similar phases together on the limited space of these genomes is aided by the frequent occurrence of overlapping open reading frames (ORF), first discovered in T4. Such situations, also found in phages with small genomes (single-stranded DNA and RNA), are now known to be fairly frequent, at least among all kinds of extrachromosomal DNAs.

Although phages do not usually code for the synthesis of the precursors of their macromolecules, the elimination of the host genome (leading to the impossibility of synthesizing new molecules of host mRNAs) does not limit phage metabolism but indeed contributes to the provision of necessary base material. Ample precursor-synthesizing capacities are present in the cells. The ribosomes, tRNAs and enzymes are stable enough not to need fast renewal, and permanent RNA and protein turnover also furnishes a pool of precursors. The genomes of large phages, such as T4, are 1/10 to 1/20 that of the *E. coli* genome, and their capsids made up of a total of 1500–2000 proteins. The mere degradation of the host chromosome(s) and mRNA molecules present at the time of infection can account for 20–40% of the precursors necessary for phage DNA multiplication, assuming a yield of 200 particles per infected cell. The *c.* $3-5 \times 10^5$ proteins representing all molecules needed for a phage burst amount to no more than 10% of those forming a host cell, not taking into account the pool of free amino acids.

4.4.1.3 Transcription during the lytic cycle of temperate phages The events taking place during the lytic cycle of this category of phages are only well known in the cases of a few phages, such as λ, Mu and P1 of *E. coli*. The mechanisms involved include very elaborate transcription regulatory controls, aiming at equilibrating the lytic versus lysogenic responses, as well as providing appropriate timing of expression as seen for virulent phages.

The λ model. λ DNA is linear in the phage head, is injected linearly and circularizes upon entry, thanks to the presence of complementary cohesive (*cos*) ends (Fig. 4.10a). Ligation into a closed circular form is probably performed by the host ligase (it is not established whether λ codes for a ligase).

A simplified map of the genome is given in Fig. 4.10a. It consists of

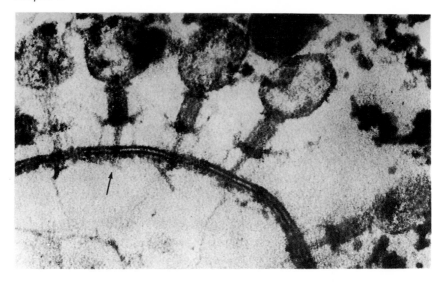

(a)(i)

(ii)   **Fixation**   **DNA Injection**

(b)

Fig. 4.8 (*Above and opposite.*) T-even phage reproduction characteristics. (a) Adsorption of T2. (i) Electron micrograph (from Freifelder, 1983); (ii) diagram to show adsorption. (b) Genetic map and transcription pattern of T4. (c) Control of transcription during development. (d) Replication. (e) Maturation (adapted from Girard & Hirth, 1989).

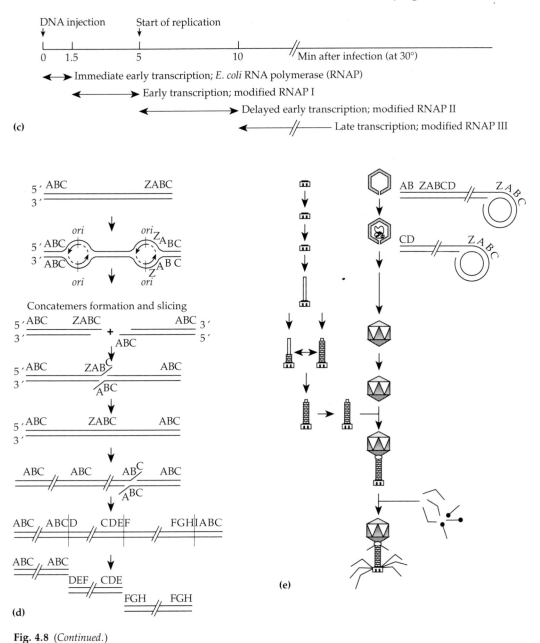

Fig. 4.8 (*Continued.*)

several main clusters of functions. The immediate early genes include two, *N* and *cI*, which code for regulatory proteins. *N* is an anti-terminator of transcription (Chapter 15), and is necessary to allow a read-through of transcription into the next group of genes-delayed early genes. This group comprises *O* and *P* which control replication;

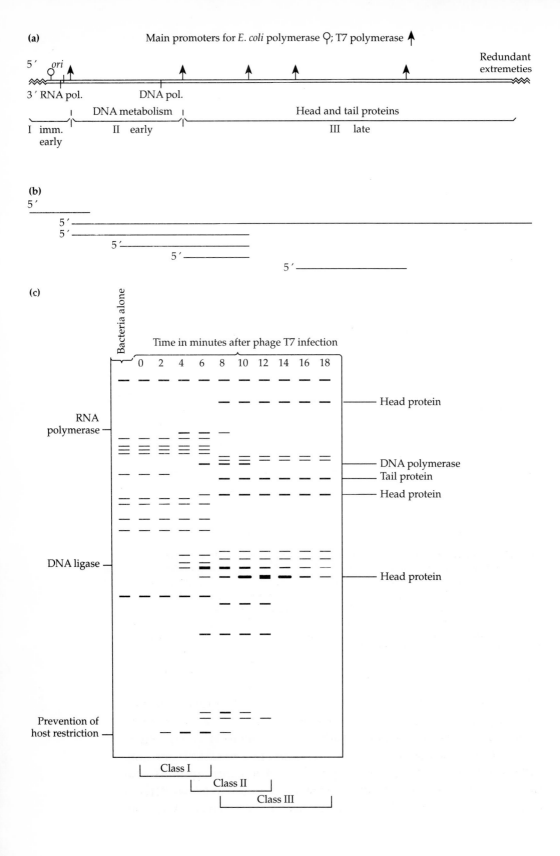

(a)   Main promoters for *E. coli* polymerase ♀; T7 polymerase ↑

5′ *ori* ↑   ↑   ↑   ↑   ↑   Redundant
                                                                                          extremeties
3′ RNA pol.   DNA pol.

        DNA metabolism   Head and tail proteins

I imm.   II   early   III   late
   early

(b)
5′

(c)

Bacteria alone

Time in minutes after phage T7 infection

0   2   4   6   8   10   12   14   16   18

RNA
polymerase

                                          Head protein

                                          DNA polymerase
                                          Tail protein

                                          Head protein

DNA ligase

                                          Head protein

Prevention of
host restriction

        Class I

            Class II

                Class III

(d)

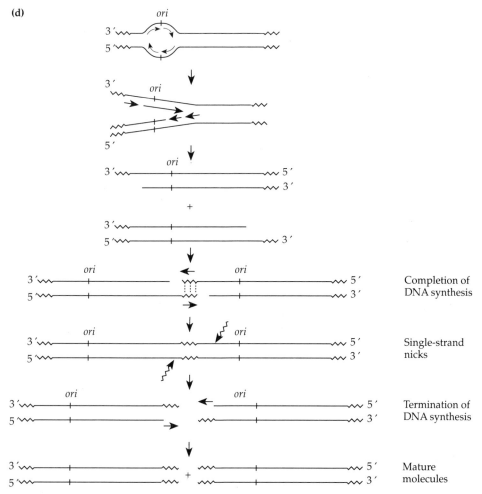

Fig. 4.9 (*Above and opposite.*) Phage T7 reproduction characteristics. (a) Genetic map. (b) Transcription pattern: main mRNAs synthesized. (c) Time after infection of appearance of various phage proteins. Crude preparations obtained by breaking open cells infected with T7, at different times (0–18 min) after infection, were centrifuged to remove large debris, and samples deposited on a polyacrylamide gel and electrophoresed. Proteins were then visualized by staining with Coomassie Blue. Many bacterial proteins are still made during the first minutes after infection (adapted from Dunn & Studier, 1983). (d) (*above*) Replication.

and *Q*, which is necessary for transcription of the late functions; and a divergent set of genes which covers excision and recombination functions. The other immediate early gene, *cI*, codes for a repressor, involved in the temperate response of the phage (Section 5). The late genes code for the capsid and tail proteins, the assembly proteins and cell lysis enzymes. A last group, located in the *N–cI* region, comprises the regulatory genes *cII* and *cro*, which are also involved in the

(a)

Mature DNA

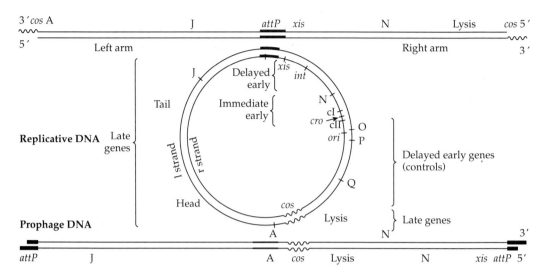

(b)

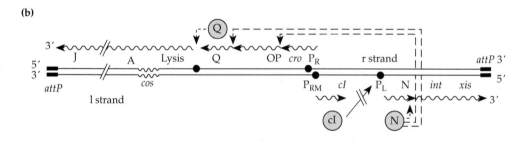

(c)

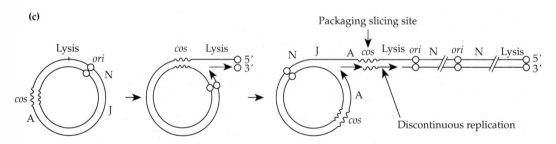

Fig. 4.10 Phage λ reproduction characteristics. (a) Genome maps. (b) Transcriptional controls (simplified). The lytic versus lysogenic evolutions. (c) Replication. The rolling-circle mode.

lysogenic response. A small site, *attP* (phage-coded attachment site), has the characteristic of being formed of two inversely oriented, symmetrical sequences. A corresponding region exists in the bacterial host chromosome, called the *attB* site. These are the sites at which crossing-over occurs, resulting in the integration of the phage genome into the host chromosome when a lysogenic state is established (Section 5).

Transcription is performed by the host polymerase according to a cascade of controls performed by the phage regulatory proteins N, O, P and Q (Table 4.3). A schematic representation of the actual events and levels of regulation is shown in Fig. 4.10b. The process starts with the expression of gene *N*, the product of which is necessary for the initiation of transcription of the *O*, *P* and *Q* genes. The O and P proteins will, in conjunction with cellular replicating enzymes, allow replication of the phage genome. The Q protein allows transcription of all the late functions.

The recombination and excision genes, as well as those controlling lysogeny, are not necessary during the lytic cycle.

4.4.1.4 Genome expression in the single-stranded RNA-containing phages
These phages raise other problems because of the chemical (RNA) and structural (single-stranded) nature and size (coding capacity for four proteins) of their genomes. Although a fairly large number have been isolated, they form a rather homogeneous group.

The RNA strand contained in the capsid is the direct template for translation (it is called the + strand) and is thus equivalent to mRNA, not needing any intermediary form of modification (Table 4.3). Transcription, the synthesis of new mRNA molecules, corresponds in fact to the synthesis of + RNA strands copied from the intermediary − ones. This is done by the same phage-coded enzyme, the replicase, that is responsible for the synthesis of the minus ones. Translation is performed by the host system. An example of an RNA phage is MS2, with a genome size of only 3.6 kb, coding for four proteins. The capacity to code for these four proteins results from maximal use of the genetic material, through overlapping genes (Fig. 4.11a). These proteins are a unique coat protein, a maturase, a cell wall-lysing protein and the replicase. An important gradient in the levels of translation of these proteins is observed during the development of the phage: 20 coat proteins are synthesized for five copies of the replicase and one maturase. This regulation is accounted for mainly by two processes:

1   Differential accessibility of the initiation codons is obtained through folding of the RNA molecule (Fig. 4.11b), the replicase and maturase coding regions being reached less frequently by the ribosomal complex.
2   Each of the proteins exerts an inhibitory action on the initiation

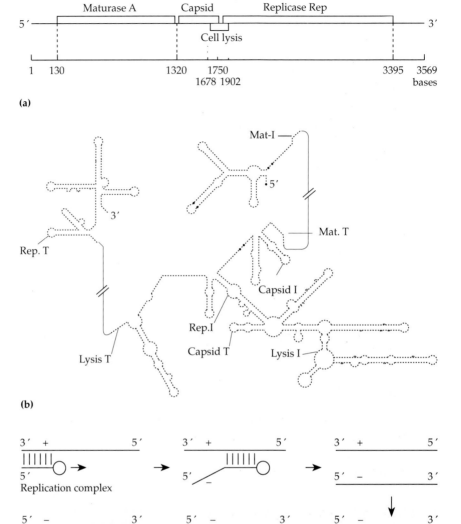

(a)

(b)

(c)

Fig. 4.11 (*Above and opposite.*) Phage MS2 reproduction characteristics. (a) Genetic map.
(b) Secondary structure of the RNA molecule during expression (hypothetical) Mat,
maturase; Rep, replicase; I and T indicate initiation and end of ORF (Watson, 1977).
(c) Replication process; (d) MS2 (1 cm = 250 nm) attached to the side of an F-pilus (a
filamentous fi phage is also attached to the tip of the pilus) (Hayes, 1968; courtesy of
L.G. Caro & D.P. Allison).

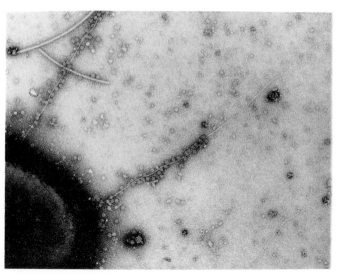

(d)

Fig. 4.11 (*Continued.*)

of translation of the others, thus enhancing the initial differences of transcription levels.

4.5 Genome replication

Vegetative replication of phage genomes starts before or simultaneously with the synthesis of the capsid proteins. The rate of replication increases rapidly, since the initial limiting factor is the number of available templates, which obviously soon become numerous.

4.5.1 *The DNA synthesizing mechanism*

This is essentially similar to that described for bacterial chromosomes, and uses the same or similar enzymes and procedures (an RNA primer, continuous and discontinuous synthesis on the complementary strands, etc.). Depending on the genetic complexity of the phage, one to several of the enzymes involved in replication are coded by the phage genome (Tables 4.4 and 4.5).

Different modes of replication are observed. They may be phage-specific and may show a correlation with the structure of the genome. Tables 4.4 and 4.5 summarize the various models known, with their principal specificities.

Many phages use the standard circular bidirectional replication mode (Chapter 2), starting from a single, fixed, origin (λ, P1, P22 from *Salmonella*). This mode is used only during the first rounds of replica-

Table 4.4 Modes of replication and enzymes involved of some *E. coli* virulent phages.

Phage	Replication mode(s)	Replicating enzymes coded by	
		The phage	The host
T4	Linear bidirect. Recombinant concatemers Several *ori* Rolling-circle?	12 directly involved (DNA polymerase) 12 indir. involved HMC synthesis, nucleases, etc.	None
T7	Linear bidirect. Recombinant concatemers 1 *ori*	Ligase Primase DNA polymerase Nucleases	DNA-binding protein Thioredoxin in DNA polymerase complex Ligase, PolI
φX174	Continuous circular for RFI and RFII several *ori* Rolling circle for ss concatemers 1 *ori*	Endonuclease RFII replication DNA maturation	At least 13
M13	Continuous circular for RFI and RFII Rolling circle for ss contatemers 1 *ori*	2 ss-DNA-binding proteins Endonuclease (initiation of rolling-circle replic.) Maturase (infection, init. replic.)	For RFI synthesis: DNA PolI and PolIII RNA polymerase Gyrase, ligase, ssb
MS2	Linear, unidirect. 1 *ori*	1 replicase	2 translation elongation factors 2 ribosomal proteins

tion, to provide template molecules for the next mode of replication used, the rolling-circle mode.

The rolling-circle mode necessitates double-stranded circular molecules as templates. A specific nick is created on one strand of the circle, at the *ori* region (Fig. 4.10c). The strand bearing the 5′-P extremity thus formed is liberated from its complementary circular strand and replicated in a discontinuous way. A complementary continuous synthesis starts simultaneously from the 3′-OH extremity, using the circular template liberated. This leads to the replacement of the double-stranded structure and allows unwinding of a continuous multimeric single strand, by a mechanism similar to the unwinding of a thread from a spool — hence the name given to the process. The linear double-stranded molecule initiated from the 5′-P extremity is thus synthesized as a long concatemer representing several (up to 100) phage genome units. Slicing down to unit sizes by a specific endonuclease takes place either during the synthesis process or simultaneously with the packaging of the DNA into the phage heads (Section

Table 4.5 Modes of replication and enzymes involved in some *E. coli* temperate phages.

Phage	Replication mode(s)	Replicating enzymes coded by: The phage	The host
λ phage	Circular, bidirectional Rolling circle 1 *ori*	2 regulatory proteins O, P	All other enzymes
Prophage	Integrates at unique site as a linear DNA, map permuted compared with phage map		All enzymes
Mu Phage	Integrates in host chromosome at multiple sites	2 control proteins Transposase A	All other enzymes
Prophage	Integrates at multiple (random) sites, as linear, non-permuted DNA		All enzymes
P1 Phage	Circular bidirect. Rolling circle *oriL*	All replicating and regulatory enzymes	All other enzymes
Prophage	Circular bi-direct. Autonomous *oriR*	Regulatory proteins Partitioning enzymes	All other enzymes

4.6). The nicking sites on the concatemer may be specific (λ, T1) or random (P22). In the latter case, the nicking sites are separated by lengths that are longer than the phage genome, a system which creates the terminal redundancies (Section 2.3) and which ensures the presence of at least a whole genome in each phage particle.

A variation of the rolling-circle model is also used by single-stranded DNA phages (Figs 4.6b & 4.7b). The DNA must first be modified to render it adequate for this type of replication. In fact, the entry of the single-stranded form seems to be dependent on the concomitant synthesis of its complementary strand. The phage maturase protein, also involved in fixation of the infectious particle to its receptor and in injection of the DNA, participates in the initiation of this synthesis. The synthesis itself is performed by the bacterial enzymic machinery. Differences appear among phages of this category as to the enzyme (RNA polymerase or primase) that initiates the process. Initiation starts at one (filamentous Ff phage group) or several (ϕX174) fixed origins on the molecule. Synthesis of the complementary strand takes place in a continuous process. Elimination of the RNA primer leads to the formation of a double-stranded structure showing a single-strand nick at the point of initiation (RFII form), converted to a covalently closed circle (RFI) by the host ligase. A unidirectional replication then takes place from a unique origin in a continuous way, from the 3'-OH extremity created as a result of a single-strand nick made on the initial + strand. The maturase protein is also involved in this initiation. A + single-stranded concatemer is formed, and maintained as such by simultaneous coating with single-strand-binding proteins coded by the phage. In fact, a few − strands are synthesized by a similar mechanism, and used either as additional templates for replication or as templates for transcription.

A linear bidirectional mode of replication is also encountered in some phages. The mechanism of replication is a classical one, reminiscent of replication of eukaryotic chromosomes. Either one (T7) or several (T4, P1) fixed origins are present on the parental molecule (Figs 4.8d & 4.9d). This type of replication applies only for phages whose mature genome possesses terminal redundancies. During the process of replication, genome units are associated and ligated together into dimers (T7) or concatemers (T4), by hybridization of the terminal redundancies. These structures play a double role: they allow the filling up of the single-strand 3′ extremities, which could not otherwise be replicated for lack of a primer from which the DNA polymerase could start, and they ensure efficient packaging, by providing molecules from which fragments longer than the genome unit length can be cut down. Also, they provide a means for genetic cross-over to occur at the same time as replication, forming the 'mating pool'. Cutting of the concatemers, which takes place during the packaging process (Section 4.6), is performed either at specific sites, forming uniform molecules (T7) or randomly, yielding molecules with circularly permuted, redundant extremities (T4 (Fig. 4.8d), P1).

Although formation of the progeny T4 genomes mostly results from linear replication, the phage probably also replicates through a rolling-circle mechanism (at the end of its cycle), which also yields concatenate structures. Replication of T4 DNA is totally independent of host replicating enzymes.

4.5.2 *Replication of RNA genomes*

A common mode of replication seems to be used by all these phages (Fig. 4.11d). The infecting + strand is copied into a complementary − RNA strand, which in turn serves as template for the synthesis of new mature + units. Only single-stranded molecules are ever found in host cells. Since double-stranded intermediate structures must exist, it has been postulated that these complexes are very unstable and dissociate rapidly to the single-strand units.

RNA phages code for their replicase, or RNA-dependent RNA polymerase, one of the four proteins necessary for their replication, the others being proteins involved in the *E. coli* translation complex (Table

4.4). The mechanism of initiation of synthesis differs from that of a classical DNA-dependent RNA polymerase (that used for bacterial transcription) since it does not require a promoter-like sequence to start from. The specificity of recognition of the template by the enzyme complex is probably ensured by the elaborate secondary structure of the RNA molecule (Fig. 4.11c). The same enzyme is used to make the new genome (+) strands and the intermediary – strands.

It is interesting to note that similar strategies are used by a number of animal single-stranded RNA viruses, with the exception of the retroviruses, in which a reverse transcriptase (an RNA-dependent DNA polymerase) copies the infecting + strand into a double-stranded DNA molecule, which is itself used as a template for either transcription or replication into single-stranded RNA molecules, depending on which strand is copied.

4.6 Phage assembly and maturation

All the constituents of the mature phage particles start being synthesized during the second half of the latent period, and continue to be accumulated from then on in the host cell until cell lysis occurs. As soon as a large enough number of the capsid proteins and genome molecules are made, assembly of mature particles starts and will cease only with cell lysis. Two main assembly processes are known, in rough correlation with the genetic and morphological complexity of the phage.

4.6.1 *Assisted self-assembly*

This process concerns all known double-stranded DNA phages and also, although by a different mechanism, the single-stranded DNA phages with icosahedral capsids.

The process is very well known in two cases of double-stranded DNA phages, namely λ and T4. Analysis of the structures obtained from numerous morphology mutants, in parallel with the genetic determination of the loci involved and characterization of the proteins formed, has allowed a very thorough understanding of the sequence of steps leading to the formation of the mature particle. Both of these phages possess a head and a tail. These two substructures are assembled separately, as a result of several precisely sequenced events (Fig. 4.8e). Preheads, with a slightly different size and structure from the mature ones, are formed first. Their transformation into the final form, which involves modification of some of the constituent proteins, is performed concomitantly with the entry of the DNA molecule. A number of phage-encoded (and possibly host-encoded) proteins are responsible for the efficiency of the process, but are not found as such in the mature particles. They play the role of assisting proteins. At least some of them do not seem to be absolutely required to ensure total assembly.

The process of DNA entry is still poorly understood. The DNA is probably pulled into the head of the phage, in which it is folded into a structure reminiscent of that of bacterial nucleoids. This probably necessitates the presence of basic proteins, although their existence has not been ascertained. Since in most cases the DNA is synthesized as concatemers, it needs cutting into correct unit lengths. At least for phages encapsulating genomes with redundant extremities, it is usually assumed (this has been demonstrated in mutants making larger capsids) that at least one element controlling the cutting is the maximal amount (the headful) of DNA that can be contained inside the heads. Similarly, a minimal length of DNA is required to ensure packaging and to prevent the mature particles from collapsing when liberated from the cell.

The DNA is cut by a specific endonuclease. The cutting sites may be variable compared with the genetic map of the phage (this is the case with T4 and P1, which encapsulate more than a genome unit (Fig. 4.8e)), or precise and fixed, as in the case of T7 (Fig. 4.9d) and λ (Fig. 4.10c). The DNA of phage λ is known to be packaged in only one orientation, a fact which means that there exists a recognition mechanism which directs the early steps of packaging. Its DNA is packaged into precisely cut genome-size units through the action of a specific nuclease which recognizes and cleaves the *cos* regions. The last step of the maturation, the closing up of the head, consists in the joining-on of the preformed tails.

The mechanisms of these processes, associated with the physiological characteristics of the phages, allow the possibility in some cases of encapsulating non-specific DNA, assuming that the necessary requirements of size (P1) or size plus presence of *cos* ends (λ) are fulfilled. Such events correspond to natural (transduction (Chapter 12)) or artificial (gene cloning (Chapter 16)) transfer of genetic information.

The packaging of φX174 DNA is concomitant to the cutting and circularization of the newly formed + strands (Fig. 4.6b). The specific phage protein, the maturase, associates with the extremities of the DNA units and ensures its circularization through a mechanism that is poorly understood. A number of phage and bacterial proteins liberate the DNA from its protecting single-strand-binding (ssb) proteins, and a phage protein encapsulates it by folding into the capsid, where it also remains. The maturase remains associated with the capsid and will serve as the recognition element for further infection.

There is now plenty of evidence that, besides those involved in the folding of the DNA, a few other proteins are included inside the heads of a number of phages. These proteins are also injected together with the DNA during the infection process. Some of them have well-known functions involved with the early steps of infection.

4.6.2 *Genome-coating process*

An apparently simpler mode of maturation is found in the case of RNA and filamentous single-stranded DNA phages.

The single-strand-binding proteins protecting the single-stranded DNA genomes are displaced by the proteins of the capsid, accumulated in the cell membrane. The formation of the mature capsid is concomitant with its extrusion through the cell membrane and envelope. Since these phages do not produce cell-wall-lysing enzymes, lysis will eventually occur passively, probably by fragilization of the cellular envelopes after a large number of particles have been ejected. This happens at a rather low frequency, since an infected population can undergo several rounds of division before it is killed. It is unknown how the concatemer precursors of the genomes are cut into unit lengths and circularized before coating. Since it is the length of the DNA molecules which directs the size of the capsid, addition of DNA into the genome will result in successful packaging. This property accounts for the use of M13, a member of this group, as a cloning vector (Chapter 16).

RNA-containing phages use a somewhat similar process for their maturation. The coat proteins bind to the RNA + strand and associate into a capsid around it. A molecule of maturase, which is necessary for the early steps of infection, is packaged with the RNA and injected with it afterwards. The release of the phages, a poorly understood process, does not involve the attachment of capsid proteins to the cell membrane.

4.7 Cell lysing functions and phage release

4.7.1 *Production of lytic enzymes*

Large phages, such as those of the T series or λ, code for one or several lytic enzymes that are capable of degrading the cell wall. Transcription of the corresponding genes — and thus synthesis of the enzymes — is precisely timed in the course of the infection cycle (Section 4.4). It usually starts after transcription of the other late genes has been initiated and then takes place continuously. Smaller phages such as φX174 also code for a lytic enzyme, although with less elaborate temporal regulation.

4.7.2 *Cell lysis*

Most lytic enzymes fall into the lysozyme category. They specifically attack the peptidoglycan constituting the cell wall, producing pores which fragilize the wall up to its rupture. The φX174 enzyme, however, interferes with the synthesis of the peptidoglycan, a process which causes lysis of all cells in the process of division. It is the accumulation of these enzymes which leads to effective lysis of the cells, an event which obviously terminates the phage cycle. Since expression of the late phage functions is never turned off, more DNA molecules and capsid proteins are usually made than can be assembled before lysis occurs.

The absence of precise regulation between the synthesis of the phage constituents and the onset of expression of lytic enzymes has been shown through artificial modifications of the time of lysis and measurement of the amount of phage particles assembled when lysis occurs. Premature lysis was obtained by means such as osmotic shock or the presence of mutations resulting in early expression of the lytic genes. Lysis could be retarded either by maintaining the cells under a higher osmotic pressure or by the introduction of phage mutations leading to the synthesis of deficient lytic enzymes. The mean number of phage particles released per cell would then be correspondingly decreased or increased. The *r* (for rapid lysis) locus of T-even phages, although not directly active in the lysing process, modifies the time of lysis and hence the burst size, depending on the allelic form present (Fig. 4.4b).

The latent period (Fig. 4.5), the start of which was defined as immediately following the adsorption step, ends at the lysis of the cell.

4.7.3 *Phage release*

The burst of each individual cell is a very rapid process. This is usually true also for the whole population, when infections are sufficiently synchronous.

In laboratory conditions, when a whole suspension is simultaneously infected, a rather synchronous evolution of the phage cycle occurs. Complete lysis of the suspension and thus an increase of the number of free infective particles take place in a very short span of time compared with the duration of the latent period or of the host generation time. Thus the increase in free phage particles in the external medium must not be considered as the measure of phage growth in the sense of cell growth. There is no *sensu stricto* phage growth, and in fact phage multiplication has already stopped when the increase in free phage particles is observed.

If only part of the cell population is infected in the first place, each phage particle released after the first cycle of infection will be able to infect a new cell and give rise to a new burst. These successive rounds of infection will take place until all cells have been lysed. Such a situation is referred to as multiple-step infections. Although theoretically synchronous, these cycles will in fact soon appear to be random, due to delays in efficient adsorptions and variations of individual lysis times reflecting differences in the physiological states of the infected cells. So, at the population level, formation of free phage particles may look like a continuous process.

4.8 Yields of production and the fluctuation test

Mean yields are reproducible for a given phage under similar conditions, but vary widely among different phages, even when infecting the same bacterial species under optimal physiological conditions (Table 4.2). No correlation exists between the yield and the natural duration of the latent period.

Individual yields among a whole population of infected bacteria show important discrepancies. This may be partly explained by differences in the physiological state of the cells at the time of infection. Individual productions have been measured in 'single-burst' experiments. A suspension of *E. coli* was infected with T4 under standard conditions at an moi of one. After a few minutes, which were sufficient to reach maximum adsorption, the suspension was diluted and distributed so as to obtain a mean of one cell per sample. The infection cycles were then continued to completion, and the total amount of phage particles released in each sample was determined. The values obtained for approximately 50 samples ranged from 30 to 400, with a mean value of 200, which agreed with that obtained by standard determinations.

This method can be applied to any measure of individual versus average productions. It is known as the fluctuation test (Chapter 7). It also allows calculation of standard variations.

It is difficult to measure yields in the case of RNA or filamentous phages. Since these phages do not disturb the cell metabolism profoundly, divisions of the infected cells continue for several generations, although at a slower rate. Phage concentrations can be increased by factors of 10^3 at the time when a large fraction of the cell population is eventually killed.

5 Lysogenization

As already defined, lysogenization is the capacity for a temperate phage to maintain its presence in a cell and its progeny without promoting their death and without producing free infectious particles. It is a latent infectious state. The structure under which the phage is maintained, called the prophage, consists of one copy (usually per host chromosome) of the phage genome in a stabilized form. The cell harbouring a prophage is lysogenic and called a lysogen, since, as we shall see, it can give rise to lysis and release of mature phages. Lysogenization refers to the establishment of the phage as a prophage in the host cell. The prophage is usually integrated into the host cell's chromosome, but in some cases may remain in the cytoplasm in a plasmid form (Section 5.2.4).

These phages are usually also capable of going through classical lytic cycles (otherwise their existence would have remained unknown). However, the presence of defective phages, stabilized in the chromosome as prophages but having lost by mutation the capacity to reproduce by a lytic cycle, has been postulated. Several such genetic structures would be present in the *E. coli* chromosome. Phages P2 and P4, which can complement each other for lytic and lysogenic functions, constitute a particular case of such partly deficient temperate phages. They will be described below (Section 5.2.2).

5.1 The general scheme

5.1.1 *Requirements for lysogenization*

The ability of a temperate phage to establish lysogeny upon infection of its host carries two implications:
1   Expression of phage functions which would normally lead to the development of the lytic cycle must be prevented.
2   Degradation of cell material or irreversible arrest of host metabolism must not occur.
Both requirements are met through the functioning of a regulatory mechanism which blocks expression of most phage functions, as in an operon (Chapter 15).

Temperate phages are always large phages, with genomes of at least 10s of kbp. This would be expected since, besides their lytic functions, they harbour a set of genes responsible for the establishment and release of lysogeny. These genes code for a series of regulatory proteins, which usually function in a regulatory cascade, that is, in a strictly hierarchical order, affecting transcription of the other phage functions. Inhibition of protein synthesis, imposed, at either transcriptional or translational level, immediately upon induction of the lytic development, affects transcription of the first set of genes which, if expressed, would initiate the lytic cycle. This results in non-expression of a second set of regulatory genes whose products are necessary for the transcription of all further functions. The whole pathway thus includes both negatively and positively acting regulatory proteins. Similarly, the establishment of the lysogenic state depends on the equilibrium between the positive and negative effects of several genes.

To maintain the lysogenic state, the initial inhibitory step must function permanently in the cell. Both the onset of inhibition of the lytic cycle of development and its maintenance are achieved by the continuous synthesis of a single regulatory protein responsible for this initial inhibition. This protein, called the repressor, is coded by the phage genome. The gene is expressed immediately upon injection of the phage DNA into the host. The mechanism of action of the repressor is a classical one: by fixation to the operator (regulator region) of the first (set of) gene(s) that should be transcribed to start the lytic cycle, it prevents, or competes with, the binding of the RNA polymerase on the promoter(s) (i.e. the site of fixation of the RNA polymerase σ subunit, the initial event in transcription (Chapter 15)) of this (these) gene(s).

A single mutation occurring in the repressor gene of a temperate phage can result in the production of a non-functional molecule, and hence in the loss of the temperate character of the phage. Modifications of the operator(s) concerned may also lead to a similar phenotype. Such mutants behave like virulent phages, even if all other functions implicated in lysogeny are present. A particular class of mutants of the repressor, the thermosensitive (ts) ones, have been isolated from the *E. coli* phages λ, Mu and P1 and from phages from other bacteria, such as *B. subtilis* and *Streptomyces* spp. They have proved very convenient for further molecular and genetic analysis of the lysogenic system and as tools for temperature-controlled expression of cloned genes (Chapter 16). The repressor of these ts mutants keeps its normal activity at low temperatures (e.g. 30°C for *E. coli*), while becoming inactive at higher temperatures (in this case around 40°C) which still allow the host to metabolize. The lytic cycle (or the expression of any gene whose transcription would be controlled by this repressor) can be induced upon transfer of the lysogenic cells to the higher temperature.

The description of specific examples (Section 5.2) will demonstrate the existence of different strategies of lysogenization and consequent genetic organizations of the lysogenic states.

5.1.2 *Probability of establishment of lysogeny*

While synthesis of the repressor prevents transcription of all other phage genes, once expression of the early lytic genes has started it prevents the inhibitory action of the repressor.

The probability of establishment of the lysogenic state thus reflects a competition between the expression of two opposite processes, both of which can take place immediately upon entry of the phage genome into the host cell. Which process is favoured for any particular phage–host association depends upon the physiological state of the cell at the time of infection. Slowing down cellular metabolism results in an increased frequency of lysogenization.

In *E. coli*, phage λ enters the prophage state in one out of 10^3–10^4 infected cells, under optimal cell-growth conditions. This figure can be shifted up or down by several units on a log scale upon action of particular physiological conditions or by the introduction of phage or bacterial mutations.

5.1.3 *Requirements for the maintenance of the lysogenic state*

Stability of the lysogenic state depends not only on the continuous production, in sufficient amounts, of repressor molecules, but also on their correct partitioning to the daughter cells at each division. Usual concentrations of repressor molecules are a few copies per cell or per genome. The excess is thus limited. Fluctuations in the rate of synthesis or unequal partitioning will result in the release of repression and hence in the onset of the lytic cycle. This process is called spontaneous induction.

Under normal conditions, the spontaneous frequencies of induction of the lytic cycle vary with the host and the prophage. Spontaneous induction of prophage λ occurs at a frequency of 10^{-5}–10^{-6} per cell per generation. Total induction can be reached upon treatment by inducing agents, such as radiations or certain mutagens.

A second prerequisite to ensure maintenance of lysogeny in a cell line is the permanent transmission of the prophage genome itself to the progeny cells. Several strategies ensuring this transmission exist, depending on the phages (Section 5.2).

5.1.4 *Specifity of immunity to superinfection*

A consequence of the presence of an excess of repressor molecules inside a lysogenic cell is the inhibition of expression of any other copy of the phage genome happening to be present in the cell. A stabilized lysogenic cell is thus insensitive to superinfection (infection of a cell by a phage of the same kind). Note that this is not due to inhibition of adsorption or DNA injection by the new phage. This phenomenon is called immunity, by analogy with the equivalent process known in eukaryotic organisms. The cell has acquired, through the presence of the prophage, a preestablished capacity to fight a new homologous infection.

This mechanism was demonstrated by superinfecting a lysogenic cell with a mutant of the initial phage modified in the operator target of the repressor. This second phage (recognizable from the first by its plaque morphology, for instance) will perform a normal lytic cycle and thus appear to be insensitive to the immunity process.

Potentially, superinfection with a large enough number of phages could overcome immunity, because there could be too few repressor molecules. High mois, however, favour establishment of lysogeny, the superinfecting phages themselves producing more repressor.

As for any immunity system, protection by the presence of the repressor is specific for phage genomes bearing a target region that is recognizable by the resident repressor. A secondary infection by any unrelated phage will end in the development of an infectious cycle, or in the establishment of a second state of lysogeny if the second phage is also temperate. Typically, only the second type of phage will be produced as a result of this lytic cycle, since only this phage's metabolism can be expressed.

The specificity of the immunity was also ascertained when super-infections were performed with recombinant phages bearing unrelated repressor target regions. Phage 434 is a temperate phage of *E. coli* K12, closely related to λ but differing in its repressor target region and its repressor molecule. No cross-immunity exists between the two phages. If the target region and repressor gene of phage λ are replaced by those of phage 434, through natural recombination or genetic manipulation, the new λ lysogens will appear to be immune to superinfection by phage 434 but not by λ.

Some related phages have retained enough homology in both their repressor structures and their target regions for cross-immunity to occur.

5.1.5 *Visualization of lysogeny*

The proportion of lysogenized cells resulting from an infection with a temperate phage is usually very low, so, if the infection is performed in liquid conditions, the amount of cells that survive the infection cannot be immediately visualized. Since, however, these cells become immune to superinfection by the progeny particles liberated from successive infections of the other cells, they will continue to divide and will outgrow the initial suspension after a delay.

A similar apparent behaviour results from the presence of cells resistant to the input phage (Section 4.2). Such mutations usually occur at lower frequencies than lysogenization and so take longer to outgrow the population.

When cells infected with a temperate phage (infectious centres) are immediately plated on to a lawn of indicator bacteria, plaques will be formed by the successive cycles of infection occurring after liberation of mature particles (Section 3.2). A small proportion of the infected bacteria will, however, be lysogenized during these successive cycles. These few cells will survive and divide at the site of their initial plating. The final plaque will thus appear turbid, as compared with clear plaques resulting from an infection by a virulent phage (or by a virulent mutant of a lysogenic phage, called a clear, or c, mutant). Surviving cells may be picked up from the centre of the turbid plaque and used as lysogenic clones.

5.2 Model temperate bacteriophages

5.2.1 *Bacteriophage λ of* E. coli *K12*

The existence of a process through which phage production could occur in some bacteria although when disrupted they did not release such particles was first mentioned in the late 1920s by Burnet and McKie. The λ system was the first one to be thoroughly studied (Lwoff, 1950s) so as to enable the notion of lysogeny to be proposed. It is now very well understood, thanks in particular to the analysis of mutants and, more recently, to molecular analysis. Isolation of the repressor has promoted this system as a model for the understanding of interactions between DNA and regulatory proteins which bind to specific sites on DNA.

An important characteristic of the lytic cycle of λ (Section 4.4, Fig. 4.10b and Table 4.3) is the organization of its regulation as a cascade. This allows an efficient timing of expression of the functions involved in the lytic response. It also controls the mechanism by which infection will result in either the lytic or the lysogenic cycle. On infection of a cell by λ, only genes *N* and *cro* (Fig. 4.10b) are switched on. The gene product N acts as an antiterminator of transcription and hence allows the expression of *cII*, situated downstream of *N* on the genetic map. The cII protein is sensitive to bacterial proteases, but, if a critical amount remains, with the help of the *cIII* gene product (provided the cell metabolism is slowed, e.g. growth in a poor medium) the action of the CII protein will switch on *cI* repressor gene expression. This switches off *cro* expression. Lysogeny will result. However, if the cII protein is reduced in concentration by the proteases, then the repressor concentration does not build up, but instead the *cro* gene product accumulates. This both prevents the expression of *cI* and allows the expression of *N*, which then stimulates the genes associated with lytic development.

During the establishment of lysogeny there is a transient expression of the *int* gene, coding for an integrase (also called transposase) that catalyses the integration of the infecting, circularized phage genome into the host chromosome. The orientation and the site of insertion of the phage genome are determined unequivocally and uniquely by recognition and pairing of the bacterial *attB* and phage *attP* regions (Fig. 4.10a). The resulting structure shows that previously adjacent genes (*xis* and *J*, a tail-protein gene) are now separated while, on the other hand, genes flanking the *cos* ends are now joined in the central part of the integrated prophage map. The mechanism of integration, which involves very small homologous DNA regions and does not require the host recombination system (Chapter 9), is that of site-specific recombination processes (Chapter 5).

Replication of the prophage is passive, as an integrated part of the host chromosome.

Release of repression by the cI protein, induction, can occur as a result of several different events. Unequal dilution of the few cI protein molecules at cell division has a certain probability of resulting in a daughter cell devoid of repressor for a duration long enough to allow expression of *N* and *cro*. A particular situation leading to induction, referred to as zygotic induction, is met when an Hfr male donor *E. coli* cell lysogenic for λ transfers its chromosome, and hence the integrated λ prophage, into a recipient non-lysogenic female cell in the course of a conjugation (Chapter 11). Since the latter cell does not possess cI proteins, competition between expressions of the *cro* and *cI* genes of the entering prophage usually ends in favour of *cro* and leads to expression of a lytic cycle and death of the F-recipient cell. Inactivation of cI, which also results in induction, will also take place as a result of specific proteolysis by the recombination-involved protein RecA, a protease which is activated by occurrence of lesions damaging the normal structure of the DNA and requiring repair (Chapter 8). Classical inducing agents are ultraviolet light and certain mutagenic agents. This process of induction by destruction of the repressor molecule by RecA is unique to phage λ.

One of the early phage proteins made upon induction, Xis, a specific endonuclease, excises the prophage genome from the host chromosome, by a specific cutting exactly symmetrical to the integration process. Circularization of the phage DNA by resealing of the two *attP* sequences thus liberated, achieved by the integrase coded by gene *int*, leads to a genomic structure identical to that formed upon infection by free phages. In particular, autonomous replication, controlled by genes *O* and *P*, which is possible only on the circular molecule (Section 4.5.1), can now take place and the lytic cycle can proceed.

Rare cases of imprecise excision occur. They result in the linkage to the progeny phage genome of host DNA, which will be transferred to another bacterium upon secondary infection. This process has been named specialized transduction (Chapter 12).

5.2.2 *Phages P2 and P4*

Phages P2 and P4 form a peculiar pair. Phage P2 can either provoke a lytic cycle or lysogenize *E. coli*, according to a scheme similar to that of λ, but its prophage cannot be induced by artificial means, because it lacks the regulatory proteins necessary to initiate the induction. Infection of *E. coli* by P4 only leads to lysogeny, because this phage does not possess structural genes for capsid proteins and for the capacity to activate transcription of its early genes. In doubly infected cells, the two phages can complement each other, since they lack different *trans*-acting proteins. The outcome of the infection by P2 functioning as a helper of an *E. coli* lysogen for P4 is the development of a lytic cycle and the production of the two phages, the P4 genomes being coated by capsids made of P2 proteins but smaller than the P2 capsids. A mixed infection or the infection of a P2 lysogen by P4 produces only P4 particles.

5.2.3 *Phage Mu*

Phage Mu is described in more detail in the chapter describing transposons (Chapter 5). It is in fact a bivalent element. It codes for capsid proteins and cell-lysing enzymes, which are synthesized when the lytic cycle is expressed, functions which allow the formation of mature infectious particles. It also behaves as a transposon, i.e. a DNA sequence capable of autonomous transposition (Chapter 5), in its chromosomal integration and vegetative replication properties, and possesses the corresponding structure and functions (Table 4.5). Its compulsory integration into the host chromosome as part of its replication process seems to be responsible for the presence of the two stretches of host DNA, from random regions of the host chromosome, forming the extremities of the viral genomes. It is also a temperate phage.

As a transposon, Mu integrates at (almost) random locations on the host chromosome, its insertion often resulting in mutations of the target genes. As a prophage, it is normally not inducible, so that the mutants produced are stable.

A peculiarity of Mu, also present in phage P1, is the possibility for a short region of its genome, G, to be present in either orientation. G bears two sets of genes which code for tail fibres involved in host-receptor recognition, readable in opposite directions but from a promoter located outside the invertible fragment. This results in variations of the host-range of the phage, switching from *E. coli* K12 (the + orientation) to *Citrobacter*, *Serratia* and *E. coli* C (the − orientation). The inversion of G is performed by a phage-coded invertase, expressed at a low level.

5.2.4 *Phage P1*

Phage P1 has many features similar to T4 (size, tail structure, regulation of the lytic cycle, packaging mechanism), although it uses the rolling-circle mode for vegetative replication and induces only limited degradation of host DNA (Tables 4.3 & 4.5). The latter feature confers transducing capacities (Chapter 12) on P1.

It can be stabilized as a prophage as a consequence of the expression of a repressor, in a way similar to that of λ. A main difference, however, is that its DNA does not insert into the host chromosome but is replicated and maintained as an autonomous, low-copy-number plasmid (Table 4.5). As such, it shows incompatibility functions against related DNAs (Chapter 3).

References

Burnet F.M. & McKie M. (1929) Observations on a permanently lysogenic strain of *B. enteritidis. Australian Journal of Experimental Biology and Medical Science*, 6, 277.

Doerman A.H. (1952) The intracellular growth of bacteriophages. I – Liberation of intracellular bacteriophage T4 by premature lysis with another phage or with cyanide. *Journal of General Physiology*, 35, 645–6.

Dunn J.J. & Studier F.W. (1983) *Journal of Molecular Biology*, 166, 447.

Eiserling F.A. (1983) Structure of the T4 virion. In: Mathews C.K. Kutter, E.M. Mosig G. & Berget P.B. (eds) *Bacteriophage T4*, pp. 11–24. American Society for Microbiology, Washington D.C.

Ellis E.L. & Delbrück M. (1939) The growth of bacteriophage. *Journal of General Physiology*, 22, 365–84.

Freifelder D. (1983) *Molecular Biology, a Comprehensive Introduction to Prokaryotes and Eukaryotes*, pp. 601, 627. Jones and Bartlett, Inc., Boston.

Hayes W. (1968) *The Genetics of Bacteria and Bacteriophages*, 2nd edn, Blackwell Scientific Publications, Oxford.

Hershey A.D. & Chase M. (1952) Independent functions of viral proteins and nucleic acid in growth of bacteriophage. *Journal of General Physiology*, 36, 39–56.

Lwoff A. (1959) Bacteriophage as a model of post-virus relationship. In: Burnet F.M. & Stanley W.M. (eds) *The Viruses*, Vol. 2, p. 187. Academic Press, New York.

Girard M. & Hirth L. (1989) *Virologie Moléculaire*. Doin, Paris.

Twort F.W. & d'Hèrelle F. (1915) Discussion on the bacteriophage (bacteriolysin). *British Medical Journal*, 2, 289–97.

Watson J.D. (1977) *Molecular Biology of the Gene*, 3rd edn, p. 540. Benjamin, Inc., Menlo Park, USA.

5: Transposable Elements

The first discovery of transposable elements in bacteria was in 1968. Interest in these genetic units increased when, in 1974, transposon-mediated antibiotic resistances were discovered and assigned an important role in the dissemination of antibiotic resistances among pathogenic bacteria. Similarities then became more and more obvious between bacterial transposons and mobile elements from eukaryotes, be they the oncogenic retroviruses or the maize transposons discovered some 40 years earlier by Barbara McClintock, who was awarded the Nobel Prize in 1981 for this discovery.

This increasing interest in transposable elements has been mainly directed towards two, non-exclusive, problems: the study of the mechanism(s) of transposition and the utilization of transposons for genetic purposes.

The larger number of such elements has been described in Gram-negative bacteria (and in their viruses and plasmids) and these have therefore become the best known. Although discovered more recently, transposable elements of Gram-positive bacteria have been shown to comprise many elements similar to their Gram-negative counterparts. Over a dozen insertion sequences have also been described, still more recently, in archaebacteria.

Bacterial transposable elements were originally regarded as forming two classes, the insertion sequences and the transposons. Analysis of their structures and the mechanisms of transposition has now led to the acceptance of a different classification.

1 Symbols and nomenclature

The large number of transposable elements described has necessitated an international agreement as to their nomenclature.

Insertion sequences are designated by the abbreviation IS followed by a number, internationally recognized, that roughly represents order of discovery, e.g. IS1, while transposons are designated by Tn, also followed by a number. In the latter case, phenotypic indication of the accessory characters they bear is denoted by using the same nomenclature as for plasmid markers, i.e. a capital letter followed by a lower case letter, e.g. Km for kanamycin resistance (Chapter 3, Table 3.6). In fact, many resistance determinants harboured by plasmids are indeed borne by transposons present in these plasmids.

The presence of any of these insertion elements in a DNA molecule is denoted by :: followed by the name of the element and preceded by what is known of the localization of the insertion; thus *galX*::IS1 means that IS1 is inserted in gene *galX*; RP4::Tn5 refers to a copy of transposon Tn5 inserted in an unknown or genetically non-determined position in plasmid RP4, whereas RP4*tet*::Tn5 means that this transposon is inserted in the *tet* gene of the plasmid.

When dealing with several different insertions of the same transposon in a chromosome that is not genetically well known, or if the insertion sites are not genetically determined, it is useful to number these insertions. The Greek letter Ω, followed by an order number, is then generally utilized.

2 Transposable elements in Gram-negative bacteria

2.1 Insertion sequences

2.1.1 *Discovery*

Bacterial insertion sequences were discovered in the late 1960s in the course of studies on the molecular basis of gene expression of the lactose and galactose operons (Chapter 15) of *E. coli* and of bacteriophage λ genes. They were identified as additional segments of DNA causing highly polar (Chapter 7) and somewhat unstable mutations. The existence of several types of such insertions was demonstrated after analysis of the structure of heteroduplex molecules generated from independent mutations occurring in the same genes, e.g. the *gal* operon, which could be isolated on transducing particles from bacteriophage λ (Chapter 12) (Fig. 5.1): each double-stranded DNA molecule was 'melted' into its single-stranded components (usually through heating), and the strands purified by virtue of their different densities (a feature of the two strands of λ DNA, referred to as + and −, or l and h, strands). Mixing appropriate single strands (one + and one −) from two different IS-containing isolates under appropriate conditions (i.e. slow decrease of the temperature) allowed formation of heteroduplex, double-stranded structures resulting from hybridization of complementary DNA sequences (a process called annealing) only if the ISs were similar. The insertion sequences thus characterized proved to belong to a few types of elements, named IS1, IS2, IS3 and IS4.

The IS sequences did not appear to confer any particular phenotypic

(a)

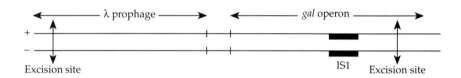

(b)

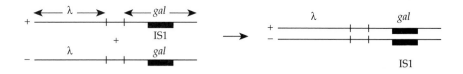

(c)

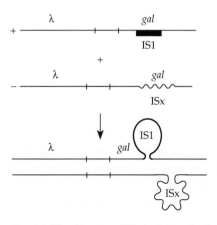

Fig. 5.1 Identification of ISs by heteroduplex formation. The *gal*::IS1 example. (a) The *gal* operon, carrying the IS1-induced Gal⁻ mutation, can be recovered on a λ genome by imprecise excision of a λ prophage from the host chromosome (Chapters 4 and 12). (b) The composite DNA fragment is separated by heating into its single-strand components, which can be separated by differential centrifugation. Reannealing of the single-strands reforms the original double-strand. (c) Annealing with a similar structure from another mutant carrying another IS forms a heteroduplex structure showing single-strand loops at the non-homologous regions.

trait on the host bacteria, apart from the strong polar mutation due to their insertion into the corresponding gene, and could be defined only by their structure, determined, at that time, by their hybridizing properties, but now also by their base sequences.

2.1.2 *Presence on the chromosomes of different species*

The presence of the insertion sequence IS1 was first examined in the *E. coli* chromosome by filter hybridization: the bacterial chromosome, fragmented, denatured and fixed on nitrocellulose filters (this enhances the efficiency of the method), was allowed to anneal with denatured IS molecules, previously radioactively labelled. This allowed both detection of the presence of the IS and measurement of the number of copies present per chromosome equivalent. It was found that all strains tested harboured five to eight copies of the IS1 element. This figure was corroborated in 1981, when a more precise technique of DNA hybridization, Southern blot analysis (Chapter 16), was used. Six to 10 copies were demonstrated, depending on the strains. IS1 was then detected in several other enterobacteria such as *Shigella dysenteriae* (40 copies per chromosome), *Klebsiella aerogenes* (one copy) and *Serratia marcescens* (two copies) (Table 5.1). However no IS1 was found in *Salmonella* species or in *Edwartsiella iorda*, in spite of their taxonomic proximity to the above species. *E. coli* was also found to possess five copies of IS3 and a single copy of IS4.

2.1.3 *Presence on other replicons*

A more systematic search was then undertaken, which showed that copies of these ISs could be found on other types of replicons (Table 5.1). Thus, one copy of IS1 is present on the coliphage P1, and two copies exist on R1 plasmids at the junction between the resistance determinants and the transfer functions.

We have already mentioned the presence of IS2 and IS3 on the F factor (Chapter 3).

2.1.4 *Other insertion sequences*

Numerous other insertion sequences, in addition to the four originally described, are continuously being discovered in new strains or species. Table 5.1 gives examples of a number of such insertion sequences. Their sizes range from several hundred bp (768 bp for IS1) to several thousand bp (5980 bp for the $\gamma\delta$ element, 7.1 kb for the IS22 originally isolated from *Pseudomonas aeruginosa*), but generally average around 1.2 kb.

Table 5.1 Some IS elements present in Gram-negative bacteria.

Designation	Host DNA and copy number	Size (in bp)	Inverted repeat[a] (in bp)	Target duplication (in bp)	Open reading frame (no.)	Special properties
IS1	Enterobacterial chromosomes, phages and plasmids (5–8 copies per strain)	768	20/23	9 (8–11)	8	Several Class I transposons are formed by inverted or direct repeats of IS1 (Tn9, Tn2350, Tn1681)
IS2	E. coli chromosomes, plasmids (F)	1327	32/41	5	2	Inverted repeats of Tn951
IS3	E. coli chromosomes (4–5 copies) Plasmid F (2 copies)	1258	39/39	3	3	Behaves as a mobile promoter
IS4	Chromosomes of E. coli K12 (1 copy at a single location)	1426	16/18	11, 12 or 14	2	1 specific insertion site
IS5	E. coli Shigella Phages λ, Mu	1195	15/16	4	3	The most abundant IS in E. coli; has a promoter activity by creation of a promoter
IS10	Tn10	1329	17/22	9	1	Tn10 inverted repeats
IS15	E. coli Salmonella Several plasmids					Tn1525 direct repeats

Element	Host/source	Size (bp)	Inverted repeat[a]			Comments
IS21	Inc P1 plasmids	2100				Mobile promoter active only when in tandem repeats
IS26	Tn2680	820	14/24	8	2	Tn2680 direct repeats
IS30	Phage P1 *E. coli*	1221	23/26	2	3	Insertion site quite specific
IS46	IncN plasmids	810				Mainly forms cointegrates
IS50R	Tn5	1534	8/9	9	2	Tn5 inverted repeats; only IS50R is active
IS52	*Pseudomonas savastanoi*	1209	9/10	4		
IS66	*Agrobacterium* plasmid pTiA66	2548				
IS102	Plasmid pSC102	1057	18	9	3	
IS136 (IS426)	*Agrobacterium tumefaciens* (pTC137)	1313	32/30	9	3	
IS200	*Salmonella typhimurium* (6–10 copies)					Only found in *Salmonella* species
IS222	*Pseudomonas aeruginosa* chromosome and phage D3	1350	40			
IS476	*Xanthomonas campestris*	1225	13	4		
IS4400	*Bacteroides fragilis*	1150				
ISRm2	*Rhizobium meliloti*	2700	24/25	8		

a, The two figures refer to the sizes of the two inverted repeats, when these are not identical.

2.2 Transposons

2.2.1 *Discovery*

In 1971, it was shown that the β-lactamase determinant conferring resistance to the β-lactam (ampicillin or penicillin) antibiotics was able to move from one plasmid on to another compatible one present in the cell. At the same time, it was also shown that this determinant could insert into the *E. coli* chromosome. The similarity existing between this element, named Tn3, and the IS elements already known was established only a few years later. Other 'antibiotic resistance' genes from numerous plasmids were then demonstrated to be able to 'jump', or transpose, on to other unrelated plasmids. These jumping elements were called 'transposons'.

2.2.2 *Other transposons*

Many more transposons have since been discovered, first in Gram-negative bacteria (Table 5.2) and then in Gram-positive bacteria (Table 5.3). Besides genes mediating resistance to antibiotics, transposons can harbour other functions. A frequently occurring one is the resistance to heavy metals, but odd ones, such as the production of β-galactosidase, have been recorded. Comparison of the known transposons led to the definition of a limited number of classes, but often similar patterns of resistances harboured by different plasmids signify the presence of identical transposons. Their size range is very large. The largest recorded example is bacteriophage Mu, of *E. coli*. The T region of plasmid Ti of *Agrobacterium tumefaciens* (Chapter 3) also has the structure of a transposon, its target being plant chromosomal DNA.

3 Structure of the transposable elements

The original distinction between IS and Tn elements was based on the presence on the latter elements of one or several genes conferring a phenotypic modification to the bacterial strain that harbours it. The presence of resistance gene(s) on the transposons that were first discovered, and in fact on most of the known ones, made their study so much easier that this distinction was not questioned for some time. However, study of their structure and other properties led to the discovery that many transposons are composed of two ISs, responsible for the transposition capacity, together with a piece of DNA responsible for the phenotype but not required for transposition (composite transposons). Several ISs occur only alone, while others have always been detected as part of composite transposons (in spite of their capacity

Table 5.2 The four classes of transposons of Gram-negative bacteria.

	A Class I	B Class II	C Class III	D Class IV
Structure of a representative example	Tn5	Tn3	Phage Mu	Tn7
mRNAs	K_m BI S_m b→ a↓	tmpA4→ tmpR→ bla→	A→ B→	$T_p S_m S_p$
DNA	IRIC IS50 L / ICIR IS50 R	IR — res — IR	S — SE Phage functions — c	IR — IR Tn7-R
Size of Tn	5.7 kbp	5 kbp	39 kbp	14 kbp
Size of target duplication	9 bp	5 bp	None	5 bp
Markers[a]	Km, Sm, Bl	Ap	Phage functions	Tp, Sm, Sp
Transposition functions	ISR: a = active transposase; b = transposase inhibitor ISL: inactive	*tnpA* transposase *tnpR* resolvase	Two proteins: A and B	Five proteins necessary
Comments	Composite Tn, with two distal, nearly identical, ISs, of which the left one is inactive	IR 39 bp	Largest known transposon	IR 28 bp
Other well-studied elements	Tn9 (IS1) Cm Tn10 (IS10) Tc	Tn1 Ap Tn501 Hg Tn21 Hg, Sm, Ap, Su Tn1000 ($\gamma\delta$) IS101 (209 bp)	Phage D108	

IR, IC, inverted repeats; res, site of co-integrate resolution; SE, c, striped ends. a, See Table 3.5 (p. 76) for explanation of the symbols.

Table 5.3 Transposable elements from Gram-positive bacteria.

Element	Hosts	Phenotype[a]	Size (kbp)	Terminal repeats	Target duplication	Class[b]
IS231	*Bacillus thuringiensis*	None	1.65	20	11	
ISS1	*Streptococcus lactis*	None	0.82	18	8	I (IS15)
IS110	*Streptomyces coelicolor*	None	1.55	10/15	ND	
IS257	*Streptococcus lactis*	None				I (IS15)
IS861	*Streptococcus*					I (IS50, IS3)
Tn4001	*Staphylococcus aureus*	Gm, Tm, Km	4.7	IS256	ND	I
Tn551	*Staphylococcus aureus*	Em	5.3	40	5	II
Tn917	*Streptococcus faecalis*	Em	5.27	38	5	II
Tn4430	*Bacillus thuringiensis*	None	4.194	38	5	II
Tn4451	*Clostridium perfringens*	Cm	6.2	12	ND	II
Tn4556	*Streptomyces fradiae*		6.62			II
Tn916	*Streptococcus faecalis*	Tc conjugative	16.4	Imperfect	0	V
Tn918	*Streptococcus faecalis*	conjugative	16	ND	ND	V
Tn919	*Streptococcus sanguis*	conjugative	16	ND	ND	V
Tn1545	*Streptococcus pneumoniae*	Tc, Em, Km conjugative	25.3	Imperfect	0	V
Minicircle	*Streptomyces coelicolor*	None	2.6	Imperfect	0	V
Tn554	*Staphylococcus aureus*	Em, Sp	6.69	0	0	V

a, See Table 3.5, p. 76 for explanation of symbols; Gm, gentamycin; ND, not determined; Tm, tobramycin.
b, The IS in brackets indicates Gram-negative elements having homologies with those described.

to transpose autonomously) and yet others are found either as in-dependent or as transposon-associated units.

Thus scientists now tend to ignore the initial distinction and speak either of 'transposable elements' or of transposons, whatever their associated phenotype. The following definition could thus cover their common, and major, traits: a 'transposable element' is a self-contained unit, encoding one or more functions which mediate(s) its transposition and which may also regulate the transposition activity. The smallest transposable elements, the ISs, encode only the genetic determinants involved in promoting (and regulating) transposition; larger elements, the Tns, encode additional genetic determinants, gratuitous to themselves, such as antibiotic resistances, bacteriophage functions or even metabolic capacities.

The insertion of a given element preserves the exact and entire base sequence of the element involved. Different insertions of a given element each contain exactly the same set of non-permuted sequences, and each insertion event involves the joining of the same nucleotides at each end of the element to a broken target molecule. It is thus easy to trace the precise identity and structure of any transposon, even though, since it is not a replicon, it can only be found inserted in a

Table 5.4 Insertion sequences from archaebacteria.

Element	Hosts	Size bp	Inverted repeats	Target duplication (bp)	ORFs
ISH1	*Halobacterium halobium*	1118	8/9	8	1
ISH2	*Halobacterium halobium*	520	19	10, 11 or 20	3
ISH25	*Halobacterium halobium*	588	none	none	
ISH50	*Halobacterium halobium*	996	23/29	none	2
ISHS1	*Halobacterium halobium*	1700	26/27	8	
ISH51	*Haloferax volcanii*	1371	15/16	3	

replicon. However, in many cases, the transposon is bracketed by a duplication of a few base pairs from the target DNA. The size of this duplication is specific for each class or group of transposons, with, however, a considerable number of exceptions. Four main classes can be distinguished in the transposable elements from Gram-negative bacteria, on the basis of their genetic organization, mechanistic properties and DNA sequence homologies. Table 5.2 presents a short list of known elements of each class, together with a few of their characteristics. Gram-positive bacteria possess a variety of transposable elements, many of which fall into the first two classes described for Gram-negative species (Table 5.3). Some are original and are ascribed to a new class. Having been more recently discovered, Gram-positive transposable elements are still less well known. This is also true for the mobile elements described from archaebacteria (Table 5.4).

3.1 Class I

Class I includes all elements in which the information necessary for transposition only is flanked by two small inverted repeats (IR). It is a very heterogenous class, which has been subdivided by taking into account characteristics which are outside the scope of this book.

Most ISs found up to now belong to this class. They consist of a short element (750–1500 bp) bordered by two inverted, more or less identical, repeats. Most of these elements contain a single coding region, beginning just inside the inverted repeat at one end of the element and extending to a few base pairs within the second inverted repeat. Usually no stop signal is present. Smaller reading frames often overlap and sometimes run in the opposite direction to the major coding segment (Fig. 5.2). In the best known instances polypeptides corresponding to all coding regions have been identified, although their functions are not yet completely elucidated. The longer one is generally the 'transposase' (Section 5). IS1 differs from all the others in

(a)

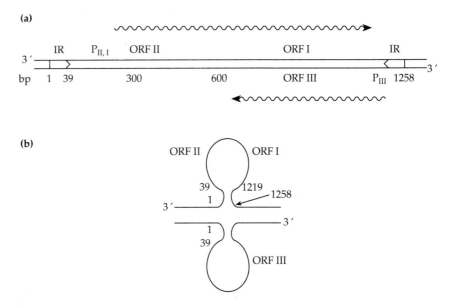

(b)

Fig. 5.2 Structure of an IS, IS3. (a) Normal linear form, showing the three coding regions (open reading frames, ORF) probably transcribed as two mRNAs, and the two terminal inverted repeats. (b) The possible stem-and-loop structure formed after local melting of the DNA.

that it has eight ORFs, a surprising situation since it is the smallest known IS. It has been demonstrated that only two of these ORFs, called *InsA* and *InsB*, code for proteins essential for transposition, whilst two other ORFs may be involved in this process but are not essential.

When these elements are part of composite transposons, two ISs surround an additional piece of DNA, the two ISs being always copies of the same one, although they may differ from each other by a few base pairs. They can be in the same orientation, such as IS1 in transposon Tn9, but are more often inverted, e.g. IS1 in transposon Tn681, IS10 in Tn10, IS50 in Tn5, etc. (Table 5.2A).

Either both ISs (as in Tn9) or only one (as in Tn5) is functional. In the former case, individual efficiencies of the ISs to transpose as independent units may differ from one to the other, and their combined efficiency, active on the whole transposon, may be higher than that of each individual one.

The additional piece of DNA in composite transposons often contains a single transcriptional unit that may encompass several resistance genes, as with Tn5. By convention, the orientation of the transposon is determined from that of this unit, transcription being from left to right; the bordering ISs are thus termed ISL (left) and ISR (right), respectively, and the two IRs bracketing each IS are then noted IR_i (inner) and IR_o (outer). Composite transposons of this group are easily detected by heteroduplex analysis, since they can form palindromic structures (Fig. 5.3).

3.2 Class II: the Tn3 family

Members of Class II often differ from those of Class I by the presence of accessory genetic determinants directly adjacent to the transposition genes, which are thus not situated in an IS (Tables 5.2B & 5.3). They do not form composite structures, and the similarities in structure and functions between all members of this group suggest a common evolutionary origin. Tn3 is the best characterized. The whole elements may be long (from 5 to 20 kb), due to the presence of accessory determinants (the *bla* gene coding for β-lactamase in the case of Tn3) in addition to the transposition functions. IS101, from *Salmonella*, a 209-bp-long element belonging to this class, possesses functional termini and a *res* site (Section 5), whereas no detectable phenotype has been detected for γδ, a 5-kb-long element present on the F plasmid (Chapter 3), which has been intensively studied.

The terminal inverted repeats have an average length of some 38–40 base pairs. They are very conserved among different elements, although a few bp differences may occur in the pair of a given element (Tables 5.2, column B & 5.3). A high degree of conservation also exists within the base sequences of each individual determinant involved in the transposition process (*tnpA, tnpR, res*). However, their order within the element may differ.

No such relationship exists between the different accessory determinants.

3.3 Class III: transposing bacteriophages

Two related *E. coli* phages, Mu and D108, use transposition as their normal mechanism to promote vegetative replication of their genome during lytic development and prophage insertion during lysogenization. They are the largest known transposons, with a genome size of 39 kb (Table 5.2C). The two phages share 95% homology of their DNA sequences but are heteroimmune. There are no homologous terminal repeats. However, the first 14-nucleotide sequence from the c (right) end is also found at position 79–92 on the S (left) end, and the next 17 nucleotides from the c end are also repeated among the first 20 nucleotides from the S end. Two Mu adjacent functions, genes *A* and *B*, are directly involved in Mu-mediated transposition.

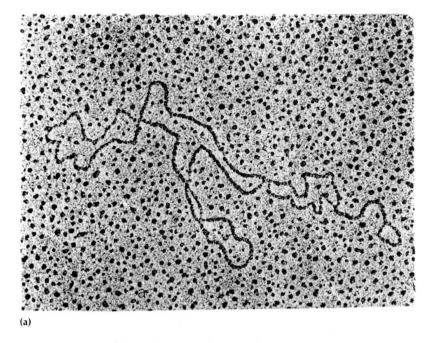

(a)

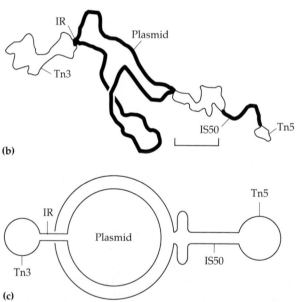

(b)

(c)

Fig. 5.3 (a) Electron microscopic determination of the palindromic structure of Class I transposons. (b) Two identical plasmids, carrying either a copy of Tn5 (right) or a copy of Tn3 (left), inserted at different locations, were denatured and annealed after mixing. The single-stranded structures formed by each transposon are clearly visible, as interpreted in the diagram (c) (Courtesy of J. Brevet, Institute des Sciences Vegétales, CNRS). The central double-stranded structure corresponds to the reannealed plasmid DNA. The small double-stranded part next to each transposon indicates the annealing of the inverted repeats of each transposon.

3.4 Class IV: transposon Tn7

Transposon Tn7 (Table 5.2D) differs from all other known transposons in several characteristics. It is a large transposon (14 kbp) and inserts in a unique chromosomal hotspot in *E. coli*, *Pseudomonas aeruginosa*, *Caulobacter crescentus* and *Vibrio* species, although in numerous sites in most plasmids. Most insertions in a given replicon are in one orientation. It codes for resistances to trimethoprim (Tp), streptomycin (Sm) and spectinomycin (Sp). Like the other transposable elements, Tn7 carries terminal inverted repeats, in this case of 28 bp, but in addition each end contains regions of 22 bp that occur several times in the same orientation and which are highly conserved (75% homology). These direct repeats are contiguous on the right-hand end but are separated from each other on the left-hand extremity. There are five genes coding for functions involved in transposition and located on a large Tn7-R segment. One of these is involved only in transposition outside the hotspots.

3.5 Class V: new transposons from Gram-positive bacteria

Some transposons from Gram-positive bacteria share a few unusual characteristics which distinguish them from those of all other classes, a situation which justifies their description as a separate group. They lack terminal inverted repeats and variable termini, and fail to produce target duplication during the transposition process.

Tn554 (Em, Sp) shows extreme site specificity on its host chromosome, *S. aureus*, inserting generally at a unique site and most often in the same orientation. Its sequence has been determined. It contains six ORFs, three of which, *tnpA*, *B* and *C*, are required for transposition. Proteins TnpA and TnpB are related to other recombinases such as the λ integrase (Chapter 4).

Numerous cases of conjugative transfer of antibiotic-resistance determinants not borne by plasmids have been described in several streptococcal species. Thus, of 82 multiply resistant strains of *Streptococcus bovis* and *Streptococcus pneumoniae*, 74 were apparently plasmid-free and 24 of these were able to transfer their resistance determinants by conjugation. In some cases these plasmid-independent conjugative transfers were shown to be due to the presence of a conjugative transposon. Tn916 (Tc) of *S. faecalis* was the first element of this group to be discovered and has been extensively studied. Tn918 and Tn919 are homologous to Tn916. Tn1545 (Em, Km) was found in *Streptococcus pneumoniae* as part of a larger (Em, Km, Tc, Cm) non-plasmid conjugative element. It is noticeable that resistance to tetracycline is the most frequent resistance determinant found among these conjugative elements. Their host-range comprises most Gram-positive bacteria, with the exception of the actinomycetes and related bacteria.

3.6 Mobile elements from archaebacteria

A few elements, all being insertion sequences, have been reported up to now in archaebacteria (Table 5.4).

Seven insertion sequences, named ISH1 to 7, whose sizes range between 300 and 1500 kb, have been found in either a single region, the bacterioopsin (*bop*) gene region, of the archaebacterium *Halobacterium halobium* chromosome or on its plasmids. Curiously, the *bop* region is approximately 10% poorer in G+C than the rest of the genomic (chromosomic and plasmidic) DNA (68% G+C).

An insertion-like sequence has also been detected in *Methanobrevibacter*.

4 Genetic features associated with transposition

4.1 Definition of the transposition process

Transposition means the insertion, into a replicon, of a transposable element from the same or another replicon present in the same cell. This recombination event does not involve DNA homology between the element and the recipient replicon, nor does it require the RecA recombination system (Chapter 9). It was often referred to as site-specific recombination, alluding to the fact that the recombination event takes place precisely at the sites defined by the extremities of the inverted repeats of the element, although it should be named transposition whilst site-specific recombination refers to processes between two molecules of DNA involving totally defined short sequences.

4.2 Mutagenic effects of transposition

Transposition may result in the formation of mutations when the insertion occurs inside a gene. The mutagenic effect is often strongly polar, as is the case in the *gal⁻* mutants which led to the discovery of IS1 (Section 2.1.1). However, secondary, or specific, structural or functional characteristics of the mobile elements may lead to apparent oddities in the manifestation of the polar, or the mutagenic, effects.

Mutations resulting from the insertion of elements such as IS12 and IS3 are polar only when the insertion is in one orientation. Polarity may also depend on the site of insertion. This is believed to be due to the presence of rho-dependent transcription stop signals (Chapter 15) belonging either to the element (as was demonstrated for IS12) or to a gene distal to the element. Many termination codons are generally found in the different reading frames of the elements, a situation which obviously favours polarity. In some ISs, read-through transcription is prevented by the presence of a particular base sequence, proximal to the promoter for the major coding region, which acts as a transcription termination signal (Chapter 15).

Insertion of some ISs or transposons can increase expression of the adjacent bacterial genes in either orientation. This phenomenon seems particularly frequent in some pseudomonads such as *Pseudomonas cepacia*, in which it might explain the versatility of their metabolism by allowing expression of genes which are otherwise not read in the wild-type chromosomal organization. Two mechanisms have been described to account for this activation.

Expression might be obtained from a complete promoter present in the element, which may be the normal promoter of a gene in the element allowing read-through due to the lack of a terminator, as has been shown with Tn3 (Table 5.2B), for instance; or, as in some ISs such as IS3 (Fig. 5.4a), a promoter situated at the end of the element, but with no particular role for the element, activates the host-distal genes. Such a promoter has been referred to as a mobile promoter.

However, in some cases, the IS possesses only part of a promoter, and can activate transcription of distal genes only if the other part of the promoter is provided at the insertion point. This seems to be the case with IS21, which can transpose only when it inserts in tandem duplication: it possesses a -35 promoter region on its right end and a -10 part on its left end, so that the presence of two adjacent copies leads to the reconstitution of a complete promoter (Fig. 5.4b).

4.3 Specificity of insertion sites

Being RecA-independent, the event involved in transposition belongs to a non-homologous recognition process. Because of this absence of homology between the transposon and its target, the question arises as to whether there is complete randomness of the insertion sites inside the target, or whether there are discrete insertion sites comparable to the *att* site of phage λ. An answer to this question involves the determination of the number and location of insertion sites of a given transposon in a given DNA molecule. Such an analysis is best performed by choosing as host molecule a small, well-known region of a replicon, so as to obtain clear and easily interpretable results. A study of the distribution of these sites may reveal the existence of hotspots, the analysis of which has helped, in some cases, to reveal the precise structure of the recognition region.

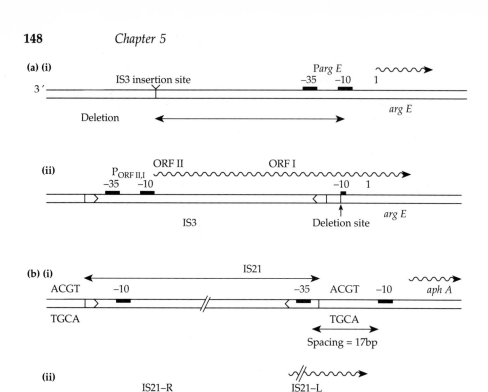

Fig. 5.4 Role of insertion sequences as mobile promoters. (a) IS3 as a mobile promoter for *argE* in *E. coli* (i). IS3 can insert upstream of *argE*. A deletion, as indicated, removes the -35 part of the *argE* promoter (ii). Transcription of *argE* is still performed, but starts from a promoter site inside IS3. (b) IS21 as a mobile -35 promoter region. (i) Insertion of one copy of IS21 in plasmid RP4. Its -35 region can complete the -10 region of *aphA* (Km^r), allowing its transcription. (Standard spacing between the two regions for the *E. coli* σ factor of RNA polymerase: 16–18 base pairs). The IS21 is not transcribed. (ii) Two tandem direct copies of IS21 into plasmid R6845. Formation of a complete and functional promoter using the -35 region of IS21-R and the -10 region of IS21-L allows transcription of IS21. Transcription of *aphA* is still possible.

Most transposons have numerous insertion sites even on small plasmids, but Tn7 and IS4 have only one site in the *E. coli* chromosome, although Tn7 can transpose at numerous sites on many plasmids. Bacteriophage Mu seems to make completely random insertions. In the case of Tn10, there exists a 6-bp homology between a sequence at the termini of the transposon and a sequence in the target 18 bp upstream of the insertion site. Mutations in the transposase have been found which decrease target specificity. IS2 inserts between two particular nucleotides. No nucleotide specificity has been demonstrated for Tn3, which transposes 100 times more efficiently into a plasmid than into the *E. coli* chromosome and may prefer AT-rich regions.

Tn5 has so far resisted elucidation of site specificity. Tc^s derivatives resulting from its insertions into the *tet^r* gene of plasmid pBR322 were isolated. Location of Tn5 in each clone was mapped along the sequence of the *tet* gene. The frequencies of occurrence of insertions per site were recorded, leading to the definition of five hotspots. Base composition along the map, determined by the A+T content, did not indicate any special trait which could be correlated with these hotspots.

Tn917 (from *S. faecalis*) and Tn551 (from *S. aureus*) show different target specificities, Tn551 transposing preferentially on chromosomes while Tn917 transposes equally on chromosomes and plasmids. Although it shows little target sequence specificity, Tn551 displays clear regional affinities on the *S. aureus* chromosome.

The susceptibility of sites to transposition may be influenced by the degree of superhelicity of the target DNA.

4.4 Duplication of a few base pairs of the target

While investigating the specificity of insertion sites, research workers came upon another important characteristic of transposition. Most known transposons, with the exception of Mu, IS3, IS4, IS5 and the new Gram-positive transposons, are flanked by two direct repeats, a few bases long, of the host target. The precise length of the repeat is often specific to the class of the transposon: 9 bp for Class I (with a possible variability for some ISs such as IS1), 5 bp for Class II transposons and for Tn7. This duplication arises as a consequence of the transposition process (Section 5).

4.5 Control of transposition frequencies

The frequency of transposition depends on the efficiency of the process and on its regulation. These frequencies can be measured by the proportion of molecules, in a population of a given replicon, that receive a given transposon. Usual values range from 10^{-4} to 10^{-7} per recipient replicon. They can also be estimated as the average length of DNA interspacing two identical insertions on a given replicon. The frequencies are then expressed in size (megadaltons) of the target DNA.

Transposition requires the integrity of the inverted repeats, since mutations in these sequences prevent transposition in all cases. In composite transposons, in which limited base-pair differences may occur between the four IRs, it has been possible to establish that the frequency of transposition conferred by the outer IRs, IR$_o$s, is often higher than that conferred by the inner, IR_i, ones, or by an inversion of one of the pairs of IRs.

Due to the mutagenic effect of transposition, it is obvious that high levels of transposition could rapidly result in non-viability of the recipient cell, as occurs during the lytic cycle of phage Mu (Section 5.3). All transposable elements except Mu possess regulatory mechanisms which limit their frequency of transposition. These regulatory mechanisms are diverse.

4.5.1 *Regulation in Class I elements*

Transposition of Tn5 is negatively regulated. When Tn5 is introduced into a cell that does not already contain a copy of it, transposition is activated for about 2 h (at 37 °C), during which time the majority of the transposition events take place. A *trans*-acting inhibitor is by then synthesized, which is likely to be the small protein coded by IS50 (Table 5.2A). This protein does not regulate the synthesis of the transposase, but may inhibit transposition by interacting with either the transposase, the IR or a host component.

Class I transposases are generally more active in *cis*, and the gene is poorly transcribed.

The *tnp* promoters of Tn10 and Tn5 are sensitive to Dam methylation (Chapter 6); transposition is stimulated in *dam⁻* mutants. Transient hemimethylated states induce bursts of transposition, while fully methylated DNA prevents it.

Several processes seem to hamper active transcription of transposase genes through activity of external promoters. For instance, through-transcription over an IS50 extremity interferes with the ability of this extremity to participate in transposition.

In Tn10, translation of the transposase mRNA is negatively regulated by an antisense RNA.

4.5.2 *Regulation in Class II transposons*

Class II transposons have a different kind of regulation. On the one hand, a protein, coded for by gene *tnpR* (Table 5.2B), causes repression of the transposase, at least in Tn3, and also represses its own synthesis. In addition, a phenomenon called transposon immunity prevents transposition of this class of element on a replicon on which a copy of the transposon is already present. This phenomenon is highly specific and does not apply to a related transposon.

Signals from the environment may also be involved in transposition regulation. Tn917 transposition is induced by the substrate of one of the resistance genes it codes for. This might be due to increased transcriptional read-through of the *tnp* gene after induction of transcription of the upstream resistance gene (both genes are transcribed from the same strand). Similarly, in Tn501 (Hg^r), the two transposition genes *tnpA* and *tnpR* are induced by mercury. Tn3 transposition is temperature-sensitive, whereas that of Tn10 (Class I) is temperature-dependent. Storage for long periods has been reported in several instances to activate transposition.

4.6 Stability

Maintenance of a transposon may depend on the transposon and on the recipient strain. Two principal modes of instability have been shown. In some cases, and generally with very low frequencies, the transposon is able to excise precisely. Loss of the resistance(s) coded by the transposon is then associated with total recovery of the function of the gene in which the transposon was inserted. This excision, mainly detectable in Class I transposons, is generally considered as independent from the transposition mechanism itself. It is *recA*-independent in Tn5 and in Tn10, and is assumed to be due to a replication error (Fig. 5.5a). It sometimes happens that the loss of the transposon is not accompanied by the recovery of the mutated function, in which case an imperfect excision must have taken place (Fig. 5.5b).

Other types of instabilities have been demonstrated in several instances, but not thoroughly studied. In these cases the transposon remains but the mutation which was initially caused by insertion of the transposon is lost with reversion to wild type, as if the localization of the inserted element has changed.

4.7 DNA rearrangements associated with transposition

Rearrangements of DNA are part of the transposition process of most Class I transposons, although the various elements differ in the extent to which they promote the diverse types of rearrangements.

4.7.1 *Replicon fusion*

When a plasmid carrying a transposon is brought into a cell that does not yet have this sequence, replication of the transposon may lead to the formation of a cointegrate, i.e. a fusion between the plasmid bearing the original transposon and another replicon present in the cell, the two molecules being linked through two copies of the transposon (Fig. 5.6). The two copies of the transposable element are direct repeats. In Rec$^+$ strains this cointegrate may undergo Rec-dependent recom-

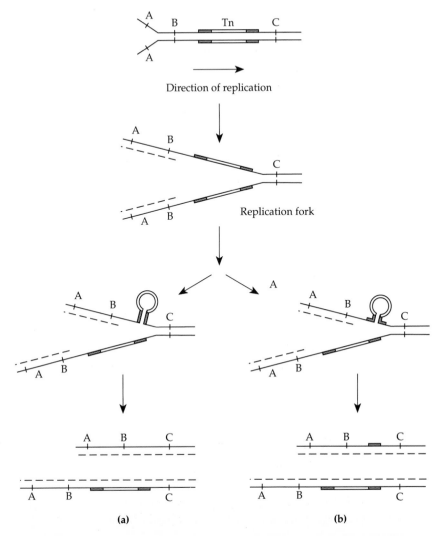

Fig. 5.5 Excision of a transposable element during replication of host DNA. Excision takes place by formation of a stem-and-loop structure, during the transient single-strand form of the DNA at the replication fork. Excision is shown in one strand only (although it may occur in both strands), involving the whole element (precise excision, a) or part of it (imperfect excision, b).

bination, resulting in the presence of the mobile element on the other replicon, at the position of junction of the fused structure.

Not all transposons can produce such stable fusions, while only Rec⁻ mutants maintain stable cointegrates. Class I transposons can mediate either mere transposition or cointegrate formation, the ratio of either event being characteristic of the transposon and possibly of the recipient strain (Fig. 5.6b). Class II transposons mediate fusions as an intermediate step in their transposition (Fig. 5.6c).

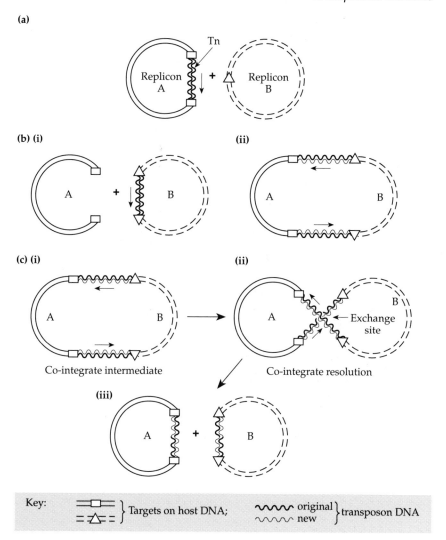

(a)

(b) (i) (ii)

(c) (i) (ii)

Co-integrate intermediate   Co-integrate resolution

(iii)

Key: } Targets on host DNA; original } transposon DNA
new

Fig. 5.6 Diagram of the structures resulting from the different modes of transposition. (a) represents the initial state of replicon A, bearing the transposon, and replicon B, a molecule possessing a possible target for the transposon. (b) Class I transposons. Transposition occurs either as conservative, with the formation of a transient circular intermediate form (i), or through formation of a co-integrate involving replication of the transposon (ii). (c) Class II transposons. Transposition comprises the formation of a co-integrate (i) which is then resolved by the transposon-encoded resolvase (ii) into two transposon-bearing replicons (iii).

If the fusion takes place between two plasmids, a larger, hybrid plasmid (or cointegrate) is obtained, whereas, if one of the replicons is the chromosome, the result is the integration of the plasmid in the chromosome. This mechanism has been used for the formation of chromosomal transfer devices (Chapter 11). Any plasmid may therefore be integrated into a chromosome provided a suitable transposable element is present on the plasmid or the chromosome. Due to negative regulation of most transposition processes, integration induced by a mobile element present on the chromosome is obviously quite an infrequent event.

4.7.2 *Inversions and deletions*

Many transposons can provoke the deletion or the inversion of sequences adjacent to their localization (Fig. 5.7). Deletions often run from the extremity of the transposon and include the host duplicated target. In many cases deletions and inversions happen in Rec⁻ cells and are manifestations of the transposition process. They can be regarded as equivalent to the formation of a cointegrate inside a single replicon. In other cases the Rec machinery is required and the deletions or inversions may stem from a recombination between two copies of the same element.

4.7.3 *Inverse transposition*

When a Class I transposon is inserted into a small plasmid, it can promote transposition of the plasmid instead of its own sequence (Fig. 5.8). This phenomenon is termed inverse transposition.

5 Transposition mechanisms

While all transposition events seem to result from a similar type of recombinational process, there are specific mechanisms in individual transposons. Only a few are well understood at present. The first to be elucidated was that of Tn3, a representative of Class II transposons. All the other transposons display definitely different mechanisms.

5.1 Transposition of Tn3 and of other Class II elements

As the easiest way to study the transposition mechanism, the classical approach of isolating mutants impaired for transposition and analysing their complementation capacities has been used. The general strategy that was used for Tn3 is described in Fig. 5.9.

Three plasmids of different incompatibility groups (Chapter 3), bearing different resistance markers, were introduced into the same cell. One of them, plasmid A, devoid of transposon, was designated to be the target on which to measure transposition ability, and was chosen for its conjugative capacities. The second one, plasmid B, had a mutated Tn3 unable to transpose, whilst the third one, plasmid C,

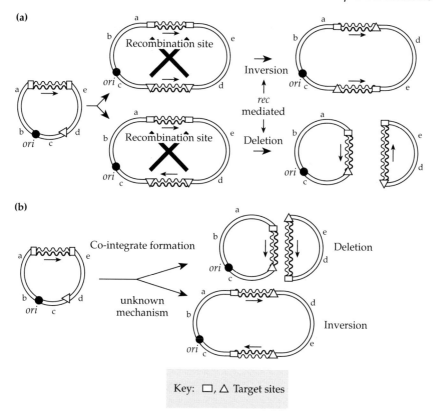

Fig. 5.7 DNA rearrangements involving transposon replication in the same replicon. (a) Rearrangements may result from a *rec*-dependent recombination between the two copies of the same transposon in the same replicon. This may happen whatever the process which brought the two transposon copies into the replicon. (b) Rearrangments may result from the formation of an intra-replicon co-integrate.

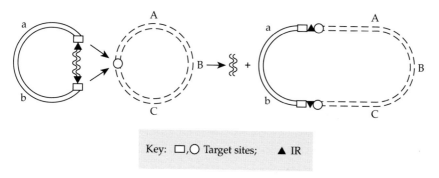

Fig. 5.8 Inverse transposition. The replicon carrying the mobile element is inserted into the second replicon instead of the mobile element.

(a)

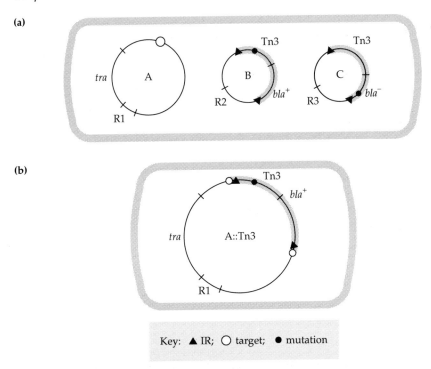

(b)

Key: ▲ IR; ○ target; ● mutation

Fig. 5.9 Complementation analysis of Tn3 transposition functions. (a) Three plasmids, A, B and C (see text for details) are introduced in the first cell. R1, R2 and R3 are three different non-transposable resistance markers. Plasmid C bears a mutation in the Tn3 Ap resistance gene *bla*, and plasmid B a mutation in a Tn3 transposition function. Plasmids B and C cannot be mobilized by A. (b) If complementation in trans of the mutation in the plasmid B-borne Tn3 by the plasmid C-copy of Tn3 is possible, some A plasmids will receive the mutated Tn3 copy of B (*bla*+). These can be detected, after conjugation to a plasmid-free cell, by the simultaneous acquisition of R1 and β-lactamase resistances.

bore a Tn3 carrying a β-lactamase-deficient gene; it could transpose normally, but its transposition could not be monitored directly, since no selection was available. If transposition of the mutated Tn3 from plasmid B could be obtained when plasmid C was present, then some A plasmids would acquire the β-lactamase determinant and thus bear resistance to ampicillin. This could be shown after transfer by conjugation into a plasmid-free strain.

Two 'complementable' groups of mutations were obtained that corresponded to two open reading frames of the transposon sequence. The function deficient in the first group could be fully complemented, and corresponded to gene *tnpA*. Complementation of mutants from the other group yielded only fused replicons; they were located in the *tnpR* gene. Transposition could occur only if both IR extremities were present. The following mechanism has been proposed. The product of gene *tnpA*, a 120-kD protein termed transposase, mediates splicing of the donor and recipient molecules and replication of the transposon, yielding a cointegrate structure. The product of gene *tnpR*, a resolvase, acts on a small specific sequence, *res*, close to each direct repeat of the transposon to efficiently mediate recombination, and thus resolution of the cointegrate. The resolvase is only active on direct repeats of the sequence present on the same replicon. The TnpR protein also acts as a repressor on both its own promoter and that of *tnpA*.

All Class II transposons possess a resolvase, which has not been found so far in any other transposon. Relative localizations of *tnpA*, *tnpR* and the resolution sequence, *res*, on the transposon map may differ significantly from one member of the family to another. Homologies have been found between this resolvase and invertases involved in several examples of phase variation in different bacteria.

Structural analyses of the complexes formed between the enzymes coded by the γδ IS and its DNA sequence are now progressing rapidly. It should be possible in the near future to understand the molecular mechanisms involved in these reactions.

5.2 Transposition of Class I transposons

No resolvase has been demonstrated in any of these transposons. Thus a distinction must be made between simple insertion and cointegrate formation. The latter (as well as inversions and deletions) obviously requires replication of the transposon. Simple insertions, on the contrary, do not necessitate duplication of the element, and are said to be conservative, i.e. excision of the transposon from its replicon and insertion into the new target.

This conservative transposition is performed in Tn10 in two steps (Fig. 5.10): (i) the element is excised from its resident replicon, which cannot recircularize, whereas the transposon is in a transient circular form; (ii) the excised transposon is then integrated into another replicon through asymmetric nicking at the new target site. This mode of transposition, which leads to target duplication, only requires a small amount of DNA synthesis, which might consist of mere DNA repair synthesis at the junction sites. The direct observation of the excised transposon is undoubtedly the best proof of the reality of this model. Due to the small number of such non-replicative transient circles, their observation is difficult to achieve, and has so far been performed only in the case of Tn10.

This conservative model has now also been conclusively demonstrated for IS10 and IS50, and also for transposon Tn7, but was not accepted for a long time. This resulted from the fact that replicative transposition was initially conclusively demonstrated for Class II transposons, while cointegrate formation does necessitate the replica-

(a)

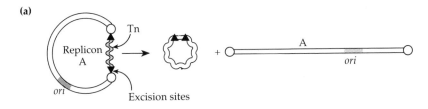

(b)

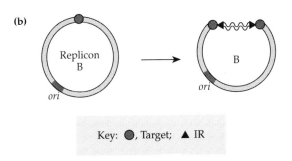

Key: ●, Target; ▲ IR

Fig. 5.10 Current model for transposition of Tn10, a Class I transposon. (a) Step 1: Formation of a transient circular form of the mobile element. (b) Step 2: Integration into a second replicon.

tion of the transposon since there are always two copies of the element between the two fused replicons. Also breakage of the donor replicon is not easily detectable. On the one hand, when the transposon is borne by a multicopy plasmid, loss of one copy is negligible, and quickly balanced by the copy-number regulation system, whilst on the other hand, Tn10 and, possibly, Tn5 are able to transpose only when the replication fork is passing through. Thus, if one copy of the newly replicated transposon excises and destroys its resident replicon, the other copy is still present on the other replicating strand.

Thus transposons of Class I are able either to perform a simple non-replicative transposition or to form a replicative cointegrate. The possibility has not been ruled out, however, that a replicative simple transposition may also exist, either in the case of the best studied transposons or for others of this class.

Thus a single model (Fig. 5.11) is likely to explain in a unified manner both simple conservative transposition for Class I transposons and replicative cointegrate formation for elements of both Class I and Class II. The proteins necessary in this transposition model comprise the transposon-encoded transposase (and sometimes several other proteins, as for IS1), and also several host proteins, among which are

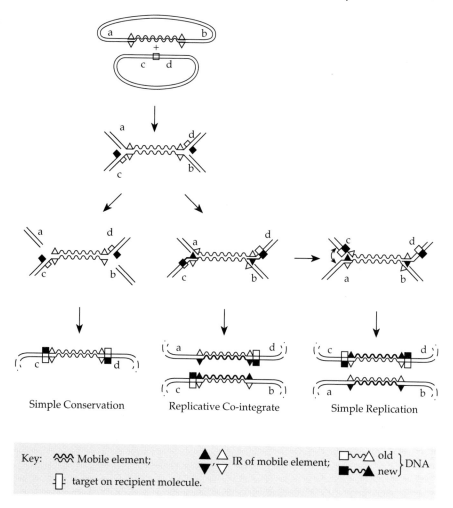

Fig. 5.11 Unifying transposition model for Classes I and II transposons.

DNA polymerase I, the DnaA and IHF (integration host factor) proteins.

5.3 Phage Mu

As seen in Chapter 4, Mu is a temperate phage which, when undergoing a lytic cycle, leads to the release of 50–100 particles per infected cell. When in the lysogenic state, the prophage synthesizes a repressor that blocks the expression of most viral functions, and the viral DNA forms a stable association with the host DNA.

A striking difference between Mu and other temperate phages is that its DNA integrates at random in the host chromosome. This is true

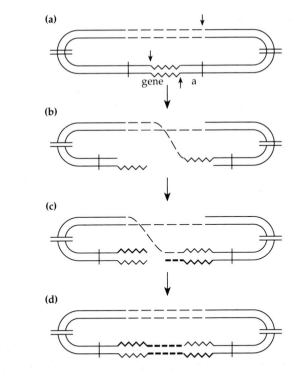

(a)

(b)

(c)

(d)

gene   a

Key: ≡≡≡≡ Original Mu DNA;   ⌇⌇⌇ original target DNA;

      ▬▬▬▬ replicated Mu DNA;   ⌇⌇⌇ replicated target DNA;

      ———— host DNA;   ↓ or ↑ single-stranded nick

Fig. 5.12 Mutagenic action and replication of Mu. Insertion of Mu, by replication, resulting in mutations and reproduction of the phage DNA. (a) Initial state of the replicon bearing Mu inserted at one site and another possible insertion site; a single-strand nick occurs at the extremity of Mu. (b) The free extremity of Mu moves and is linked to the appropriate strand of the other target. (c) The single-strand region (the host target and Mu) is replicated. (d) The original strand of Mu is resealed at its original target, and a second complementary synthesis takes place from the newly synthesized strand at the second target site. The single-strand nicks are sealed, producing a host molecule containing two copies of Mu, and a new mutation at the second site.

not only for the establishment of lysogeny, but also during the development of the lytic cycle, when Mu transposes to new locations and generates chromosome rearrangements as a result of replication of its genome (Fig. 5.12).

The Mu genome has been extensively analysed (Table 5.2C). In the phage particle, the DNA is linear and proteins are linked to its extremities. The prophage has the same genome structure and map as the infectious phage. It comprises 22 genes, most of which are essential for replication, capsid formation and lysogenization, but a few code for dispensable functions. In addition, a variable DNA sequence, of average size 1.7 kb, extends at one of its extremities (the S end), which are visualized as single-stranded regions (striped ends, SE) upon denaturation–renaturation of a population of phage DNAs. Short SEs are also attached to the other end (c end). These variable ends correspond to random host-chromosomal regions. If the length of the phage genome is increased by the presence of an insertion, the SEs are shorter at the S end, but the c end is not affected. This confirms that Mu DNA packaging involves copies randomly integrated in the host genome and starts at the c end, constant lengths being encapsulated (Chapter 4). The transposition leading to lysogenization seems to be a conservative one, whereas those involved in the lytic cycle have to be replicative. Thus Mu behaves as a giant transposon which, upon induction of its phage functions, transposes with an extremely high frequency.

Experimentally, a distinction between phage behaviour and transposon behaviour has been shown by the isolation of the 'mini-Mu', carrying a large central deletion covering the genes required for lysis. With this it was conclusively demonstrated that transposition and replication cannot be separated, since those mini-Mu which possess the two ends and the two genes *A* and *B* are able to promote their own transposition and are thus effective transposons. The possibility of achieving *in vitro* transposition with such minitransposons has prompted extensive studies, at the molecular level, of the mechanism of action of the A and B proteins. The need for some host proteins, such as HU (Chapter 2), has also been demonstrated.

5.4 Transposition of new transposons from Gram-positive bacteria

In Tn916 the 10 nucleotides of the right end are similar to the 10 nucleotides of the target situated just before the left end. This suggests a non-replicative insertion mechanism through a recombination event involving a very small homologous target. After excision, the remaining 10-bp sequence of the target is identical to either the original one or that from the right end of the transposon. Thus both decamers can be lost with the same efficiency. The same is true for the six terminal base pairs of Tn1545.

Regions of the host chromosome showing high degrees of homology with the transposon may enhance transposition efficiency. Transposition and conjugative transfer are coregulated and seem to share a common step. When a plasmid bearing Tn916 is introduced into *E. coli*, excision happens at an extremely high frequency (>90%). This led to the proposal of a model, later confirmed for both Tn916 and Tn1545, according to which excision is the first step for both transposition and conjugative transfer. In contrast to the case of Class I transposons from Gram-negative bacteria, this excision does not destroy the replicon from which the transposon originates, and can be either precise, since expression of the inactivated gene is restored after excision, or not. A covalently closed circular, non-replicative, intermediate form has been observed, which can undergo transposition into the same or another replicon, or can be conjugatively transferred into another host as a result of the transposition event.

Two transposon-encoded proteins, which share homology with the phage λ Xis and Int proteins, are necessary for excision, while only one is needed for integration. If the transposon is introduced into a transposon-free cell (on a conjugative plasmid for instance), the frequency of transposition is increased. This is reminiscent of zygotic induction (Chapter 4), and suggests the existence of a negative regulation of the transposition process.

A circular molecule, or minicircle, observed in *Streptomyces coelicolor* is also thought to be a transposition intermediate.

The transposon has only two chromosomal target sites, the specificity of which does not result from homology of these sites with the element. Transposition does not appear to cause a duplication of the target site.

6 Role of mobile elements in evolution and the 'selfish DNA' theory

The large number of transposable elements and their widespread presence raise the question of whether they have been favoured by a selective advantage conferred to their host cells. Transposons that code for a resistance or some other function may be selectively useful to the host bacterium. Such a situation is probably exemplified by the observed tremendous increase in their frequency since antibiotics have been widely used. This has been established by comparing the resistances borne by similar strains from ancient collections to those of more recent ones (Chapter 3).

However, such a hypothesis does not account for the maintenance of the ISs, which may be even more frequent than transposons on bacterial chromosomes, since they do not bring any additional function to the cell. Their existence has been explained by two of their properties, mutagenicity and self-reproduction.

On the one hand, they may confer a great genetic variability to the strains that harbour them, the mobile elements behaving as mutator genes. This is the selectionist theory, which would have the effect of speeding up Darwinian selection and could thus be an answer to criticism of the Darwinian theory on the grounds of the slowness implied. As an argument in favour of it, competition experiments have been performed between two isogenic strains differing only in the presence or absence of IS10. These strains were maintained together for numerous generations under non-selective growth conditions in a chemostat. The possession of IS10 indeed gave a selective advantage to the strain harbouring it, although the exact nature of this advantage in the conditions used was not determined. IS50 has also been shown to contribute to bacterial fitness by increasing the rate of adaptation to a new environment.

Another theory considers any DNA as capable of having a selective advantage as long as it has developed adequate multiplication devices. Mobile elements would be 'selfish' DNA elements, behaving as parasites, like phages for instance.

Whatever the theory, and both may be true depending on the external conditions, mobile elements have obviously played a great role in the evolution of bacteria. We have already seen that many plasmid functions are indeed transposon-borne. Much more information may have been transported by transposons which may have been subsequently lost or degenerated. This may explain the origin of most resistances to antibiotics from a few antibiotic-producing strains (although this is still a non-proven theory), or the 'horizontal' propagation of functions such as toxin production. It has already been shown that the β-lactamase-producing plasmid of *Neisseria gonorrhoeae* has been derived from that carried by *Haemophilus influenzae* and is presumed to have been transferred by neighbourly association in the pharyngeal environment. Horizontal propagation was first thought to occur only between strictly related bacteria, Gram-negative ones on the one hand and two separate groups of Gram-positive ones on the other hand, one of which was actinomycetes. However, several striking similarities between elements found in non-related bacteria suggest that wider horizontal transfer must occur, even if at extremely low frequency. Thus transposons of Class II sharing a high degree of homology have been observed in both Gram-negative bacteria and streptococci. Recently, Tn4556, a transposon from *Streptomyces fradiae*, has been shown to possess an ORF coding for a putative peptide sharing 61% identity with the Tn3 transposase. In contrast, their resolvases show no homology. Similarly, 30% homology, mostly in the transposase genes, exists between the *Streptococcus* IS861 and the Class I IS3 and IS50 from Gram-negative bacteria. IS493, from *Streptomyces lividans*, has the structure of a Class I element. IS15 is considered to be the archetype for a large family of elements from both Gram-negative and Gram-positive species.

In addition, more or less stable associations of transposons, forming structures which would then acquire multiple resistances, may also arise. The Class II transposon Tn21 for instance, has four resistance genes (Hg, Sm, Ap, Su). Some elements showing close relationships to Tn21 possess even more resistance characters, and may have resulted from recombination events taking place at favourable hotspots present on the transposons. Such structures are referred to as integrons. Transposons in which other different transposons have inserted represent very important factors in the spread of antibiotic resistances among pathogenic prokaryotes. Each constituent transposon has its own IRs and therefore can transpose individually, or the whole structure may transpose.

A very special example of interspecific propagation of a transposon is the case of the T element present on the Ti plasmids of *Agrobacteria* (Chapter 3).

6: Protection of DNA Integrity: DNA Methylation and the Restriction–Modification Systems

1 General scheme of the restriction–modification phenomenon

There exists a mechanism that is designed to protect cells from the invasion of foreign DNA, i.e. DNA not closely related to the genotype and sometimes even from different but related species. These protective systems usually involve the activities of two sets of enzymes, the methylases, whose action results in 'modification' of the DNA, and the specific endonucleases, causing 'restriction' of the DNA. The methylases, some of which at least seem to be coded by most living organisms (with the exception of small bacteriophages and viruses), add a methyl group to specific targets on the DNA. This secondary information spread along the DNA molecule is strain-specific, since the pattern of distribution of these sequences is specific for each particular DNA, i.e. for each bacterial strain. The specific endonucleases introduce double-strand nicks, causing subsequent DNA degradation at locations related to these targets, by recognizing a foreign DNA by its difference of methylation at these targets.

This general process has thus been named the restriction–modification (R–M) of DNA. In fact, this general designation covers several systems involved in either methylation or restriction, or both, but displaying different mechanisms of function and purposes. One strictly methylating process, the Dam system, an adenine-methylating pathway, is implicated in other cellular processes that are now well documented although not well understood as yet.

2 Discovery

The historical development of the understanding of the R–M phenomenon, from its chance discovery in the 1950s to the key role it has played in opening the way to molecular genetic technology, is worth a short survey. It is interesting in that it forms a typical example of the unpredictable importance of an apparently trivial observation. The experiments performed in the 1950–1960s also constitute a remarkable example of classical genetic and physiological elucidation of a new and, in the first place, odd phenomenon.

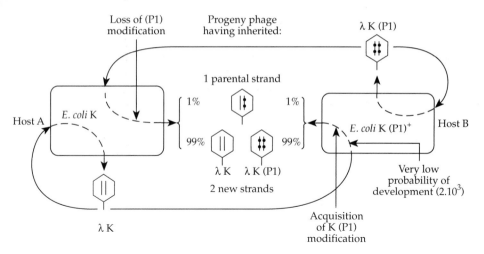

Fig. 6.1 Host-controlled modification and restriction. Variation of the efficiencies of plating of phage λ depending on which derivative of *E. coli* it originates from. Figures in progeny phage indicate percentage of the two different genome structures obtained, having inherited one or no parental DNA strand.

The initial observations, performed by Luria's and Arber's groups, highlighted a phenomenon whereby a single cycle of phage growth in a particular bacterial host may alter the host-range of virtually all the progeny. This is not a mutational event, since the property concerns the whole population, and is not heritable (as soon as the phages are grown in some other host, the modification can be lost). Figure 6.1 gives an illustration: phage λ will infect *E. coli* K12 (host A) with an efficiency of 100%, i.e. each infecting particle will propagate (Chapter 4). The progeny phage, designated λK, will infect *E. coli* K12 with the same 100% efficiency, but will infect *E. coli* cells that are lysogenic for phage P1, *E. coli* K (P1)$^+$ (host B), with an efficiency of only 0.002% (this represents the few infecting particles which by chance have escaped restriction or have infected a restriction-deficient mutant of host B). The λK is said to have acquired a restricted range. Host A is said to be permissive and host B non-permissive (or restrictive) for the λK particles. However, the rare phage progeny emanating from host B consist of particles that will not only plate on host B with an efficiency of 100%, but will also plate on host A with an efficiency of 100%, i.e. their host-range has been extended. These few phage particles have been said to have undergone modification in host B to give λK(P1) particles, whilst the majority of the original λK is said to have undergone restriction in host B. After a single cycle of development in host A, about 99% of the λK(P1) particles return to the original restricted host-range with only about 1 in 100 particles retaining the ability to form a plaque in host B.

Two important features should be noted:

1   The phenomenon is not due to differences in the ability of the phage to adsorb to the non-permissive host; nor is it due to absence of DNA injection, but the DNA is unable to replicate.

2   The rare phage particles arising from a single burst of host A which can form plaques in host B show a random distribution, not the clonal one expected if due to mutant particles.

Also, the number of plaques arising on host B is affected by the physiological state of the bacteria, e.g. previous irradiation with ultraviolet light or heating at 49 °C of the *E. coli* cells greatly increases the rate of acceptance (non-restriction) of the otherwise restricted phage.

The phenomenon was therefore interpreted as the acquisition by the phage of a reversible phenotypic modification responsible for variation of its host-range specificity. Since both restriction and modification processes were controlled by the bacterial hosts, the phenomenon was called phenotypic, or host-controlled restriction and modification. In fact, in the above case, the genetic information leading to both the restriction and the modification capacities were present on the P1 genome present in *E. coli*(P1)$^+$ cells.

It was afterwards established that the capacity to both modify and restrict is not a privilege of phage P1, but is a widespread property. Most organisms can confer their own type of modification, and thus have their own range of acceptance or restriction of invading DNA.

A series of experiments led to the molecular interpretation of the phenomenon:

1   Restriction, resulting in non-permissive infections, is due to a rapid degradation of the infecting phage DNA. The restrictive degradation produces double-stranded breaks (Section 4), thus creating substrate molecules for non-specific nucleases, which continue to degrade the fragments.

2   Modification is related to a state of the DNA. Density labelling of the DNA of a λK(P1) phage (obtained by growth on *E. coli* K12(P1)$^+$ in the presence of deuterium, provided as D_2O) allowed the fate of the original and the newly made DNA molecules during infection of *E. coli* K12 cells to be followed. Only the minority of progeny which retained original, D-labelled, DNA strands were able to reinfect *E. coli* K12(P1)$^+$. Thus alterations to the DNA imposed by the host concerned only the newly synthesized DNA molecules, and alterations imposed by a previous host would not be repeated in the new DNA. Preservation of a single original strand, as is the case in the hybrid molecules composed of an original and a new strand, was sufficient for maintenance of the original host-range.

3   Modification can be imposed on non-replicating DNA, and the functions of modification and of breakdown of DNA are separable. Infection of *E. coli* K12 cells with D-labelled phage λK at a high multiplicity, which results in partial prevention of replication, yields a proportion of progeny in which the injected genomes reappear unchanged, that is, they contain heavy-density DNA. If, during the latent period, the infected cells are superinfected with phage P1, all the progeny, including those carrying non-replicated genomes, are found to be modified to the λK(P1) pattern. Hence P1 can exert its methylating activity quickly, fast enough for the λ phage to finish their cycle before being prevented from doing so by the effect of the P1-encoded restriction system.

4   Modification is due to methylation of adenine and/or cytosine bases of DNA (Fig. 6.2a), using *S*-adenosylmethionine (SAM) (Fig. 6.2b) as methyl donor. This was in fact the first demonstration of the biological importance of methylation. Comparative infections, performed in the absence of their essential amino acid, into *E. coli* K12 strains auxotrophic for various amino acids showed that Met⁻ strains, and only these, failed to impose their host-range on the few progeny phage that could be formed under these starvation conditions. Similarly, a mixed infection with the λ test phage and phage T3, which happens to code, as an early gene, for a SAM 'degradase', leads to an absence of methylation of all newly made nucleic acid and a parallel absence of modification of the phage host-range.

5   The dual aspect of the phenomenon, modification of DNA through methylation and degradation of non-modified molecules, was corroborated by the isolation of mutants. Two types of mutants, r^-m^+ (for restriction and modification, respectively) and r^-m^- ones, were obtained. With the transfer into *E. coli* K12(P1)⁺ *lac*⁻ cells of an F' *lac*⁺ plasmid, the acceptance of the plasmid could be visualized through the acquisition of the Lac⁺ phenotype. Analysis of such (P1)⁺Lac⁺ clones, obtained at a low frequency, showed that a proportion was altered in the ability to restrict, and these belonged to two types: (i) one type had lost the ability to restrict but could modify, i.e. r^-m^+; and (ii) the other type could neither restrict nor modify, i.e. r^-m^-. The mutations were shown to be borne by the P1 phage genome itself. The fact that the third expected phenotype, r^+m^-, could never be recovered was interpreted by the hypothesis, since verified, that such clones were non-viable because of their failure to protect their own DNA against their endogenous restricting nuclease. Similar phenotypes resulting from chromosomal mutations were also isolated in *E. coli*, and contributed to the understanding of the genetic organization of these functions (Section 4).

All the fundamental characteristics of the phenomenon were thus elucidated in the mid-1960s. It later appeared, though, that the general

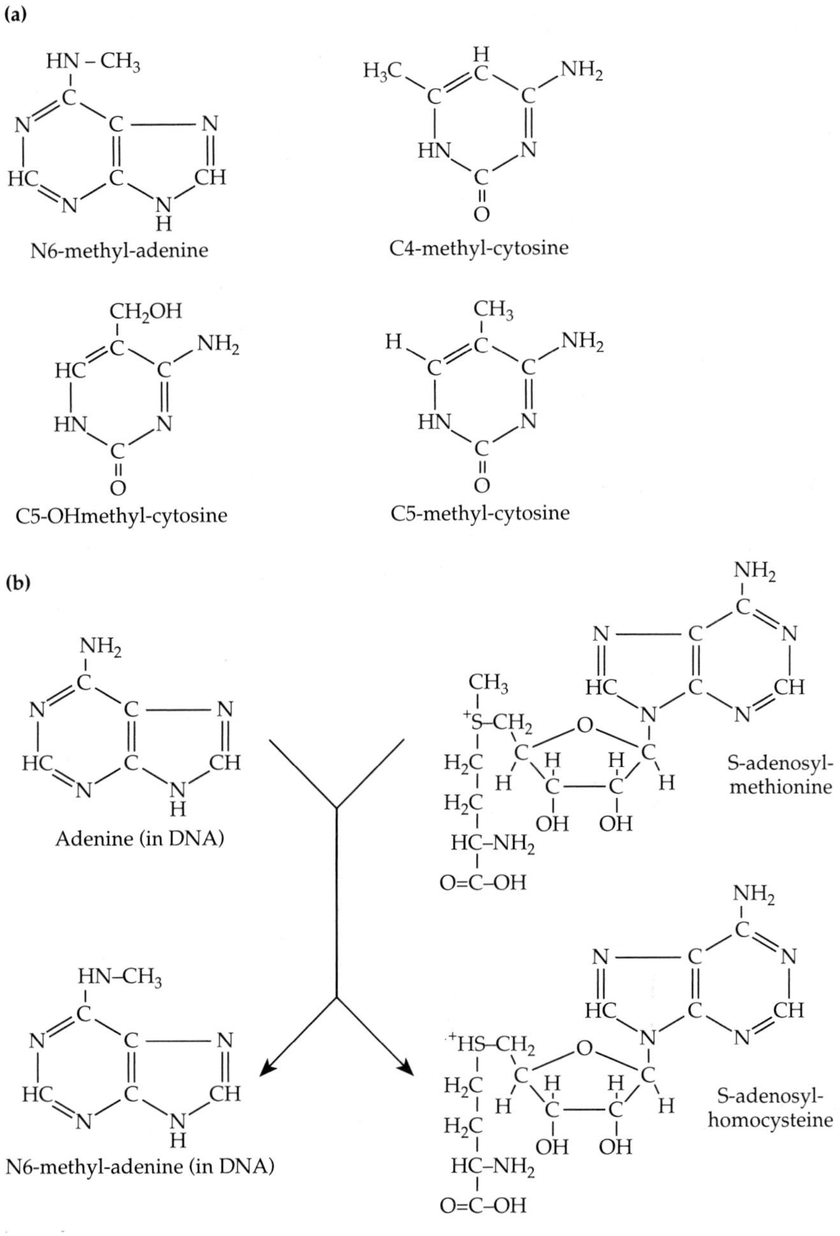

(a)

N6-methyl-adenine

C4-methyl-cytosine

C5-OHmethyl-cytosine

C5-methyl-cytosine

(b)

Adenine (in DNA)

S-adenosyl-methionine

N6-methyl-adenine (in DNA)

S-adenosyl-homocysteine

Fig. 6.2 Base methylation. (a) Adenine methylation by S-adenosylmethionine (SAM). (b) The known methylated bases.

concept of R–M thus defined in fact covers several independent, non-exclusive but different processes, which can be grouped into three types of systems: (i) the Hsd (host specificity of DNA) systems, which includes the one originally described; (ii) the Mar (methylated-adenine restriction) and Mcr (methylated-cytosine restriction) systems; and

(iii) the Dam (DNA adenine methylation) and Dcm (DNA cytosine methylation) systems.

A practical implication of the understanding of R–M systems was soon put into application for *Salmonella typhi* phage-typing, an important tool for the epidemiology of typhoid fever. It was essential to make the distinction between host specificities for a range of test phages due to host-range mutations and those resulting from phenotypic modification, since only the former are related to the pathogenic typing of the bacteria.

The main application derived from the discovery of R–M processes, however, had to await the further elucidation of the unique characteristics of some Hsd-type restriction nucleases, which took place in the early 1970s (Section 4.5).

3 General features of DNA methylation

DNA methylation has been shown to occur in many organisms, both prokaryotes and eukaryotes, for several decades, and is thought to be universal. Exceptions are known only in a few eukaryotes such as *Drosophila* and yeast. In all known cases it concerns only a small proportion of the bases (about 1%), and only adenine and cytosine (or their modified equivalents, such as hydroxymethylcytosine in *E. coli* T-even phages) are affected. These features indicate that there must exist a genetic control of the methylation process(es), and hence a (several?) biological implication(s) attached to the consequent state of the DNA. This conclusion was supported by the understanding of other aspects of the phenomenon. At least in prokaryotes, the choice of which adenines or cytosines are methylated is strictly reproducible in a given organism and dependent on local base sequences. The methylases possess a target-specific recognition mechanism. All possible target sites determined by these recognition sequences are methylated. Thus the methylases show base specificity and site specificity. The methylase of one strain recognizes one set of bases in a DNA molecule, whereas the enzyme of another strain will affect another set of bases; e.g. the methylases of *E. coli* B will methylate all the bases they can act on in a sample of DNA while those of *E. coli* K12 will still methylate other bases in the same DNA. Two sites at which the methyl group is bound to the base are by far the most frequent: nitrogen-6 for adenine and carbon-5 for cytosine (Fig. 6.2a), but others, such as carbon-4 of cytosine, are encountered.

As demonstrated above (Section 2, **3**), methylation is not concomitant with DNA replication. It usually occurs after a span of double-stranded structure has been completed. The methylating enzymes thus do not belong to the DNA replication complex. It follows that a hemimethylated state, showing the history of the two strands (the parental one fully methylated, the daughter one not), may exist, for a given region of DNA, during a significant period of the life cycle of the cell (0.5 to 3 min for *E. coli* growing in about 50 min at 30 °C). This has very important consequences for maintenance of DNA integrity and cellular metabolism (Section 7; Chapters 8 and 10) and possibly in control of DNA replication cycles (Chapter 2).

An interesting and puzzling characteristic of the methylation capacity of bacteria and phages is that it is non-essential. Mutants can be isolated, either spontaneously (for instance, the r⁻m⁻ ones mentioned above) or by gene inactivation of up to all the known methylases in a given strain. These mutants are normally viable, at least with certain genetic backgrounds for some of them (e.g. the r⁻m⁻ double mutant). Only particular multimutated combinations involving supplement deficiencies for genes functioning in mismatch repair (Chapter 8) are non-viable. This holds true for the three known (and probably only) methylating systems present in most strains of *E. coli* (Sections 4, 5 and 7). The consequent phenotypes, besides loss of the corresponding methylating capacity, can be either not distinguishable from the wild-type or pleiotropic, depending on what system has been inactivated. In the latter case, these mutants have been key tools for deciphering the roles of DNA methylation in cellular processes.

4 The host specificity of DNA (Hsd) systems

Hsd systems are the genuine restriction–modification systems. The association of the two complementary functions is matched by a linked organization of the corresponding genes in all the cases analysed up to now. They represent four classes, defined according to the mode of cleavage of the endonuclease involved. The molecular structure and mechanism of action of the enzymes, for both activities, are also homogeneous within each class (Table 6.1).

A clearly defined nomenclature rule allows the identification of the bacterial R and M enzymes: endonucleases are designated by the first and first two starting letters of the respective genus and species names of the organism by which they are coded. This acronym is italicized and followed by the appropriate strain denomination. A roman order number is given if several enzymes are present in the same strain (e.g. *Ava*I refers to the first restriction endonuclease isolated from *Anabaena variabilis*) (Table 6.2). Cognate methylases are identified by the same acronym preceded by a capital M.

4.1 Class (or Type) I systems

Three adjacent genes, *hsdR, M* and *S* (for restriction, modification and specificity, respectively), in this order on the chromosome, code for

Table 6.1 Characteristics of the four classes of Hsd systems.

Properties	Class I	Class II	Class III[a]	Class IV[a]
R and M activities	Trimeric complex	Single enzymes	Trimeric complex	Single protein + 2nd methylase
Genetic organization	Two transcripts: *hsdR, hsdSM*	Usually two transcripts: *hsdR, hsdM*	Two transcripts: *hsdR, hsdSM*	Not known
Recognition sequence	Non-symmetric, hyphenated	Usually palindromic	Non-symmetric	Palindromic, hyphenated
Requirements for restriction methylation	SAM[b] SAM, Mg^{2+}, ATP	Mg^{2+} SAM	SAM, ATP, Mg^{++} SAM, ATP, Mg^{2+}	Mg^{2+}
Location of cleavage site	Random, at least 10^3 bases away from recognition site	Usually within the recognition sequence	24–26 bases 3' of recognition site	14 bases 3' of recognition site
Restriction versus methylation	Mutually exclusive	Separate	Simultaneous	± simultaneous
Recycling of endonuclease	No	Yes	Yes	Yes
Model systems	*Eco*B and *Eco*K	*Eco*RI	SP1	*Eco*57I + M-*Eco*57I

a, Only five Class III and one Class IV cases have been studied. b, SAM: *S*-adenosylmethionine.

Table 6.2 Recognition sequences of some restriction enzymes.

Enzyme	Name	Organism	Recognition sequence	Observations
Class I	*Eco*K	*E. coli* K12	AACN$_6$GTGC	Cleavage $\geq 10^3$ bases away
Class II	*Eco*RI	*E. coli* RY13	G/AATTC	Palindromic sequence
	M-*Eco*RI	*E. coli* RY13	GAA*TTC	Cognate methylase
	*Rsr*I	*Rhodopseudomonas sphaeroides*	G/AATTC	Isoschizomer of *Eco*RI
	*Ava*I	*Anabaena variabilis*	C/PyCGPuG	Palindromic degenerate sequence
	*Dpn*I	*Diplococcus pneumoniae*	GmeA/TC	Necessitates me-DNA
	*Bam*HI	*Bacillus amyloliquefaciens*	G/GATCC	} Compatible enzymes
	*Mbo*I	*Moraxella bovis*	GATC	
Class III	SP1	Phage P1	AGACC	Cleavage 24 bases from 3' end
Class IV	*Eco*57I	*E. coli* RFL57	CTGGAG	Cleavage 14 bases from 3' end

By convention, only the 5' to 3' strand is shown. N, nucleotide; Py, pyrimidine; Pu, purine; meA, methyl-Ade; * represents the methylating site; / indicates the cleavage site.

proteins which function as a trimeric complex. Each monomer, however, plays a specific role, as has been demonstrated by the introduction of mutations in each of the three genes, singly or in association. This system is illustrated in Fig. 6.3. HsdS, by binding to the DNA, is

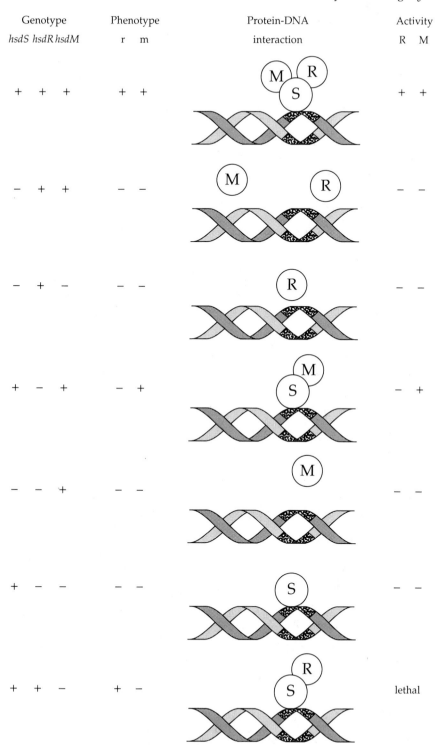

Fig. 6.3 The Class I Hsd system.

necessary for efficient and specific recognition of the target sequence. HsdM, in association with HsdS, is responsible for methylation. HsdR, active only in the trimeric complex, causes DNA cleavage. Binding of the trimeric complex to the DNA is irreversible, each complex perform-ing only one cleaving event.

The recogition sequence, 14 bases long overall, is hyphenated, with two specific external domains separated by a non-specific spacer, 6 nucleotides long (Table 6.2). Methylation takes place inside the target sequence, on a single Ade residue on each strand. The cleavage sites are always located outside the recognition sequence, at distances 1000–5000 nucleotides away, but the actual nicking takes place at random sites. The functioning of the enzymic complex renders the two catalytic activities mutually exclusive and prevents any risk that the endonuclease will be active on DNA which has been specifically methylated.

The *hsd*I genes are expressed as two transcripts, read in the same orientation, *hsd*R on the one hand and *hsd*M and *hsd*S on the other. This organization gives an opportunity for the cell to synthesize only the methylating part of the complex, and avoids the exclusive synthesis of an active nuclease, which would obviously be suicidal.

Class I R–M systems have been described in several bacteria, although it is mostly enteric species that have been screened. The first systems recognized were those from several *E. coli* and *Salmonella* strains. The relatedness of some of these systems is seen in the similar organization of their *hsd* coding region and its location close to the *serB* gene in both species. Reciprocal complementing capacities of the *hsd*M and *hsd*R genes, immunological cross-reactivities and DNA sequence homologies have resulted in the classification into families of *hsd* systems among several enteric bacteria (Section 8.1.4). The target specificities, however, are strictly host-defined. New specificity pat-terns can be created through intragenic recombinations between the *hsd*S genes from two *Salmonella* strains. The HsdS proteins all show a common structure, composed of alternate sequences of conserved and non-conserved domains (Fig. 6.4a). Analysis of the nucleotide sequences and of the target specificities of the recombinants (Fig. 6.4b) suggests that each non-conserved domain plays a role in recognition of (binding to) each of the defined regions of the target site.

Besides those encoded by the bacterial genomes, some plasmid-encoded Class I Hsd systems have also been found.

4.2 Class (Type) II systems

These have simpler protein and genetic structures than the Class I types. Each system comprises two genes, *hsd*M and *hsd*R, responsible

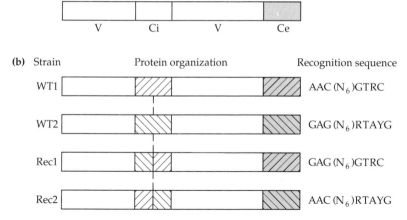

Fig. 6.4 The organization of Class I HsdS proteins in *Salmonella* species. (a) Structure of a HsdS protein. V: variable regions; Ci, Ce: internal and external conserved regions. (b) The Hsd polypeptides and their recognition sequences on the DNA, of two *S. typhimurium* wild-type strains (WT1 and WT2) and two derivatives (Rec1 and Rec2) resulting from *in vivo* intragenic recombination.

for modification and restriction, respectively. Although some enzymes are active as homo-oligomers, they never require the formation of an HsdM–HsdR heterodimer for activity. Their sites of action, consequently, must be more strictly defined and coordinated. The two enzymes must be antagonistically active on the same DNA sequence, otherwise the system would be lethal for the host cell itself. The information for target recognition is present in both enzyme molecules, the regions recognizing the target sequence being either identical or partly homologous. These features led to hypotheses suggesting that they arose either from a common ancestral gene by subsequent duplication or as a result of convergence of the two genes.

The target generally consists of a sequence of 4–7 bp, but may be as large as 9–10 bp. In most cases, this sequence shows a rotational symmetry about a central point and is a palindrome, i.e. the nucleotide sequences in both strands are identical but reversed (Table 6.2, Fig. 6.5). Thus the sequence read on one strand, e.g. in the 5′ → 3′ direction, is identical to that read in the same orientation on the other strand. Both enzyme activities take place at strictly defined sites, but not necessarily the same ones. Methylases are either Ade- or Cyt-specific, and modify only one residue per recognition sequence on each strand if more than one are present. Endonucleases usually cleave within the sequence, although some do so outside the sequence but never further than a few bases off. Control of the endonuclease activity by methylation most frequently acts negatively: addition of the methyl

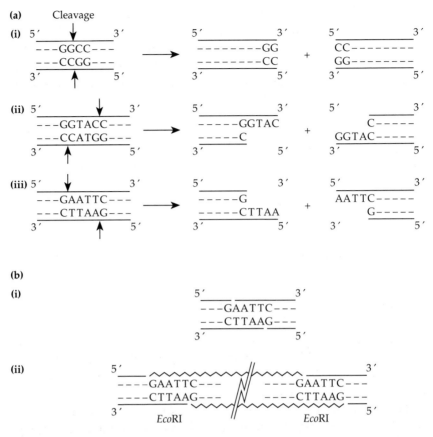

Fig. 6.5 Cleavage and reannealing of DNA extremities cut by Class II endonucleases. (a) Cleavage: (i) Blunt extremities. Ex: *Hae*III (from *Haemophilus aegyptus*); (ii) 3′-protruding extremities. Ex: *Kpn*I (from *Klebsiella pneumoniae*); (iii) 5′-protruding extremities. Ex: *Eco*RI (from *Escherichia coli*R). (b) Annealing. Ex: *Eco*RI-cut DNAs: (i) Intramolecular; (ii) Intermolecular. Insertion of a DNA fragment.

group prevents efficient binding, and thus cleavage, by the restriction enzyme.

A consequence of the palindromic nature of the target is that the nuclease, given the opportunity, automatically cleaves both DNA strands. The cuts take place symmetrically every time they are situated within (or at an extremity of) the recognition sequence. In these cases, three configurations of cleaved extremities can be produced (Fig. 6.5). If the sequence consists of an even number of bases and if the enzyme cleaves at its centre, the two cuts separate exactly the two adjacent and inverse pairs of bases, forming blunt cleaved extremities (Fig. 6.5a). In all other cases, identical single-stranded extremities, the length of which depends on the target size and the site of cleavage, are formed on each DNA extremity. The location of the nicking site *vis-à-vis* the axis of symmetry of the target determines whether the extremities are 3'- or 5'-protruding (Fig. 6.5a). In both cases, however, their antiparallel complementarity allows them to anneal and regenerate the original DNA sequence. Thus, these single-stranded ends are known as sticky, or cohesive, ends (Fig. 6.5b). Endonucleases of this class are frequently used in genetic engineering (Section 4.5; Chapter 16).

Because of the small size of the target and the specific sites of action of the enzymes within this sequence, enzymes from different strains or species showing exactly identical specificities are fairly frequent. These are referred to as isoschizomers (Table 6.2). Enzymes that recognize partially different targets that produce identical single-stranded ex-tremities, i.e. sticky or cohesive ends, are said to be compatible (Table 6.2). Methylation of similar sites on identical sequences leads to an inhibition of cleavage by heterologous isoschizomeric enzymes, and thus protection of foreign DNA.

Class II restriction–modification systems are widespread, and may indeed be the most frequent ones. Surveys published in 1986 and 1987 listed more than 700 of them, among approximately 800 restriction enzymes and 100 methylases recorded in the literature from about half as many strains. The first discovered, and the most extensively studied, is the *Eco*RI/M-*Eco*RI endonuclease/methylase system found in an *E. coli* R strain (Table 6.2).

The genes for a fairly large number of Class II systems have been cloned and often sequenced. Among about 50 of them, belonging to different species, the two genes are close to each other, but not organized in a transcriptional unit. In general, the methylase and the endonuclease genes occur together in a given strain, although some strains seem to code only for a methylase (this is not *sensu stricto* a Class II system, except that these methylases display the characteristics of the other, endonuclease-associated ones).

Comparisons, though still limited, of the structure of several isoschizomeric enzymes have failed to reveal common homologous regions which might be candidates for target recognition. Certain domains, however, are conserved in various methylases, within either the N6-Ade or C5-Cyt enzymes, and are also found in the equivalent methylases from other types of modification systems. For instance, the *E. coli* and T4 phage Dam methylases (Section 7), the Type II *S. pneumoniae* M-*Dpn*II, which all add a methyl to the Ade of GATC targets, and the Type II M-*Eco*RV, which methylates the 5'-Ade of a GATATC sequence, have either high or significant amino acid sequence homologies. The presence of weakly conserved regions shared by Cyt-methylases and Ade-methylases has been taken as suggestive of a common origin.

4.3 Class (Type) III systems

Class III systems are intermediary, in several of their characteristics, between the two preceding ones. Their genetic structure is very similar to that of Class I systems: three genes are organized in two, closely located, transcription units, *hsdM*, *S* and *hsdR*. Each polypeptide has a role similar to that of its equivalent one in the Class I complexes. The precise mechanisms of action of the two complexes, however, clearly differ (Table 6.1), one major difference being that Class III enzymes can simultaneously methylate and cleave the DNA on the same sequence. The efficiency of cleavage is usually not maximal, probably because the possible simultaneous methylation prevents further cleavage.

The recognition target is precisely defined and extends over five to six bases (Table 6.2). It does not show any symmetry, and sometimes can be methylated on only one of the strands, due to the absence of an appropriate methylating site on the other. Cleavage by the restriction enzyme takes place outside the recognition region, 24–27 bases away on the 3′ side.

Only a few systems belonging to this class have been described. The first and best known is that coded by phage P1, which happens to be also the first case of restriction discovered (Section 2).

4.4 Class (Type) IV systems

The characteristics of an R–M system recently discovered in an *E. coli* strain led to its classification as a new type. Another system, though still poorly characterized, from a *Gluconobacter suboxydans* strain, might also belong to this new Class IV. The *E. coli* system, *Eco*57I, comprises two enzymes: a methylase and a methylase–endonuclease, with both methylating activities having the same specificity. Cleavage takes place 14 nucleotides 3′ to the recognition site. The target shows a hyphenated type of sequence, with a twofold symmetry of its extremities.

The methylase–endonuclease requires Mg^{2+}, and is stimulated by S-adenosylmethionine. Complete cleavage is never achieved, probably because of the joint methylase activity of the protein. No information is available as to the genetic organization of this system, or the possible homologies of the proteins with their equivalent from other systems.

4.5 The use of restriction enzymes in molecular genetics technology

A theoretical DNA molecule possesses, on the average, a given four-base sequence every 256 nucleotides (this is exactly correct only for a duplex target sequence/DNA molecule having the same base ratio, for instance a GATC sequence in a 50% G+C chromosome) and a six-base one approximately every 4×10^3 nucleotides. An average-length bacterial chromosome (such as that of *E. coli*, approx. $4.75.10^9$ bp long, representing 3000–5000 genes (Chapter 14)) would thus be cut down to approximately 1000 fragments (assuming all restriction events would occur on the two DNA strands and lead to denaturation of the intervening sequence) by an endonuclease specific for a six-base recognition region, and to 16-fold as many by an enzyme specific for a four-base region. (It is worth remembering that the length of a gene coding for an average-size protein, 30 kD, is approximately 10^3 nucleotides.) Obviously, reality differs from these theoretical estimations:

1 Base composition along the DNA is not homogeneous.

2 G+C contents among prokaryotes show a large range of variation (Chapter 2).

3 Most bacterial strains possess (usually no more than two or three) restriction–modification systems, which implies a certain degree, and consequently a certain pattern, of methylation.

4 Discrepancies between expected and observed frequencies of particular target sequences among bacterial or phage chromosomes have led to the speculation that selective evolutionary processes might have introduced orientated biases in localized DNA sequences (Section 8.1.3).

All these features result in interference with predictions of both the frequency of occurrence and the distribution of a given sequence, and hence introduce large variations in the length of the fragments created by a given restriction enzyme. The frequency of cleavage of the whole chromosomal DNA of a given strain by a particular restriction enzyme may vary from an (almost) total absence of action to a very drastic degradation (Fig. 6.6). Enzymes with average frequencies yield fragments ranging in the size scale of common genes. This makes it worth considering the use of restriction enzymes for cutting down chromosomes to fragments close to the size of the genetic units needed by geneticists.

Two essential characteristics of the mode of action of Class II restriction enzymes explain their primary role in present-day DNA molecular technology. The possible reversible annealing and ligation of the cleaved extremities, at least for enzymes yielding cohesive ends, allows complete recovery of the initial DNA sequence (Fig. 6.5b). It also renders possible the controlled ligation of unrelated DNA molecules, providing they have been cut by the same (or isoschizomeric or compatible) enzyme(s). Since the reannealing operation restores the initial target sequence (not always true for pairs of compatible enzymes, e.g. *Bam*H1 and *Mbo*I, Table 6.2), the whole cleavage–ligation scheme allows for the insertion of a DNA sequence with duplication of the original target sequence on each side (Fig. 6.5b). This takes place with total reproducibility of location and structural configuration and can be

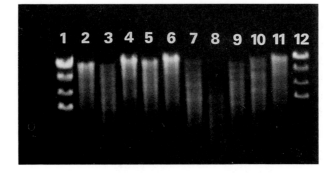

Fig. 6.6 Range of fragments sizes obtained by cleavage of chromosomal DNA by several restriction endonucleases. Total DNA from the cyanobacterium *Synechocystis* PCC6803 was purified and treated with several restriction endonucleases (lanes 1: size scale. λ phage. DNA digested by *BclI*; 2: *AccI*; 3: *AvaI*; 4: *Bam*HI; 5: *Eco*RI; 6: *Eco*RV; 7: *Hae*II; 8: *Hinc*II; 9: *Hind*III; 10: *Hpa*I; 11: *Kpn*I; 12: size scale). Fragments were resolved according to size by electrophoresis on an agarose gel (Chapters 3, 16). Smears indicate large numbers of fragments with, statistically, very small size increments. Enzymes such as *Bam*HI, *Eco*RI and *Eco*RV cut the DNA very poorly because their possible restriction sites are protected, since replication of fragments of this DNA in an m⁻ *E. coli* leads to cleavage by these enzymes (Courtesy of L. Beuf, Université Aix-Marselle II).

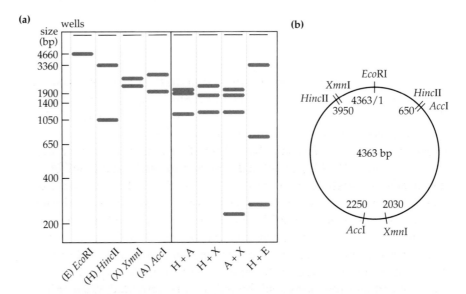

Fig. 6.7 Construction of a restriction map. A partial restriction map of the artificial *E. coli* plasmid pBR322 (total length 4363 bp) is shown. The DNA was digested with four restriction enzymes, either alone or as pairs, and the resulting fragments separated by electrophoresis on an agarose gel. (a) Comparison of the sizes of the fragments obtained after double versus single digestions allows the placing of the restriction sites on the DNA molecule. (b) The data shown here leave an ambiguity regarding the position of the *Eco*RI site inside the small *Hinc*II fragment. It has been placed on the map at its correct position, taken arbitrarily as 1.

performed as many times as needed. These properties have obviously opened the field to possible transfers, and thus cloning, of chosen DNA sequences (Chapter 16).

The fixed and specific locations of recognition sequences along each DNA molecule have been used to construct restriction (also called physical) maps of either particular regions, small molecules or plasmids (Fig. 6.7) or whole chromosomes (Chapter 14).

5 The methylated-adenine (Mar or Mrr) and methylated-cytosine (Mcr) restriction systems of *E. coli*

Studies (1975–1980) concerned with the origins of the low transformation efficiencies exhibited by various *E. coli* strains led to the identification of three new restriction systems, which function as the counterparts of the Hsd systems. Equivalent systems have not yet been seriously looked for in other species. The common trait of their recognition sequences is the obligate presence of a modified base (Ade or Cyt) included in the sequence. Since the unrestricted, thus protected, form is the non-modified DNA, which is completely opposite to the systems previously described, these systems do not comprise cognate methylases.

5.1 The Mar system

The first methyladenine restriction system (*Dpn*I) was reported in 1975 in *S. pneumoniae* (the name *Dpn* instead of *Spn* originates from the former name, *Diplococcus pneumoniae*, given to this strain). The endonuclease cleaves a GmeATC sequence. Other restriction enzymes showing the same specificity have been found in a few other bacteria (Section 6.1). Extensive analyses have been carried out on a methyladenine restriction system present in *E. coli* K12, named Mrr (*m*ethyladenine *r*ecognition and *r*estriction) or Mar (*m*ethyladenine *r*estriction), discovered in 1987. The approach which led to the identification of the Mrr system as a restriction one (implying a specific endonuclease) and of its probable recognition site is a typical one. It consisted in the observation that the introduction into an Hsd⁻ *E. coli* strain of certain exogenous Hsd methylase genes resulted in either low transformation efficiencies, cell death or induction of the broad-range SOS rescue process and DNA repair (Chapter 8). These deleterious effects could be attributed to DNA cleavage, specifically triggered as a consequence of the establishment of patterns of DNA methylation normally not present in the *E. coli* host and corresponding to the action of the M-*hsd* cloned gene. A consensus sequence (GmeAC or CmeAG) was elaborated from comparisons of the individual recognition sequences of several M-Hsd enzymes incompatible with survival of the host strain. Both the induction of DNA cleavage and the necessity for a specific recognition sequence were taken as evidence that Mar was a standard restriction system.

The *mar* locus maps close to the *hsdR, M, S* ones in *E. coli* K12. It is absent from *E. coli* B.

5.2 The Mcr systems

The Mcr$^+$ phenotype is a complex of two genetically distinct and sequence-specific systems, McrA and McrBC.

5.2.1 *The McrBC system*

In *E. coli* K12, the *mcrBC* region also maps close to the *hsd* cluster and comprises two or maybe three genes. This clustering of three restriction systems out of the four presently known in *E. coli* K12 is noteworthy and may have a functional rationale. Tight linkage of these loci might limit the risk of the strain divulging its DNA identification. It might also ensure global transfer of an equilibrated system of protection for both the potentially transferred DNA and the recipient strain (Section 8.1.4). Whether such a situation is also the rule in other species is still not known.

Comparative sensitivities to the McrBC system of DNAs bearing different patterns of methylated cytosine produced by appropriate M-Hsd enzymes led to a hypothetical GmeCG minimal recognition sequence. The nature of the modification on the cytosine is rather flexible, as methyl groups on the C5-carbon or the N4-nitrogen atoms and hydroxymethyl groups on the C5-carbon mediate McrBC restriction. The capacity to modulate the recognition specificity results from the activity of the McrC protein, a basic protein capable of binding to the DNA. Modulation of the restriction specificity was observed by comparing *mcrC*$^+$ and *mcrC*$^-$ mutants.

The McrBC$^+$ phenotype is frequent. Among about 30 common *E. coli* laboratory strains, mostly K12 derivatives, approximately half were positive.

Genetic mapping and phenotype comparisons have established that the McrBC system is identical to one of the two Rgl (restriction of glucoseless phage DNA) ones, namely RglB, described in 1952. Wild-type T-even phages, with glucosylhydroxymethylcytosine in their DNA, and mutants deficient for the addition of the hydroxymethyl group to their cytosines were found able to develop, while others which lacked the glucosylating capacity were restricted, in the same *E. coli* K12 host (Section 6.2). Since these former observations cover only one aspect of the McrBC phenotype, the more accurate mnemonic McrBC rather than RglB has been maintained. The nucleotide sequence of the *mcrBC* region itself is poorer in G + C bases than the average *E. coli* chromosome. This might indicate either a recent acquisition by this genus (or by a species from it) of DNA from an exogenous origin, or a locally biased use of codons so as to limit the chance occurrence of an McrBC cleavage sequence inside the region.

5.2.2 *The McrA system*

Less information is available concerning the McrA restriction system. In *E. coli* K12, the *mcrA* locus maps near *purB*, i.e. clearly away from the cluster of restriction–modification genes lying close to *serB*. Comparisons of mutant phenotypes and complementation and recombination studies have also led to identification of the *mcrA* locus with that of the previously known *rglA* one. The recognition sequence is still not completely elucidated. Similarly to that of McrBC, it includes meCG, but it is clearly distinct.

The McrA$^+$ phenotype was found somewhat less frequently than that of McrB$^+$ among the same set of *E. coli* laboratory strains.

6 Other modification and/or restriction systems

6.1 Other methyl-DNA restriction systems

Restriction enzymes requiring a methylated sequence have been described in *Caulobacter* and several *Neisseria* strains. Like the *Dpn*I enzyme, they are active on the GATC sequence.

Two original systems have been described in bacterial species considered to be phylogenically distant, not only between themselves but also distant from both *E. coli* and *S. pneumoniae*. A strain of mycoplasma, *Acholeplasma laidlawii*, was shown in 1986 to be able to restrict DNA containing C5-meCyt. Its originality resides in the absence of sequence stringency necessary for restriction, since 5-meCyt by itself would serve as a recognition site.

Methyl-DNA restriction is frequent in *Streptomyces* species, since seven strains out of nine tested showed either meAde- or meCyt-specific restriction. The capacity of *Streptomyces avermitilis* to restrict DNA containing either N6-meAde or C5-meCyt might reflect the presence of two meDNA restriction systems, with tight specificity for one or the other methylated base, respectively.

The small number of such systems described and, in all cases but the *S. pneumoniae* one, their rather recent discovery, might reflect a low frequency of occurrence of this type of protection strategy. It may also, however, correspond to a bias introduced by the screening procedures. Because of the ubiquitous presence of Hsd-type systems, search for restriction capacities has often been limited to the use of non-modified DNA as possible substrate, thus preventing the discovery of enzymes active on methylated DNA.

6.2 Glucosylation as a modification system

Another type of modification utilized in the T-even phages of *E. coli* is based on glucosylation of the 5-hydroxymethylcytosine (HMC) bases. Normal T2, T4 and T6 phages are unusual in carrying HMC instead of cytosine, some of which (all in T4) carry glucose residues (Chapter 4). This is dependent on the presence of uridine disphosphoglucose (UDPG), an intermediate in galactose metabolism also involved in cell wall synthesis. *E. coli* B/4, a strain resistant to T4, lacks the UDPG pyrophosphorylase, and thus can neither ferment galactose nor glucosylate phage DNA. When *E. coli* B/4 becomes infected with T2 or T6, the phage can develop and give a normal yield, but these progeny are unable to undergo another cycle of growth in B/4 bacteria. Their DNA, not being glucosylated, is broken down by the bacterial restriction endonuclease(s). These progeny can nevertheless develop on hosts such as *Shigella* which do not restrict them. The phage-encoded restriction endonuclease(s) do not cut their own non-glucosylated DNA provided it is still hydroxymethylated and thus different from host DNA.

7 The DNA adenine-methylation (Dam) and DNA cytosine-methylation (Dcm) systems

Since they have no accompanying restriction endonucleases in *E. coli*, the strictly methylating systems, Dam and Dcm, do not belong, *sensu stricto*, to the restriction–modification processes, but their presence obviously has consequences on the restriction profiles of the corresponding DNAs.

7.1 The Dam system

Dam methylases add a methyl group originating from an *S*-adenosylmethionine to the Ade residue present in GATC sequences arranged in a palindrome. It can exist at three levels of methylation: null, hemi- or full methylation, and Dam-mediated GATC modification is involved in a large range of cellular processes.

The methylated state of the GATC sequences and their over-representation in DNA regions corresponding to origins of replication have been shown to participate in the control of replication initiation. Modification of the DNA structure consecutive to the presence of the methyl group controls the efficiency of interaction with the DNA of the proteins forming the initiation complex.

Similar mechanisms are offered to explain the implication of GATC methylation in the expression (at the transcriptional level) of a number of genes or operons. Either the -35 or the -10 boxes of the corresponding promoters contain a GATC sequence. The influence of methylation on the expression rate of different genes in the same *E. coli* K12 strain is either positive or negative, but the rationale for this is still unexplained.

Repair of small mismatches along DNA duplexes is also dependent on methylation of GATC sequences for recognition of both the erroneous strand and the site of initiation of repair (Chapter 8). The endonuclease involved in this process (MutH), triggered by the presence of a mismatch, requires hemimethylated (and non-methylated) DNA for recognition and cuts only one, non-methylated strand.

A crucial question which still awaits an explanation concerning Dam methylation arose when *dam⁻* mutants were isolated in *E. coli*. These clones, although they show a pleiotropic phenotype corroborating the influence of the level of GATC methylation on the various processes mentioned above, are otherwise normally viable. This methylation process thus might not play an essential role, unless the cell is saved by the presence of bypasses or subsidiary mechanisms.

Several *dam* genes have been sequenced, revealing that the proteins share regions of homology, probably involved in DNA sequence recognition. No strong conservation of these enzymes seems to have taken place.

While the *E. coli* Dam enzyme has no cognate restriction nuclease, GATC methylases functioning as part of Class II Hsd systems exist in a number of bacteria.

Restriction systems compatible with this methylation profile may have been preferentially selected during evolution. The large proportion of restriction enzymes using as recognition target either GATC or larger sequences including GATC (15% of all listed Class II systems) supports this idea. The figure approaches 50% if only systems using four-base targets are considered. A random occurrence would correspond to around 6% (one of the 16 possible four-base palindromic sequences). Strains from three species, however, have developed restriction systems specific for GmeATC sequences (Section 5.1).

7.2 The Dcm system

C5-cytosine methylation is also a widespread capacity among both prokaryotes and eukaryotes. The role and the specificity of this process in eukaryotes are very poorly understood. Among prokaryotes, the best-known case is that of *E. coli* K12. The sequence recognized by the *E. coli* Dcm methylase is CC(A/T)GG, in which only the internal cytosine is affected. Although the mechanism has not been elucidated, Dcm methylation is necessary in mismatch repair, in which the very-short-patch (VSP) mismatch correction is concerned (Chapter 8). While the two proteins implicated in initiation of Dam-dependent mismatch repair are also required for VSP repair, the endonucleases involved in the two processes are different. No other cellular role is known for the Dcm system.

8 Restriction–modification and evolution

The two reciprocal aspects of the possible interrelationship between restriction–modification processes and evolution should be considered.

8.1 Taxonomic analysis of the restriction–modification systems themselves

8.1.1 *Distribution of restriction–modification systems*

Although a large number of restriction–modification systems have been characterized, and an even larger number of strains screened, the information available is still too scarce, and mostly too unsystematic, to allow conclusions to be drawn regarding the distribution of this function among prokaryotes, allowing only a few tentative generalizations to be put forward:

1 Not all strains possess restriction–modification systems. The proportion of those devoid of detected systems is impossible to evaluate, since negative information is usually not published. An example of a systematic study performed among a group of bacteria is described below (Section 8.1.4).

2 In addition to the presence of the strictly methylating systems, some bacterial strains (and phages) possess several restriction systems, of similar or different nature.

3 Occurrence of the various types of restriction and/or modification systems is widely varied. The bias introduced by many screening procedures on the discovery of methyl-DNA-restricting enzymes may have led to a large underestimation of the representation of these systems among prokaryotes. A discussion on comparative efficiency versus risks between the two protective strategies, i.e. restriction of methylated DNA versus restriction of non-methylated DNA compulsorily associated with cognate methylation, obviously has to remain open.

4 Among the three classes of Hsd systems, Class II enzymes are by far the most frequently found (95%), even though this figure should be taken with caution, since at least in recent screening procedures the search for Class II enzymes is favoured because they are of interest in molecular biology technology (Part 3). A possible argument explaining a selection in this direction during evolution is that the tighter definition of the restriction site may limit the frequency and the span of uncontrolled or erroneous cuttings. These systems are also less costly to the cells.

5 The ubiquitous presence of methylation in prokaryotes and eukaryotes and its involvement in a large range of cellular processes argue in favour of the hypothesis that a general function was the primordial role of this system. Its protective function might have developed secondarily.

8.1.2 *Conservation of the enzyme molecules*

The information available indicates that conservation of primary or secondary structures of either the endonucleases or the methylases may not have been an essential requirement. It was not possible to identify ancestral active core molecules, but only conserved domains probably in charge of recognition of, and binding to, the DNA targets for a few methylases. Even isoschizomers do not seem to form families of molecules.

Conservation of primary structure among Dam enzymes seems unpredictable. Thus, only limited regions of homology are shared by several methylases recognizing the GATC sequence, namely the *E. coli* and the *E. coli* T2 and T4 phage Dam methylases and the M-*Dpn*I enzyme. In contrast, the two phage enzymes are highly homologous (the two phages themselves are closely related (Chapter 4)). The methylase of the *Eco*RV Hsd system, specific for the GATATC sequence, also possesses the homologous regions common to the Dam enzymes.

8.1.3 *Limited variability of target specificities*

No oriented selection leading to preferential occurrence of particular target types within bacterial species seems to have taken place. Thus isoschizomers are found in widely varying taxonomical groups. This diversity, however, is in contrast with the marked limitation of the actual number of site specificities encountered as compared with the range of possible ones. (The palindrome-forming sites composed of four to seven bases can give 160 different sequences.)

A recent survey covering 700 restriction enzymes identified only 120 different targets, taking into account all sizes and structures possible.

Among the 256 four-base sites theoretically possible, the 16 combinations corresponding to palindromic structures have been almost the only ones encountered. Of these, GATC, the site for Dam methylation, is also the most frequent one for restriction, and the only known one for which a few me-DNA Hsd restriction enzymes have been found. On the other hand, the presence of GmeATC restriction enzymes must indicate an absence of Dam methylase in the corresponding strains (unless the restriction enzyme is compartmentalized, a rather improbable situation). It will be interesting to know whether equivalent systems replace it in its various control functions. The GATC sequence has not been found as a restriction target in any of the enterobacteria screened (Section 8.1.4).

The above survey has clearly pointed out the existence of limiting processes in the choice of targets within a species and a family. A taxon specificity, corresponding to the selection of only a subset of all possible targets inside each taxonomic group, has been postulated. Two arguments support the idea that this process consists of a negative selection relying on methylation rather than restriction specificity:

1 Methylations incompatible with gene expression or the structural stability of the DNA tend to be eliminated, since deleterious effects resulting from methylation, such as enhancement of mutation frequencies or the induction of SOS repair processes (Chapter 8) have been described.

2 The existence of methyl-DNA-restricting enzymes suggests a control of the specificity of methylation.

This proposed scheme is corroborated by the frequency of occurrence of four-base sites compared with larger recognition sequences among all known restriction systems. Their proportion (about 14% among 750 Class II systems) is much lower than that expected (approx. 50%) considering all four- to seven-base palindromic sequences and the frequency of a four-base (1/250) as compared with a six-base ($1/10^3$) sequence on a DNA molecule. That such a selection may not have worked against the GATC methylases could result from the general (and possibly more ancient) role associated with this function. Selection may, however, have discriminated against the GATC sequences themselves. Thus the *E. coli* chromosome is relatively poor in such sequences. Their high density in the *oriC* region does not contradict this, since Dam methylation, primordial in controlling initiation of replication, must have been maintained and even amplified there.

8.1.4 *The example of the Enterobacteriaceae*

A systematic survey of the known R–M characteristics of more than 1000 strains belonging to three species of Enterobacteriaceae led to a

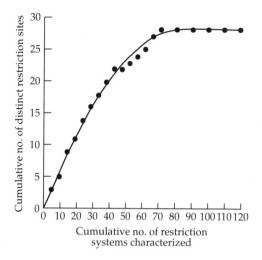

Fig. 6.8 Previsible variability of *E. coli* strains DNA restriction sites. The cumulative numbers of distinct restriction recognition specificities determined in the course of characterization of newly discovered restriction systems in a group of *E. coli* strains are plotted against the total number of systems tested (adapted from Janulaitis *et al.*, 1988).

number of interesting observations. The restricting capacities were determined by testing cell extracts against DNA sequences containing recognition sites for a very large number of known enzymes. The proportion of strains possessing R–M capabilities was low, since only about 30% of the strains screened showed activity.

Recognition sequences have been determined for 170 of the 348 R–M systems detected. The most interesting conclusion reached is that the variability of specificities is limited, as proved by two observations:

1   Only 33 different recognition sites were covered by the 170 systems.
2   The number of new specificities discovered among the subset of *E. coli* strains decreased during the course of the study (Fig. 6.8), suggesting that the 28 sequences used by this group of strains represent the quasi-totality of those used by the whole *E. coli* species.

The genetic relatedness shown by the 24–60% DNA homology existing among three species is reflected in the proportion of their common restriction specificities: three out of the four sites used by the *K. pneumoniae* systems and seven out of the 11 sites from the *Citrobacter freundii* ones are also used by the *E. coli* systems, one site being common to all. Although the numbers dealt with are small, and should thus be taken with caution, they fit in the range of genetic relatedness exhibited by these species.

No four-base recognition site has been found, as is the case for all known enterobacterial restriction enzymes. The only exception is in a strain of *Providencia*, a genus which appears to be the most distantly related member of the family, on the basis of having the lowest level of DNA homology with the others.

8.2 Possible roles of restriction–modification in bacterial evolution

Besides a role tending to eliminate the mutations which create, in a given strain, sites that are recognizable by the restriction system(s) present in that strain, R–M processes may also have influenced the evolutionary role of genetic exchanges among strains or species. The normal substrate for a restriction enzyme is invasive foreign DNA, the presence of which results from the functioning of any of the natural DNA transfer processes existing among prokaryotes (Chapters 10–12), including phage infection itself (Chapter 4). A capacity for the recipient strain to select the DNA at entry has been described only in the case of transformation by *Haemophilus influenzae* (Chapter 10). The selection, based on the recognition of a particular sequence, is not very stringent, since at least the various strains of the same species have a high probability of carrying that sequence. Even these related strains, however, may differ in their R–M profiles. In all other cases of DNA transfer, obviously including the artificial ones developed for molecular genetics studies, the most frequent situation will be for the invasive DNA to be sensitive to the endogenous restriction of the host cell. This explains why a strict control of the R–M capacities of the strains used for genetic manipulation purposes is necessary (Chapter 16).

The ubiquitous presence of R–M systems is thus considered to be the safest means for the cells to protect their genetic integrity. The process of protection, i.e. selective destruction of the invasive DNA, has been likened to a primitive immune process, in which the neutralizing system does not differentiate one from among various infective agents, but classifies them as foreign through their absence of a recognition (protective) signal.

A consequence of the functioning of these protective processes is the establishment of barriers between independent cells, and thus the construction of stricly defined clonal lines. It is a well-known rule in genetics that such genetic isolation is deleterious for the long-term survival of the clone considered. It is also generally accepted that exchange systems in general, and in prokaryotes in particular, have indeed played a decisive role in evolution. It has thus been suggested that the understanding of the influence of R–M systems in evolution had been oversimplified. The very distinct characteristics of the three Hsd restriction types, the most frequent modes of restriction, led to the suggestion that their presence may result in opposite effects on DNA integrity.

Type I restriction systems are costly to the cells. The R–M system consists of three proteins, of which the endonuclease is not recycled after enzymic action but remains bound to the DNA and is thus used only once. The protective effect is not specific because the cleavage site is distant from the recognition site and randomly chosen. The massive DNA degradation that follows cleavage utilizes large amounts of ATP. Between 1% and 0.01% of a phage population infecting a cell harbouring only a Class I R–M system escape restriction, a figure decreasing to about 10^{-6} in Class II or Class III hosts (Section 2). It has been argued that the Class I R–M systems are in fact important (and have therefore been maintained through evolution) in promoting occasions for general recombination, by creating DNA fragments with random single-strand extremities. If such were the case, these systems would favour rather than limit genetic exchanges.

The characteristics of Class II R–M systems, strictly defining both their recognition sequences and the site of cleavage, are more efficient for protection and less likely to provide chance substrates for recombination. Their overwhelming proportion among Hsd systems probably results from their greater efficiency and lower cost in performing the vital task of maintaining the integrity of genetic equilibrium.

Class III restriction enzymes occupy an intermediary position between those of Class I and Class II. The more limited randomness of their actual site of DNA cleavage, however, favours a protective role.

The genetic organization of most R–M systems as clusters of genes with a precise expression scheme, which is constant among individual Hsd classes, must have favoured efficient transfers of R–M modes present on invasive DNA molecules (bacterial and phage genomes or plasmids) to other DNA entities. Such stabilizations of new R–M systems in a cell obviously reorientates its relationship with the other organisms, both positively and negatively, thus contributing to more varied evolutionary possibilities.

References

Janulaitis A., Kazlauskiené R., Gilvonauskaité *et al.* (1988) Taxonomic specificity of restriction-modification enzymes. *Gene*, 74, 229–32.

PART 2
MODIFICATION OF THE
GENETIC MATERIAL

7: Mutations and Mutants

Mutants have already been defined, and extensively employed, in the previous part of this book, as tools to elucidate various processes. Indeed, the demonstration of the genetic organization controlling a biological process cannot be achieved without the isolation and study of mutants displaying changes to the wild-type phenotype. In fact, mutants arise continuously in cell populations, not only as organisms exhibiting a metabolic block, but also as organisms that show wide ranges of modifications for any genetic trait of the strain. The reversion of a mutant to the initial state may also happen, and this constitutes another mutagenic event. With the accumulation of knowledge about the molecular nature of the modifications that result in mutants and the mechanisms responsible for these changes, it is now possible, in many cases, not only to isolate spontaneously formed mutants, but also to induce, at random or sometimes specifically, a required modification. In the latter case, a high level of knowledge of the biological process and, more precisely, of the DNA region involved is necessary (Chapter 16).

The aim of this chapter is to analyse the nature of mutational events, their mechanisms and the conditions of their appearance. In parallel, it is of importance to delineate the possible correlations between the categories of mutations, as defined at the molecular level of the genetic material, and their consequences at the functional level for the active molecules. Emphasis will also be put on the means and strategies available for the recognition and isolation of mutants, since this is one of the key problems of any genetic work.

1 Definitions

Operationally, a mutant is defined as a variant of a given strain or species in which usually one but sometimes several characteristics are modified and which transmits the new trait(s) to its progeny. In order to be heritable, the modification at the origin of the formation of the mutant, referred to as a mutation, can only have taken place in the genetic material of that organism. A mutation is thus a modification of the genome, and hence of the chemical organization of the genetic material.

Mutations appear suddenly, without any transitional stage between the initial and the final states of the organism. When established, they will be permanently present, whether or not the conditions of development of the mutated organism allow their detection. Thus a bacterium that was originally sensitive to an antibiotic can undergo a mutation rendering it resistant to the drug. The new form of response to the drug, however, may remain undetected. Only chance or deliberate assays in the presence of the drug, which can take place any length of time (i.e. any number of generations) after the mutational event, will reveal the difference.

It is important to differentiate a mutation from an adaptation process. Adaptation is a consequence of a related modification in the environment, needs time for its expression and is reversible, since the disappearance of the new environment factor results in the consequent arrest of expression of the adaptative trait (Chapters 15 and 16). Only the capacity to adapt is transmitted to the progeny organisms.

The initial isolate of an organism, often from the 'normal' naturally occurring population, is referred to as the wild type, from which many different mutants may arise for each genetic character of the organism. It is clear, however, that the definition of the wild type is a purely arbitrary one, and is often imprecise, depending on the individual specimens which were first isolated and purified. In principle, it is characterized by the genetic form of each of its genes. However, obviously only a fraction of these are described or anticipated, even in the case of a well-known species such as *E. coli*. Hence, it is often better to describe a particular organism as wild-type for the particular gene or trait in question. Deviations, that is mutations, for unknown characters, which may occur during successive subculturings of a strain, may well happen without being noticed.

It is thus important to keep in its initial state, or as close to it as possible, the isolate considered as the reference one. It is advisable, also, to go back to it every so often, if comparisons are to be made with other strains, isolates or mutants. Genetic deviations of 'lab' strains have been mentioned in a number of instances, and indeed can be expected (Section 4.3). Most of the time, if not always, the growth conditions imposed in laboratories differ from natural ones. In particular, it is a frequent custom among scientists to look for 'optimal' conditions, defined as those leading to rapid growth. These may in fact be equivalent to a pressure tending to favour faster-growing deviants (that is mutants) of the original organism.

The sum of the genetic traits present in the genome (the total amount of genetic material, whether chromosomal or plasmidic) of an organism, i.e. all the functions it can express in the form in which each is present, constitutes its genotype. Two strains differing by a single mutation in a given gene involved in function A will have the same overall genome (the same total functions or genetic determinants), but their genotypes will differ by this single trait. They are said to be isogenic except for function A. Similarly to eukaryotes, different forms of a gene are called alleles. Any gene can potentially exist in a large number of allelic forms (Section 3).

The genotype of an organism can thus be considered as a more or less stable property, modified only by mutations (we will consider their frequencies of appearance in Section 4.3) or by gain or loss of supernumerary genetic elements, such as plasmids (Chapter 3), phages (Chapter 4) or mobile elements (Chapter 5), and can be used to define the organism, as stated above.

Of all the existing genes of an organism, only a fraction may need to be expressed under a chosen set of conditions. This sum results in the phenotype of the organism under these conditions. The word phenotype refers to the visible, or expressed, traits of the organism. It is usually defined at the functional level, as one of several possible results of the expression of the genotype. Changing the environment may lead to a different phenotype, due to changes in the expression level of some genes (Chapters 15 and 16).

The phenotype is thus an intrinsically variable characteristic, for which the conditions of observation must always be specified. For example, a wild-type cell of *E. coli* may seem not to synthesize β-galactosidase, and will indeed not do so in the absence of lactose, the inducer of the lactose metabolic functions (i.e. the lactose operon (Chapter 15)). Its phenotype would then be described as 'β-galactosidase-minus', as would that of a mutant deficient for either the synthesis or the activity of the enzyme. Upon addition of the inducer, however, the two phenotypes would then exactly reflect the respective genotypes of the two strains. In this example, the lactose operon specifies a coordinate synthesis of two more enzymes, the β-galactoside permease and acetylase (Chapter 15). As described below, this situation holds further consequences for the phenotype.

A number of circumstances add to the ambiguity of definition of the phenotype and to its inaccuracy as representing the genotype. Most biosynthetic pathways involve several successive enzymic reactions that ensure the elaboration of the end-product, such as a metabolite, a vitamin, a cofactor, or the catabolism of a substrate. Obviously, an impairment in any of the enzymic activities involved will result in an impairment of the whole metabolic chain. The apparent deficiency, that is the phenotype, of such cells will at first sight be defined as the lack of production or catabolism of the molecule which constitutes the key of the whole process, usually either the end-product or the initial substrate of a metabolic chain. Only a more precise definition of the phenotype will distinguish which of the various possible points in the chain is defective, and thus the various genotypes, in a series of, at first sight, phenotypically identical mutants.

To illustrate this in the case of the lactose utilization genes in *E. coli*, two genetically different mutations can lead to a 'lactose-minus' phenotype: an inactive system for lactose transport into the cell (permease) or an inactive β-galactosidase. Since it is technically easier, and less time-consuming, to determine an incapacity to grow on lactose than to measure each of the two specific enzymic activities responsible for this catabolism, the first definition of the phenotype will usually be the imprecise, lactose-minus one.

This remark is also valid for any metabolic pathway or multistep process that participates in a common physiological event. Chromosome replication and all known cases of adaptation involving cellular differentiation (sporulation, N_2 fixation (Chapter 16), morphogenesis related to cell-cycle periods) are examples.

2 Nomenclature

Rules have been elaborated to make as clear as possible the level at which a strain characteristic is defined. Recommendations initially proposed by Demerec and collaborators in 1966 have since been generally followed and should still be used as a guide.

A genotypic trait is properly designated with italic letters, usually three, in lower case, often an acronym more or less reminiscent of the function implied, followed, if necessary, by an upper-case letter to identify each gene out of a set controlling a particular phenotype, e.g. *trpA*, *trpB* (for genes implicated in the synthesis of tryptophan). Whenever a particular mutation needs to be indicated, a number is finally added, so that different alleles can be described in the same gene, e.g. *trpA1*, *trpA2*. Identical phenotypic characters may sometimes result from different metabolic processes. A case is the resistance to a number of antibiotics, which can be coded by either chromosomal or plasmidic genetic determinants (Chapters 3 and 5). A nomenclature allowing easy attribution to either genetic structure has been described (Chapter 3).

A phenotype is usually represented by a three-letter non-italicized symbol, starting with a capital letter. The symbol is not always the same as that of the genotype. For example, a *Rhizobium* strain deficient for N_2 fixation (Nif$^-$) may be so because of a mutation in the N_2-reducing system (*nifX*$^-$) or because it can no longer form the nodules necessary in the plant host for N_2 reduction (*nodX*$^-$). No further indication is applicable since by definition the phenotype represents the expression resulting from the genotype, be it well defined or not, e.g. Trp⁻.

Deficiencies are indicated by a minus (−) superscript following the symbol. Abilities, when their specification is important, are similarly shown by a + superscript. In order to simplify the reading, however, it is accepted that an absence of mention means a wild-type allele (usually proficiency) and the mention of the character with no further indication means a mutation of the character (Table 7.1). Sensitivity and resistance are represented by small s and r superscripts, respectively. Here again, the wild-type (usually sensitivity) character may be omitted if no ambiguity results.

Applications of these definitions are shown in Table 7.1. Wild types and possible mutants are described for the *E. coli* lactose catabolic pathway, controlled by two genes coding for β-galactosidase and lactose permease. Bacteriophages adsorb to specific cell receptors (Chapter 4), and mutations in these complexes will render the cells resistant to the virus. This is exemplified by the case of *E. coli* and bacteriophage T1.

This second example illustrates a characteristic of all rules, that they have exceptions. In this case (and this seems to be generally true for responses to bacteriophages), the phenotype is directly represented by the name of the phage, which is very short (also a general situation).

The previous rules have been widely respected since they were proposed by Demerec *et al.* in 1966. Some previously studied characters have been renamed since in order to fit the rules. A fairly large number of exceptions remain, however, and this is particularly true of processes for which little is known of their genomic organization. Historical

Table 7.1 Examples from nomenclature of genes, mutations and mutants.

	Phenotype	Genotype
Lactose metabolism		
Wild-type lactose utilizers	Lac⁺(β-galactosidase⁺, permease⁺)	*lacZ⁺*, *lacY⁺*
Possible lactose-utilization-deficient mutants	Lac⁻ (either β-galactosidase⁻ or permease⁻)	*lacZ⁻*, *lacY⁺*, or *lacZ* *lacZ⁺*, *lacY⁻*, or *lacY* (*lacY00* or *lacY11* if information available)
Response to phage T1 infection		
Sensitivity (wild-type state)	T1ˢ	*tonA⁺*, *tonB⁺*, etc., i.e. phage T1 sensitivity
Resistance	T1ʳ	*tonA⁻*, *tonB⁻*, etc., i.e. phage T1 resistance

reasons may also have determined the maintenance of non-orthodox symbolism. Actual lists of functions, genes and symbols are available for a number of species, including *E. coli*, *Bacillus subtilis*, *Salmonella typhimurium* and *Streptomyces*. Homogeneity of nomenclature has, unfortunately, not always been the rule among different organisms. Phage gene nomenclature often also constitutes an exception.

3 Types of mutations

Mutational events can lead to several types of modifications on a DNA molecule. The genetic and functional consequences may depend both on the nature of the change and on its location within the genetic organization of the DNA, and also on the role of the genomic region affected. Mutations can be classified in more than one way, e.g. with regard to: (i) the molecular nature of the change; or (ii) its physiological effect. Although the consequences of mutations will be considered mostly for DNA regions coding for proteins (i.e. open reading frames, Chapter 15), it is obvious that modifications in a DNA region either expressed only as RNA (e.g. ribosomal or transfer RNAs) or non-expressed but used as recognition signals for expression (e.g. promoters) can also occur, and will also have consequences which depend on the nature and the location of the change. The nature of possible changes can be characterized as follows.

3.1 Base subsitutions

Base substitutions are mutations in which one base has been changed for another and are for this reason also referred to as point mutations. Point mutations can exist either as transitions, in which a change to the same chemical type of base has occurred, e.g. a purine to another purine or a pyrimidine to another pyrimidine, or as transversions, in which a change to the other type of base has occurred, e.g. a purine to a pyrimidine, or vice versa. In recent years, the addition or deletion of a single base has incorrectly been referred to as a point mutation; these were originally named frameshift muations, a description which is much more accurate and informative.

The following situations may occur with point mutations in regions actually coding for proteins:

1   A change to a related codon (the unit specifying an amino acid), i.e. one of a redundant class coding for the same amino acid, will allow the composition of the resulting polypeptide chain to remain the same. The mutation is said to be silent or conservative. However, the protein specified may well be produced at a lower rate. A particular organism usually has favoured codons with a greater quantity of the correspond-

ing tRNAs present so that a related codon, although still coding for the same amino acid, may limit expression.

2 A change to an unrelated codon leading to an amino acid change, is known as a missense mutation. Its effect will depend on the position of the change in the protein. If the change is at an active site, it could have a profound effect, ranging from partial to total inactivity. The consequences will depend on the importance of the system to the life of the cell and will range from a relatively subtle phenotypic change to non-viability. If the change occurs in a position of the protein structure which has no functional group, and has no deleterious effect on the tertiary structure of the protein regarding the availability of active sites, then no or little (the mutation is then called leaky) change may be manifested.

3 A change of a sense to a nonsense codon will have the effect of terminating the synthesis of the polypeptide chain at that point. The consequences to the life of the cell would again depend on the position of the change and on the essentialness of the polypeptide concerned.

A convenient tool used to elucidate whether a deficient phenotype is caused by the absence of synthesis of a protein or by the formation of an abnormal protein is immunoassay, i.e. the immunological detection of the 'cross-reacting material' (CRM) with antibodies raised against the wild-type protein. A lack of reaction, however, should be treated with caution, since the mutation may have eliminated the immunoreactive site(s) of the protein or caused the product to be degraded by proteases.

When several adjacent genes are expressed as a single transcription unit (Chapter 15), a polar effect on the gene(s) downstream from the mutated one may be observed if the mutation is in a DNA region necessary for expression (e.g. a promoter (Chapter 15)) or in the case of some nonsense or frameshift (see below) mutations in that gene.

3.2 Local modification of the number of bases: additions and deletions

Additions and deletions of one or two bases at adjacent or close locations usually provoke mutations which result in an incorrect reading frame and lead to the synthesis of a completely different distal extremity of the protein from downstream of the mutational site. They are known as frameshift mutations. Such mutations often result in the appearance of a stop codon downstream of the frameshift, thus shortening the size of the resulting polypeptide. As seen above, if the mutation occurs upstream of a transcriptional unit involving more than one gene, it may have a polar effect, affecting one or all the proteins coded downstream of the unit (Fig. 7.1). Transcription may be affected

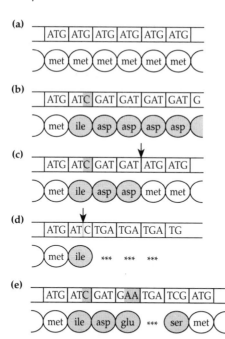

Fig. 7.1 Frameshift mutations and their consequences if located in a polypeptide-coding region. Three stars *** represent a nonsense codon. (a) Wild-type sequence. (b) Addition of 1 base. (c) Addition of 1 base followed by deletion of 1 base. (d) Deletion of 1 base. (e) Addition of 3 bases.

by insertion or deletion mutations in the promoter region. The existence of polar mutations is of particular importance (and was even more important when no direct analysis of mRNAs was possible) in deciphering the processes of gene expression and the organization of operons (Chapter 15).

It should be noted that the addition or deletion of three bases within a short sequence of coding DNA restores the sense downstream from the third modification to the original coding sequence and may often cause no phenotypic change.

3.3 Conditional mutations

The term 'conditional mutation' does not refer to a specific type of molecular modification but to the ability of the mutated protein to show different levels of activity upon modification of its environmental conditions, e.g. sensitivity/resistance to temperature, light, ions, etc. Two extreme sets of conditions are usually defined which correspond to total loss (non-permissive, or restrictive) and maximal maintenance (permissive) of the activity. The mutation generates a change in the structure of the protein such that its activity is usually only slightly impaired under permissive conditions, but which renders it more sensitive than the wild-type form to inactivation by the critical conditions. For example, heat-sensitive mutations are conditionally lethal mutations in which the mutant may grow at a low temperature, in *E.*

coli, say, 30 °C (the permissive temperature) but not at 40 °C (the non-permissive temperature) although the wild type could still do so. Such a mutant is known as ts (temperature-sensitive). These mutations must, to a large degree, leave intact the integrity of the protein, so that it can keep enough (if not all) its activity under permissive conditions. It is thought that ts phenotypes are due to an unfolding of a part of the tertiary structure of the protein because of the interchange of a particular amino acid. Conditional mutations constitute a very convenient tool since proficiency and deficiency can be compared in the same organism. They are also of interest for the definition of the relations of protein structure to function.

A class of conditional mutants known as conditionally dependent is also known. One such mutant described in *E. coli* was obtained after prolonged growth in medium containing streptomycin, giving a conditionally streptomycin-dependent mutant that will not grow in the absence of the antibiotic. In streptomycin-sensitive strains, the antibiotic, which binds to a ribosomal protein, favours misreading of the mRNA, i.e. the introduction of erroneous amino acids into the polypeptide chains synthesized by these ribosomes. A mutation in the streptomycin-binding protein may occur in which the molecule of streptomycin has become necessary in order for this protein to recover its normally active form. The antibiotic has then become essential to the organism.

Another type of conditional mutants, which are in fact suppressible mutants, will be described below (Section 6.2.2).

3.4 Rearrangements

Rearrangements of the DNA in the form of inversions or translocations are mutational events which will most often manifest themselves in impairments of at least one of the genes involved in the modification. These may be accompanied by deletions of varying sizes.

4 Occurrence of spontaneous mutations

The occurrence of genetic alterations may conveniently be placed into two categories, those said to be spontaneous, i.e. caused by uncontrolled natural agents, and those that are deliberately induced by known factors, i.e. mutagens.

4.1 Probable causes of spontaneous mutations

Mutations arise all the time, but at low frequencies, even in cells developing under perfectly favourable conditions. The causes are of various types.

4.1.1 *Misfunctioning of natural processes*

A large proportion of spontaneous mutations result from errors occurring during the normal vital processes that directly implicate the genetic material: replication (Chapter 2), adenine methylation by the Dam system (Chapter 6), recombination between homologous chromosomes (Chapter 9), repair of damage occurring to these molecules (Chapter 8). In fact, these processes include a number of similar events and share some of the enzymes involved.

The introduction of erroneous bases during replication or errors arising from the operations (cutting, resealing or gap-filling) leading to recombination or repair processes engenders discrepancies in the nucleic acid molecule. Protection mechanisms (the error-editing capacity of the DNA polymerases, the repair systems themselves) erase most of the errors. However, a small proportion escapes correction. If one recalls the rate of synthesis due to the *E. coli* DNA polymerases (several hundreds to thousands of bases added per second), it is amazing that so few errors eventually remain.

Mutation rates can be increased by the presence of mutator genes, a number of these having been identified as genes coding for proteins involved in DNA synthesis, genes which themselves have mutated toward loss of accuracy. Thus *mut* alleles are known in the *E. coli* polymerases PolI and PolIII. Conversely, antimutator DNA polymerase mutants giving decreased mutation rates are also known. Mutator alleles of other genes not involved in replication are also known.

4.1.2 *Mobility of transposon elements*

Another class of mutations arises from the presence, which new techniques have shown to be widespread, of transposing elements (Chapter 5). Their particular mode of mobility makes them capable of disturbing the normal gene sequences and hence of inducing mutations. The relative importance of this cause of mutations varies depending on the organism and the genes considered.

4.1.3 *Other causes*

Other causes of spontaneous mutations exist. The effect of the permanent exposure of all cells to natural ionizing and ultraviolet radiations has often been mentioned. Their relative influence, however, while certainly real, is not always clear. An indirect argument confirming their mutagenic efficiency comes from the observation that photosynthetic prokaryotes exhibit very large photorepair capacities, specific for UV-induced damage to DNA (Chapter 8). Being dependent on light for their energy supply, photosynthetic prokaryotes are constantly exposed to the UV component of light. Evolution has probably favoured subclones that are more capable of maintaining the integrity of their genome. Strictly, these effects are due to external agents and are classified as spontaneous only because their occurrence is independent of deliberately created mutagenic situations.

4.2 The fluctuation test

There has been a long controversy over the problem of whether a mutation arises independently or as a result of the application to the wild-type population of conditions allowing the selection (recognition) of the mutants under study. Thus, did a mutant isolated as, for instance, streptomycin-resistant and therefore selectively able to grow in the presence of the antibiotic, exist before the addition of the drug into the growth medium? Demonstration of the prior existence of the mutants was established by Luria and Delbrück (1943) in the following way: they studied the appearance of phage T1-resistant (T1^r) mutants from a wild-type strain of *E. coli*. They established the average spontaneous frequency, measured on large populations, as close to 10^{-8}. They then determined the individual proportions of mutants present in each of a series of similar populations initiated from 'mutant-free' samples. These were ensured by using inocula smaller than 10^8 cells/assay, so that the probability of carrying a preexisting mutant in each sample was negligible. The cells were allowed to divide under non-selective conditions until reaching densities slightly larger than the inverse of the mean mutation frequency, i.e. approx. 10^8/ml, and the populations were then challenged with the selective agent. Had the selective conditions been the inducer of the mutations, equal numbers of mutants should be obtained when identical samples from each tube were analysed. However, the preexistence hypothesis implies random appearance of mutants during growth of the suspensions, so the number of generations undergone by each newly formed mutant clone would thus depend on the time of formation of the initial mutation. Fluctuations in the total number of mutants per tube larger than that expected from sampling variations should thus result, which was indeed the case (Table 7.2).

Luria and Delbrück's experiments, referred to as the fluctuation test, have since been taken as one of the basic rules of genetics. The question has recently been reopened, however, after several scientists had come upon what looked like deviations from this dogma. It has been argued that the mutational system used by the previous authors, i.e. selection on resistance/sensitivity to phage T1 by addition of the phage suspension to the bacterial samples, excluded the possibility of

Table 7.2 The fluctuation test. Spontaneous mutants resistant to phage T1 were numbered from 10 samples aliquoted from a single culture of a T1^s *E. coli* strain (column A), and from 10 independent cultures of the same strain each inoculated with a small enough number of cells to avoid inoculation of preexisting mutants (column B).

Sample no.	A	B
1	46	30
2	56	10
3	52	40
4	48	45
5	65	183
6	44	12
7	49	173
8	51	23
9	58	57
10	47	51
Average/sample	51.4	62
Variance	27	3498
Calculated mean mutation frequency	2.4×10^{-8}	–

(Data adapted from Luria and Delbrück, 1943).

observing potential 'selection-induced' mutants because the killing effect of T1 was faster than the protective effect expected after occurrence of such mutations. A repetition of Luria and Delbrück's procedure applied to the detection of a non-traumatizing mutation, the Lac$^+$/Lac$^-$ system, indeed seems to show that mutations could derive from two origins, independently of and as a response to the presence of the selective agent. Hypotheses as to the mechanism of preferential transfer of information from presence of the selective agent to fixation of the mutation in the corresponding genetic system deal with abnormal events during replication and/or transcription processes.

This important observation, confirmed for several other metabolic traits among several organisms, may necessitate a reconsideration of a number of established conclusions in bacterial (and perhaps general) genetics, e.g. the irreversibility of the 'DNA makes RNA makes protein' dogma. While synthesis of DNA from RNA has now been established (in several viruses and in prokaryotes themselves), should it also be considered that metabolic constituents not directly involved with DNA synthesis (e.g. lactose metabolism) may specifically alter the fidelity of DNA-synthesizing devices?

4.3 Frequencies

The frequency of mutation is defined as the ratio of the number of organisms bearing a mutation for a given character to the total viable population in the same sample. Since, when working with prokaryotes (and microorganisms in general), very large populations are considered, values can be measured with reasonable accuracy, and have indeed proved to be reproducible. In spite of its inaccurate genetic meaning but for the sake of simplicity, most of the time only the phenotypic character of the mutants is used. Since a unique phenotype can cover mutations in several genes or different mutations in the same gene (Section 1), comparisons of mutation frequencies should often be used with caution.

The mutation frequency gives the genetic image of a population at the time of the assay. It represents the equilibrium reached by the population as a result of forward and reverse mutations (Section 6) for the character considered.

Another concept that can be considered is the rate of mutation, which is the proportion of mutants, for a given character, appearing per generation; this value has rarely been determined, due to the difficulty of measuring it, and it is not necessarily of the same order of magnitude as the frequency.

Average spontaneous mutation frequencies among prokaryotes range between 10^{-9} and 10^{-5}, values similar to those encountered for eukaryotes. They can be considered as fairly homogeneous considering the span of characters and species for which information is available and the limit in the genetic meaning of the phenotypic criteria used. These values are considered as representing single mutational events. Except in certain cases, mutations in unrelated functions and thus in different genes can be considered as independent events. The probability of the simultaneous appearance of multiple mutations can be calculated as the product of the frequencies of appearance of each individual one. Their observation is thus very rare, even when dealing with large populations as are available with microorganisms. In fact, the argument is used in reverse: high frequencies of formation of pleiotropically mutated phenotypes, i.e. simultaneous modifications of several phenotypic traits, are taken as an indication of a correlation between the functions involved, either in the organization of the genetic material or in the expression of the functions (Chapter 16).

Single mutational events affecting up to a few per cent of a population also take place. These mutations were overlooked for a long time because of their instability (they may revert with similar frequencies), which makes the mutated clones difficult to maintain. These mutations often correspond to unstable genetic organizations, from which wild-type or pseudo-wild-type revertants emerge at high rates and overgrow the mutated population. One type of such mutations arises from the presence of mobile elements (Section 4.1; Chapter 5).

5 Mutagenesis: induced mutations

A number of physical and chemical agents can be applied experimentally to cause mutations, i.e. are mutagenic. The resulting damage is called an induced mutation. The mutagenic effect of radiations, both ionizing and ultraviolet, on bacteria has been recognized and studied since the 1950s. During the same period, a number of chemicals capable of reacting with DNA have also been shown to be mutagenic. The known chemical mutagens are basically of five types as classified by their alteration of the DNA. Most of these products, together with others more recently discovered, constitute the arsenal generally available to provoke mutations. In spite of the specificity of the reaction each of these mutagens produces in DNA, a distinction between physical and chemical mutagenic agents has proved to be insignificant when considering the final consequences in terms of mutagenesis.

It is interesting to note that some agents (chemicals, ultraviolet light) show a directed action, that is, they act on specific types of bases, while others introduce structural disturbances. The location and the molecular change of the mutation which may be engendered can be variable, since they often depend on the subsequent misfunctioning of the repair systems. This is true of radiations and cross-linking or alkylating agents.

5.1 Mutagenic agents: mode of action

5.1.1 *Ionizing radiations*

Through the ionizations they provoke, either directly on the DNA or indirectly via electronic excitation of other molecules (active radicals) which then interact with the DNA, ionizing radiations lead to breakage of the phosphate–deoxyribose backbone of the DNA helix. Depending on the energy of the radiation, and thus on the density of ionizations produced along the path of the particles emitted, either single- or double-stranded breaks are produced. The former are rather easily repaired in double-stranded DNA molecules since they only slightly alter the structure of the molecule, which is obviously not true in single-stranded DNAs.

5.1.2 *Ultraviolet light*

The best studied system is that of ultraviolet (UV) radiation at 254 nm, the wavelength absorbed by most purine and pyrimidine bases. Purines are resistant to chemical change by irradiation, but pyrimidines become hydrated, particularly cytosine. However, the damage done to DNA is

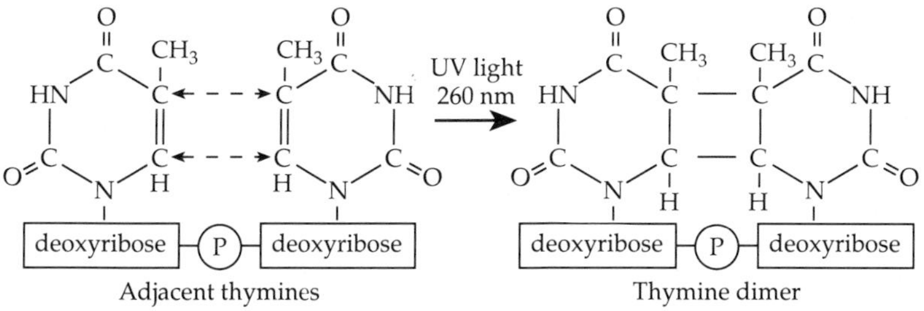

Fig. 7.2 Thymine dimer formation caused by UV irradiation.

particularly the formation of thymine dimers, i.e. adjacent thymine rings coalesce in pairs to form dimers (Fig. 7.2). Dimerization causes a shift in the absorption spectrum which allows the change to be physically detected. Although dimerization can occur between thymines on opposite DNA strands, giving cross-linkage, the reaction occurs predominantly between adjacent thymines on the same strand. A distortion of the double helix then appears which could lead to conditions allowing for dimerization between strands, making replication impossible. A number of mechanisms exist for the repair of these dimers (Chapter 8), one of which, photorestoration, specifically deals with this type of lesion. Since its action is the *in situ* disruption of the dimer bond, it is a 'clean', i.e. not error-prone, system. A mutation is left if errors are introduced by other repair processes, or if some dimers escape photorestoration. The type of mutations caused by UV irradiation are, then, mostly base substitutions, both transitions and transversions, although frameshifts can occur.

5.1.3 *Deaminating agents*

Deaminating agents chemically modify bases *in situ* inside a DNA molecule. The modified base, wrongly recognized by the DNA polymerase at the next round of replication, will thus most frequently be erroneously paired, leading to a mutation. The effect is hence not found until the DNA is either replicated or transcribed. On further replications, the mutation is preserved, and so is the modified base in the originally mutated strand. An example of a deaminating agent is nitrous acid. Adenine is deaminated to give hypoxanthine with a change of base pair from A–T to G–C, whilst cytosine is deaminated to uracil so that a G–C base pair is changed to an A–T one (Fig. 7.3a). Hypoxanthine prefers to pair with cytosine, whilst uracil pairs with adenine.

Guanine is deaminated to xanthine, but the X–C pairing is non-viable.

(a)

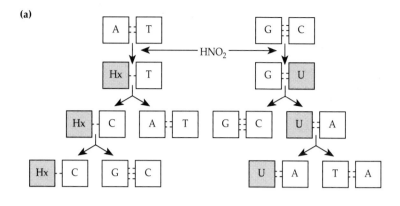

(b)

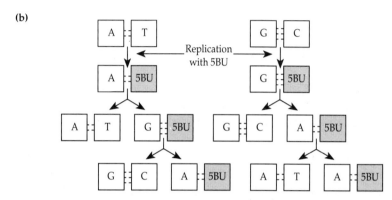

Fig. 7.3 Examples of mechanisms of mutagenesis by chemical agents. (a) Deamination by nitrous acid. Hx: hypoxanthine. (b) Base-analogue mutagenesis. 5-BU: 5-bromo-uracil.

5.1.4 *Base analogues*

Base analogues are molecules that can be erroneously integrated in the DNA by the polymerizing enzyme. Their mutagenic effect is a consequence of their ability to switch from one to another of two tautomeric forms, resulting in different pairing capacities with normal bases. 5-bromouracil (5BU) and 2-aminopurine (2AP) have structures similar to thymine and adenine, respectively. Hence, when 5BU is in the normal keto form, it pairs with adenine in the place of thymine, but the enol form will pair with guanine. Thus, an A–T pair may be changed to a G–C, or a G–C pair may be changed to an A–T (Fig. 7.3b).

5.1.5 *Alkylating agents*

Alkylating agents are a group of chemical mutagens that covalently bind alkyl residues to bases by their action on ring nitrogen and carbon groups and on phenoxy groups, thus rendering their proper

(a) CH₃—SO₃—CH₂—CH₃ Ethyl-methyl sulphonate (EMS)

(b) 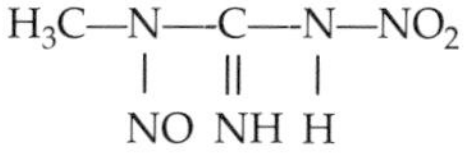 H₃C—N—C—N—NO₂
       | || |
      NO NH H
      N-methyl-N'-nitro-N-nitroso-guanidine (MNNG)

(c)

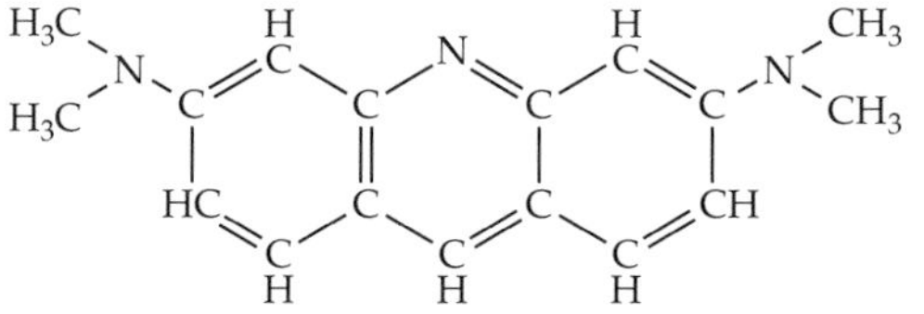

Acridine orange

Fig. 7.4 Examples of chemical mutagens. (a, b) Alkylating agents. (c) Intercalating derivatives.

recognition impossible and leading to random pairing at the following cycle of replication. Among them are some of the most powerful mutagenic agents known. They are very widely used on prokaryotes, the most frequent being the sulphonates methylmethane sulphonate (MMS) and ethylmethane sulphonate (EMS), and N-methyl-N'-nitro-N-nitrosoguanidine (MNNG) (Fig. 7.4a,b). Multiple mutations occur frequently with EMS and even more with MNNG. The latter compound is particularly reactive at the replication fork (Chapter 2), thus producing two mutations at sites symmetrically positioned with regard to the origin of replication. This property has in fact been used to demonstrate bidirectional replication (Chapter 2).

5.1.6 *Intercalating agents*

Intercalating agents are flat molecules (Fig. 7.4b) which can fit between the base pairs in the DNA, a process known as intercalation. This causes frameshift to occur when the DNA is replicated or transcribed. Although the initial event is the intercalation of the compound, the final modification can be either an addition or a deletion, depending on the structural alterations induced in the DNA molecule. Examples of these compounds are acridine derivatives (proflavine, acridine orange), nitrogen mustards and the fluorescent dye ethidium bromide, which is used as a chemical marker in DNA electrophoresis.

5.1.7 *Cross-linking agents*

Cross-links, i.e. covalent binding of two bases from opposite strands, obviously disturb the DNA double-helix structure and, by preventing helix unwinding, block DNA synthesis. Mutations probably occur as a result of the necessary action of repair mechanisms. Well-known agents of this kind are mitomycin C and various psoralens, the latter requiring UV light for activation. These agents usually have a high lethal effect.

5.1.8 *Directed, or targeted, mutagenesis*

All the mutagens discussed above will give mutations more or less randomly distributed over the genome. However, using recombinant DNA techniques, it is now possible to alter a particular site in a gene, a process referred to as site-directed mutagenesis (Chapter 16). This requires that the base sequence of the gene in question is known. If the change aims at modifying a protein, the process is known as protein engineering.

Gene inactivation, resulting from interruption by insertion of an exogenous piece of DNA (usually an easily detectable character), is also frequently used. This can be performed *in vitro* on isolated DNA fragments or *in vivo* through the use of mobile elements, mostly transposons (Chapters 13 and 16).

These techniques are limited to bacterial species in which genetic techniques, particularly genetic transfer, are well estabished.

5.2 Efficiency of mutagenesis

The laws that control the efficiency of action of a mutagen are similar to those of any chemical reaction. Efficiency, practically measured as the proportion of mutated clones obtained after application of the treatment (see below), depends on the intrinsic reactivity of the agent with its particular target on the DNA, and on the concentrations of both the target (the amount of DNA, or of a particular base, or DNA structure) and the mutagen. The dose of mutagen can be monitored either through its concentration (or emission rate for radiations) or through the duration of contact with the organism. It can be deduced, from what we have seen of the chemical nature of mutations and of the mode of action of mutagens, that in principle a single event occurring in the studied target is sufficient to induce a mutation. At the population level, the number of a given type of mutants formed should be proportional to the dose of mutagen applied. Several characteristics of the system, however, modify the conditions of applying this rule, and should be considered when setting up a mutagenic procedure.

1   A proportion of the lesions occurring on the DNA prevent the reproduction of the damaged cells. They belong to two main classes. The first class is made up of those lesions which interfere directly with replication such as, an unconditional deficiency in a function necessary

for DNA synthesis, a cross-link between bases of the two opposite DNA strands, a breakage of the sugar–phosphate backbone, or even deletions. The second class of lesions are those which do not alter the structural continuity of the DNA but induce genetic and physiological changes incompatible with the reproduction of the cell under the conditions being used – e.g. a deficiency in the synthesis of a metabolite not present in the growth medium. Lesions of the first group can be considered as true lethal events: the damaged cells cannot divide and thus do not survive. Those of the second class can be considered as conditionally lethal: the DNA of the corresponding cells can replicate, and the cells could metabolize and divide if incubation conditions were rendered appropriate. We shall see below (Section 7) the result of experimenting with the incubation conditions in order to control the effect of the latter lesions. Whatever their nature, the presence of such lesions results in the death (or non-division, which is phenomeno-logically equivalent with prokaryotic organisms) of a fraction of the population. Mutagenic agents are said to have a lethal effect intrinsically linked to their mutagenic action.

If the dose of mutagen is high enough, several lesions will occur simultaneously in the same cell. Among these, lethal lesions can be formed and hinder isolation of the hoped-for non-lethal ones, and thus the expected mutants will be lost.

Under usual conditions of application, the mutagenic agent will affect only a proportion of the cells. The actual isolation of mutants takes place only among survivors of the mutagenic treatment. The mutagenic efficiency, E_M, for a particular phenotype, is calculated as the proportion of mutants in the surviving population:

$$E_M = \frac{M/ml}{N/ml}$$

where E = efficiency, M/ml = concentration of mutants, and N/ml = concentration of viable organisms after a given dose of treatment.

A typical dose–response curve is shown in Fig. 7.5. Death of the cells (curve A) occurs at a constant rate (except for a possible lag, which will be explained below, point 3). The absolute amount of mutants in the surviving population and its proportion increase with dose up to a dose at which the probability for each mutated cell to bear a second lethal lesion reaches the mutagenic rate (curve B). The apparent mutagenic efficiency rises (curve C), but the actual number of viable mutants in the population decreases as they accumulate other (lethal) mutations.

2   When inducing mutations by use of a mutagenic agent, it is usually important to avoid the isolation of clones bearing one or several (viable) mutations in addition to that looked for, i.e. multiple mutants.

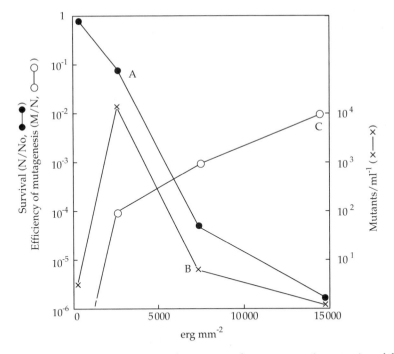

Fig. 7.5 A dose-response curve after treatment by a mutagen. A suspension of the cyanobacterium *Synechocystis* PCC6714 was irradiated with UV light for increasing doses (i.e. increasing lengths of time). Aliquots were then plated on normal medium (measure of survival, curve A) and on the same medium complemented with pF-phenylalanine, a toxic analogue of phenylalanine, to determine the induction of pF-phenylalanine resistant mutants. These are expressed as their concentration (/unit volume) in each sample (curve B) and in proportion of total viable cells in each sample (curve C). (After Astier *et al.*, 1979) N, N_0 = viable cells; M = mutants.

Statistically speaking, this depends solely on the dosage of mutagen applied, as described above for lethal lesions. The only way to minimize the probability of inducing secondary mutations is to decrease the dose of mutagen to less than one total event per organism (in practice less than 0.4, according to the Poisson distribution curve; Chapter 4, p. 93). These conditions will greatly decrease the yield of mutants since the chance that each individual event reaches the gene studied is inversely proportional to the number of genes (grossly speaking) harboured by the genome of the organism. This general rule may have exceptions: MNNG, for instance, is mostly active close to the replication forks and is considered as systematically producing double mutations, one at each fork location.

3   Under many circumstances, prokaryotic cells contain several copies of their genome (Chapter 2). Application of an optimal mutagenic treatment, as defined above, usually yields only one genome copy mutated for the function under study. Depending on the dose applied,

the others will either not bear any lesion or bear another mutation or a lethal lesion. The lethal lesion may remain cryptic, due to the multiplicity of (non-hit) genomes. This partly explains the apparent lag in the survival curves mentioned above, since the cell will not die until all the targets (genomes) have been hit. The other cause for the apparent lag is the presence of repair processes (Chapter 8). For the same reason, the induced mutations may also remain cryptic. Genotypic (and hence phenotypic) homogeneity and the isolation of mutated clones will then necessitate the segregation of the genomes (and, con-comitantly, of independent mutations on different genomic copies) under appropriate conditions. Selection conditions applied after the mutagenic treatment must be specifically defined (Section 7).

An interesting strategy that can be used, when possible, to eliminate a secondary mutation from a mutated clone prepared after mutagenesis consists in transferring the mutant gene to a wild-type cell. The prob-ability that the secondary mutation will be genetically linked to the one under study is low enough (considering the relative genetic size of the system studied in the whole genome) to enforce their separation. Obviously systems which transfer shorter pieces of DNA are more favourable.

4 Notwithstanding their direct action on DNA itself, mutagenic agents, unexpectedly, show a wide range of efficiencies in different strains or species. The presence of a large capacity of DNA repair may be responsible for the apparent low activity of certain drugs or of radiations, e.g. in *Deinococcus radiodurans*. In such instances, both lethal and mutagenic effects are similarly decreased.

It is also unclear why the relative efficiencies of various mutagens differ largely from one organism to another. Thus the alkylating agent MNNG, known as one of the most powerful agents in *E. coli*, is an almost inefficient mutagen in certain cyanobacteria (with a comparable G+C ratio) although it shows similar killing kinetics. Ultraviolet light efficiency, in contrast, differs only slightly in the two groups of bacteria, under conditions preventing photorepair of the pyrimidine dimers.

5.3 Hotspots

In many cases, certain positions on the DNA molecule are found to be more susceptible to mutagenesis than others. This led to the hypothesis of the existence of hotspots, probably resulting from the processing of unusual structures induced by the mutagenic agents. Association of the mutagen with particular base sequences on the DNA may lead to stabilization of modified structures that are less amenable to repair and hence more mutation-prone. Patterns of hotspots depend on both the mutagen and the particular DNA region considered.

6 Reversions

Just as a change from the wild-type to the mutated state occurs, the reverse process may also happen. Detailed analysis of revertants has proved that there may exist several ways through which a cell can reestablish a lost function. The study of reverse mutations is thus an important means of defining the nature of the mutation itself and/or the genetic organization of the system. All mutational processes leading to the establishment of a simple base change are able to give rise to revertants by a secondary mutational event modifying the base pair

Table 7.3 The various types of reversions.

Type of reversion	Process	Results
In situ *reversions*		
True	Return to WT DNA sequence	Complete recovery of WT activity
	Return to WT amino acid sequence, through replacement of the mutation by another one, thanks to the degeneracy of the code	Complete recovery of WT activity
	Precise elimination of a mobile element	Complete recovery of WT activity
Pseudo	Replacement of a substitution mutation by another one at the same location	Usually partial recovery of activity
Suppressor reversions		
Acquisition of a neutralizing second mutation in the same gene	Addition of a second mutation, particularly in the case of 1-base frameshifts	Usually partial recovery of activity
Acquisition of a suppressor process	A modified tRNA synthetase, which will charge a tRNA with an erroneous amino acid	Recovery of activity usually partial
	A modified tRNA, which will deliver its normal amino acid when recognizing a nonsense codon	Three classes exist, specific for the suppression of one nonsense codon each: amber (UAG), ochre (AUU) or opal (UGA), no matter what gene the codon is in. Usually partial recovery of activity
	A modified 30s ribosome subunit stabilizing an incorrect pairing between the mutated codon and a tRNA, more or less at random	Recovery of activity usually partial
Activation of a bypass metabolic pathway		
This is possible only when two routes may lead to the same final product, one of which shows no or a low activity in WT cells		No recovery of the missing function *per se*; efficiency of recovery variable

that was concerned with the initial mutation (*in situ* reversions). Frameshift mutants, in contrast, are able to give revertants most frequently by mechanisms such as a secondary, suppressor, mutation.

Table 7.3 shows the possible reversion processes.

6.1 *In situ* reversions

In most cases the molecular change at the origin of a reversion is of a similar nature to that leading to the forward mutation, and hence reversions are defined in the first place by the reacquisition of the original (wild-type) phenotype. A true *in situ* reversion corresponds to the restoration of the original amino acid sequence in the protein. This can sometimes be achieved through several different base-pair substitutions, if the code of the particular amino acid restored is degenerate. The stringency of the modification looked for restricts the probability of occurrence of true reversions to a low level. Thus *in situ* reversions in a mutated leucine codon may have a good chance of occurring since there are six codons for leucine and these codons taken pairwise share two conserved bases. In contrast, for methionine and the three 'nonsense' codons, which are represented by only one codon each, there is only one possible choice that will restore the initial triplet. Partial phenotypic reversions, in which another erroneous but acceptable amino acid replaces the mutated one and allows the protein a certain degree of activity, would be expected to occur more frequently.

6.2 Suppressor reversions

Many apparent reversions correspond, in fact, to the acquisition of a second mutation, known as a suppressor mutation, independent from the initial one and at a different site. This is more commonly located outside the original mutated gene, in which case it is known as an intergenic suppressor, but it may also be located within the original mutated gene and is then known as an intragenic suppressor. Several unrelated processes may be envisaged (Table 7.3).

6.2.1 *Intragenic suppressors*

Reversions due to intragenic suppressors occur typically, but not compulsorily, after an intercalating agent has caused a frameshift mutation by addition or deletion of one or a few bases. The reverse process, e.g. addition of one base in the close vicinity of a one-base deletion, reestablishes an in-frame reading (Fig. 7.1). The efficiency of the reversion will depend not only on the discrete changes produced by each mutation, but also on the distance between them, which corresponds to an abnormally read stretch of DNA.

Intragenic suppressors may also occur after base substitutions, if the second base change (probably also a base substitution), by interacting with the first one, is capable of restoring a sufficient level of activity, e.g. an appropriate tertiary conformation to the protein product of a gene or the proper pairing of two bases involved in the stem-and-loop structure of an RNA molecule.

Examples of frameshift reversions were studied in the rII region, responsible for host-range characteristics, of the phage T4. A modified host-range phenotype was obtained in clones harbouring a frameshift mutation resulting from either an addition or a deletion of one base. Phenotypic revertants, in fact pseudo-revertants having acquired a second mutation, could be obtained. The second mutation was a frameshift of the sign opposite to the first one and occurring within the same gene and usually close to the site of the first mutation, thus reestablishing in-frame reading. This very elegant work, totally based on genetic analyses at a period when direct sequence analysis of DNA was not possible, led to the understanding of the nature of these mutations, but also served to ascertain the three-letter (or multiple of three) organization of the genetic code.

6.2.2 *Intergenic suppressors*

The most frequent intergenic suppressions concern biases of the translational process such that initial mutations leading to erroneous amino acids in a protein sequence or to a stop codon on the messenger RNA are erased by a correcting mechanism occurring during synthesis of the protein. Three major constituents of the translational machinery may be concerned. The deciphering of how these suppression systems work has been important for the understanding of the processes and constituents involved in translation.

The first type of such suppressor reversions, bearing modified aminoacyl-tRNA synthetases, was discovered in the 1950s. Examples of these are found in the tryptophan synthesis pathway in *E. coli*: the 210th amino acid in the tryptophan synthetase A protein is glycine, but in mutant *trpA23* this has been replaced by arginine, giving an inactive enzyme. Among revertants to the wild-type phenotype is a mutant that carries a tRNA with an altered translation fidelity which recognizes the codon for arginine but actually delivers glycine. The erroneous charging of the tRNA corresponds to a mutation of the cognate aminoacyl-tRNA synthetase.

The second translation constituent that may be concerned is the tRNA itself. Several such suppressor mutations have been identified in

E. coli. They can suppress mutations conferring either a sense or a nonsense codon, the latter having been better characterized. For example, in response to the UAG codon, serine will be delivered by the mutated seryl-tRNA allele *supD* 20% of the time, glutamine by *supE* (glutamyl-tRNA locus) (14%), tyrosine by *supF* (tyrosyl-tRNA locus) (25%), tyrosine by *supC* (25%) and a basic amino acid by *supG* (lysine-tRNA locus (5%). *supC* and *supG* will also respond to UAA (12% and 6%, respectively).

More recent work has been concerned with the third type of suppression, the 'nonsense suppression', in which a ribosome will favour the binding of a charged tRNA through erroneous complementarity with a stop codon in the mRNA, and thus allow the continuous synthesis of the protein. The mutation of the ribosome (in fact of a constituent protein of the ribosome) results in a decreased accuracy of recognition or of positioning of the mRNA codon/tRNA anticodon during the synthesizing process. The first cases that were described were streptomycin-resistant mutants of *E. coli*. This antibiotic, by binding to the wild-type S12 protein of the small subunit of the ribosome, increases the level of misreading of the messenger RNAs. The resistance mutation, borne by protein S12, decreases the capacity of streptomycin to bind and simultaneously the frequency of misreadings. Analysis of this category of reversions, enlarged to include other ribosomal mutations, has led to a better understanding of the control of accuracy of function of the translating apparatus.

The availability of suppressor mutations has proved a very practical tool for geneticists. A suppressor gene, i.e. the appropriate allele of either suppressor process, can be introduced at will (using genetic transfer systems as described in Chapters 10–12 and 16) in a cell in which the genetic organization of a metabolic system is being studied. If this study includes analysis of mutations that are susceptible to suppression, the effect of these mutations can be easily turned off by transfer of the suppressor gene without alteration of the initial mutations. Conversely, it is possible to eliminate the suppressor gene at will through the use of one of several tricks; for example, the suppressor allele may bear a temperature-sensitive mutation or it may be carried by a plasmid conditionally replicated in the host cell. These systems provide flexible control of the genetic system under study.

These suppressor processes clearly involve a second mutation affecting a molecule that is functionally and genetically unrelated to the previously modified one. The fact that the nonsense-suppressor reversions are not very frequent can easily be explained: the newly acquired capacity to 'erase' a nonsense codon could a priori apply to all such codons present in a cell genome, obviously resulting in a general disturbance of protein synthesis. Only poorly efficient suppressor mutations will be compatible with normal cell viability, so that all normal essential identical codons in other genes have a chance of being read as such.

6.3 Other possible reversion processes

The reversion pathways described above are far from exhaustive. Any secondary mutation which results in the modification (for instance, the induction or increase of expression of a gene(s) or activation of an enzyme) of a metabolic pathway which can replace the missing function may be considered as a suppressive mutation.

7 Selection and enriching conditions

Mutants constitute a powerful tool for the genetic approach to biological problems. One of the difficulties of this approach, however, resides in the selection of mutants. Important technological developments over the past 30 years have indeed dealt with the elaboration of strategies improving their isolation. This section defines the problems encountered and describes the main solutions available.

As seen above, some mutations do not confer any (or any measurable) functional variation and are therefore irrelevant for most practical purposes. They can only be recognized through molecular analysis of the DNA, using techniques such as cloning and DNA sequencing (Chapter 16), or they can be induced by the insertion of an easily detected secondary marker (Chapters 13 and 16).

7.1 Direct selection

Mutants bearing phenotypic differences can, in practice, be divided into two groups, depending on whether the difference confers a selective advantage on the modified cell over the predominant number of wild-type ones. An example of an advantageous mutation is the acquisition of a prototrophic capacity by a cell from a population that is auxotrophic for a given metabolite, e.g. Leu$^+$ among a Leu$^-$ population, supposing the corresponding metabolite, i.e. leucine, is not provided. Another directly selectable phenotype is that of resistance to a drug, e.g. Str^r among an Str^s population. Utilization of their selective advantage allows direct isolation of the mutants.

Leu$^+$ mutants can be selected by applying samples from a Leu$^-$ suspension directly to a medium lacking leucine. The mutant cells will begin immediately synthesizing leucine and thus grow and form a colony, whilst the Leu$^-$ ones will undergo at most one residual division using their store of leucine if they have any. The procedure is usually carried out on solid medium. The mutants can then be isolated by picking up cells from individual colonies. The strategy for isolating an Str^r mutant from an Str^s population is the same, except that the selection consists in the addition of the selecting agent, streptomycin, to the medium. However, newly arisen Str^r mutants, e.g. cells that have just been mutagenized, still harbour a heterogeneous genotype and a totally streptomycin-sensitive phenotype, i.e. they have only streptomycin-sensitive ribosomes. Since this phenotype is recessive, an immediate addition of the antibiotic might thus kill the mutants, which should be allowed time for phenotypic expression, that is, to undergo a few divisions so as to exchange their stock of ribosomes for a stock of streptomycin-resistant ones, before application of the selecting agent.

7.2 Indirect selection: enrichment

Reciprocally, no advantageous conditions can be set to specifically favour an auxotrophic or a sensitive clone arising among a prototrophic or a resistant population, respectively. Indirect selection procedures, such as replica-plating, must be devised when looking for such mutants. Direct search would mean individual checking for the property studied among at least as many clones as the ratio of wild-type to mutant cells in the population (in practice several times more, in order to reach favourable probabilities of success). This is feasible for mutation frequencies no smaller than about 10^{-3}, a situation sometimes encountered after application of an efficient mutagenic treatment.

Otherwise the strategies used may be generally defined as finding a means to circumvent the problem. It is rather like playing a trick on the non-desired, most numerous, cells so as to disadvantage them, without harming the mutants. In the usual procedure, the suspension is maintained under conditions which do not allow division or metabolism of the mutants, but are favourable to the activity of the wild type. Selective killing of the latter is achieved through application of a treatment that is bactericidal only to actively metabolizing cells, e.g. penicillin, an antibiotic which inhibits cell-wall synthesis, will only act on actively dividing cells. The effect is an increase in the proportion of mutants in the final viable population. The magnitude of this enrichment varies with specific cases. In favourable situations, factors of 10^3 to 10^5 can be obtained. Should low killing efficiencies be the case, it is always possible to repeat the treatment until a sufficient level is obtained. Enrichment is usefully carried out until the proportion of mutants reaches 0.1–1% of the population. Direct individual screening is then performed. The treatment applied is usually not totally harmless to the mutants. This brings constraints in its utilization. Intermediate conditions are usually used, which kill only a limited portion of the population to be eliminated (e.g. 99–99.9%) but lessen the risks of losing the mutants. The enrichment thus obtained is usually still insufficient to allow easy isolation of the mutants. The application of repeated cycles is then advisable.

Enrichment was first set up by Davis (1948), who was looking for histidine auxotrophs (His$^-$) in a wild-type prototroph population of *E. coli*. The spontaneous mutation frequency was expected to be around 10^{-7}. A suspension was grown in medium containing histidine, so as to ensure equivalent growths of both wild-type and mutant cells. When at least 10^8 total cells (several times the ratio of wild-type/mutant, i.e. several mutants/sample) were obtained, the suspension was transferred to a medium lacking histidine. Incubation at 37 °C for the equivalent of two to three generation times allowed exhaustion of any histidine pool previously built up in the mutants, the division of which was then completely blocked. The enriching agent, penicillin, which interferes with cell-wall synthesis in dividing cells, was then applied. Its concentration was adapted so that it could stop division of wild-type cells but not lyse them. The latter point is important since cell lysis would release metabolites, including histidine, in the medium, and the mutants, insensitive to penicillin while not metabolizing, would then be able to reinitiate division and be killed by the antibiotic. After the equivalent of one to two generation times, the penicillin was washed away and histidine added. Under Davis's conditions a survival of approximately 10^{-3} was attained, which would correspond to 10^{-4} mutants/surviving cell. This proportion was still too small for easy direct screening. The surviving suspension was allowed a few divisions in the presence of histidine, after which a second identical penicillin treatment was performed and the final survival level determined.

Penicillin actually accumulates slightly in non-dividing cells; its dilution down to harmless concentrations as a result of division of mutant cells justified the growth step applied between the two enriching treatments. Individual clones (a few thousands) were prepared from the surviving cells, by plating on complete (histidine-complemented) medium. Direct screening for the capacity/incapacity to grow without histidine allowed final isolation of the deficient mutants.

Penicillin enrichment has proved efficient for many prokaryotes. The difficulty of the procedure resides in the exhaustion of all residual metabolizing capacities in the mutants. Pool sizes vary widely (from one to 10 generation equivalents) depending on the function being considered and the organism.

Many variations of this strategy have been used. In principle, any inhibitor or antibiotic is usable in place of penicillin. The isolation of photosynthetic-deficient mutants from *Rhodobacter capsulatus* constitutes an interesting example. The enriching agent was tetracycline, an antibiotic to which this strain is sensitive and which is known to utilize an active transport system to enter the cells. Photosynthetic-deficient mutants are phenotypically deficient for the transport of tetracycline through lack of energy when maintained under a photosynthetic regime, and thus are not killed by the antibiotic.

It is worth emphasizing that the cells which survive the enriching procedure are only phenotypically (under metabolically unfavourable conditions) insensitive to the agent. However, true genetically resistant mutants might also exist and, obviously, the lower their frequency, the more efficient will be the chosen enriching agent.

7.3 Replica-plating

An improvement of the screening procedure, known as the replica-plating technique, was achieved by J. and E.M. Lederberg (1952). The fact that this author was awarded a Nobel Prize for this achievement, which may nowadays appear trivial, as part of his pioneering work on bacterial genetics emphasizes the importance mentioned above of mutant isolation in genetic studies. As shown in Fig. 7.6, grids were prepared by transferring groups of 50 clones on to each of a series of complete-medium plates with a platinum wire (nowadays sterilized wooden toothpicks are frequently used). This step, aimed at ordering the clones, is not compulsory, but makes further analysis more convenient. The original plates can be used, so long as the colonies to be replica-plated are discrete. After growth of the colonies, the plates were printed, via a sterile piece of velvet material set on a replica drum, on to, in order, a histidine-deficient and a histidine-containing (control) plate. Appropriate marks were made to orientate the plates and to allow the comparison of individual clones. Specific incapacity to grow on histidineless plates was easily depicted in a series of fifty clones. The His⁻ mutants could be picked from either the master (grid) plate or the control one, and grown for further work.

When applied under good conditions, as many as 10–12 different selective plates can be printed from the same velvet template. This allows rapid analysis, for example, of clones issuing from a cross in which several markers are involved.

The replica-plating technique has since been widely used and applied to many microorganisms, including yeasts and fungi. Its limits reside in the possibility of having regular, independent colonies on solid medium. Variations concern the size and shape of the Petri dishes used (square ones being frequent now) and hence the number of clones per plate, or the printing system, e.g. Whatman paper or grids of platinum wires. In the latter case, the clones are suspended in the wells of an immunoassay rack and a multiple-wire grid is used as the replicator.

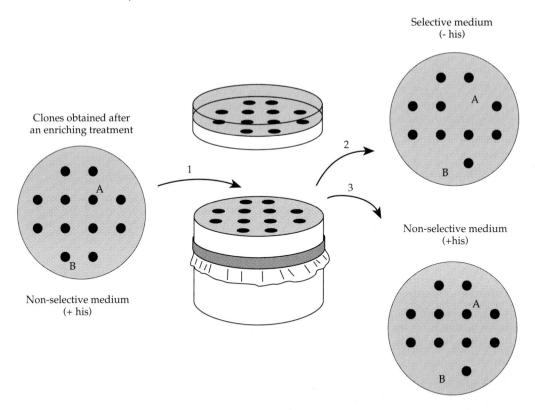

Fig. 7.6 The replica-plating method for indirect detection of mutants. Application to the isolation of His⁻ mutants. Clones obtained after an enriching treatment are plated on a non-selective medium, used as master-plate (possibly after individual transfer into an organized grid) to print a velvet set on a replica-drum (1). Two plates (containing selective and non-selective medium, respectively (2) and (3)) are then applied in this order on the velvet, and incubated. Colony A, which does not grow on the (negatively) selective medium, is His⁻, while colony B, which has not grown on both replicas, does so because of inefficient transfer from the master plate.

7.4 Other selection procedures

It may happen that a mutation that is deleterious in regard to its effect on the function directly affected may have a secondary consequence that can be favourable under particular conditions. These conditions may not be specific for the phenotype looked for, and thus may lead to the simultaneous isolation of other unrelated mutants which also respond positively. The enrichment achieved, however, is usually sufficient to allow individual analysis from a feasible number of clones.

An example is the isolation of clones that are impaired in the accuracy of function (translation fidelity) of their ribosomes. This mutation obviously brings no direct advantage to the cells, and thus cannot be selected easily. We have mentioned that at least some of these mutations simultaneously confer a streptomycin-resistant phenotype (Section 6.2.2) which, in contrast, allows easy selection. The streptomycin resistance phenotype can, however, be due to several other causes, such as an impairment of entry of the antibiotic into the cells. Screening must thus be carried out among the clones obtained.

Respiratory-deficient mutants from the versatile cyanobacterium *Synechocystis* PCC6803, which grows either photosynthetically or by respiration, have been obtained in such a way. Enrichment under chemoheterotrophic conditions (the cells being provided only with glucose, which they use as both carbon and energy source, in the absence of light) was in principle possible but in practice difficult because of the slow chemoheterotrophic growth rate of the wild type. Respiration is also used by these cells for the energization of a sodium-extrusion system when they are maintained in the presence of high salt concentrations. A secondary effect of a respiratory deficiency was expected to be an incapacity to adapt, and thus divide, under high salinity, whatever the growth regime. Penicillin enrichment under these high salt conditions proved efficient. As expected, a second type of mutant was also recovered, namely true sodium-extrusion-deficient ones. Individual screening was easily performed.

A classical method consists in the utilization of analogues, i.e. molecules that are structurally closely related to the biological one, of the constituent the metabolic pathway of which is under study. Each analogue usually interferes with a precise step (an enzyme) of the pathway, but its presence blocks its overall realization. Analogues show either bactericidal or bacteriostatic effects on the cells, both resulting in an arrest of growth of the population. Among resistant clones, isolated by classical direct techniques, mutants modified in the specific target of the analogue can be screened. The other frequently obtained resistant phenotype results from an impairment of entry of the analogue, which usually utilizes more or less specifically the transport system of either its cognate or another molecule. Sorting out the different types of mutants can easily be achieved.

A strategy widely used whenever possible is the simultaneous production of the mutation and the acquisition of another phenotypic characteristic conferring direct selection capacity. This is achieved when the mutation results from the insertion of a transposon (Chapter 5) or of a derivative of a transposon (a marker cassette) inside the gene to be mutated. Insertion of a transposon is carried out *in vivo*, while integration of a cassette must be performed *in vitro*. These techniques are described at greater length elsewhere (Chapters 13 and 16).

7.5 Selecting independent mutants

When starting a selection operation, conditions should be set so as to ensure efficient selection. The initial population should obviously contain at least one of the mutants being looked for. This necessitates a precise definition of the nature and size of the cell population from which the isolation is performed. When looking for spontaneous mutants, the culture should be pregrown under conditions which will not interfere with the growth of these mutants and, if possible, which will not slow them down either. For instance, selection of auxotrophs should be done from a culture inoculated in a medium containing the substance for which the auxotrophy is sought. The size of the sample should be large enough for at least one mutant to be present. This criterion is important to take into account when applying a mutagenic treatment, which will kill a large proportion of the cells (Section 5.2).

It is often important to obtain a collection of mutants representing all (or most) of the genes involved in the metabolic pathway under study. These mutants, however, will at first sight have the same phenotype, i.e. the incapacity to carry out the metabolic pathway. Thus the tryptophan biosynthetic pathway includes three enzymes, which transform the initial precursor, chorismic acid, into, in order, phosphoribosylanthranilate (anthranilate synthetase), CdRP (carboxyphenylaminodeoxyribulosepyrophosphate) (indole glycerol-phosphate synthetase) and tryptophan (tryptophan synthetase). By blocking the metabolic chain, a deficiency in the activity of any of these enzymes will result in the non-synthesis of tryptophan, the necessary metabolite, an impairment which can in all cases be circumvented by addition of tryptophan. In other words, selection on tryptophan-supplemented medium will allow the isolation of any of the likely mutants, but will not make it possible to distinguish them. If the aim of the operation is to ensure isolation of mutants covering all possible enzymatic steps, a large number of such Trp$^-$ mutants should be obtained and secondarily defined. It is then advisable to isolate them through independent operations, a process which increases the chances that different mutations are obtained, whilst several mutants from a unique population could be descendants of a single original one (Section 4.2). This is usually achieved by performing parallel isolation procedures from a large number of independent cultures, each started from a small enough sample (inoculum) of the original suspension for there to be a very slight chance that it contains a preexisting mutant. These samples are inoculated in a non-selecting medium, allowing growth of the mutants. These will thus appear independently in each culture. When each population has reached a large enough size, standard selection procedures are applied, and only one mutant from each culture will ultimately be retained. The mutations have thus arisen from independent events; this, however, does not exclude the possibility that several concern the same gene or, although less probably, that they bear the same mutational modification. The existence of hotspots may bias the pattern of mutants obtained.

References

Astier C., Joset-Espardellier F. & Meyer I. (1979) Conditions for mutagenesis in the cyanobacterium *Aphanocapsa* 6714. *Archives of Microbiology*, 120, 93–6.

Davis B.D. (1948) Isolation of biochemically deficient mutants of bacteria by penicillin. *Journal of the American Chemical Society*, 70, 4267.

Demerec M., Adelberg F.A., Clark A.J. & Hartman P.E. (1966) A proposal for a uniform nomenclature in bacterial genetics. *Genetics*, 54, 61–76.

Lederberg J. & Lederberg E.M. (1952) Replica plating and indirect selection of bacterial mutants. *Journal of Bacteriology*, 63, 399–406.

Luria S.E. & Delbruck M. (1943) Mutations in bacteria from virus sensitivity to virus resistance. *Genetics*, 28, 491–511.

8: DNA Repair

Although DNA is a genetically stable molecule (spontaneous errors occur at the rate of 10^{-9} per base per replication), interaction with various physical or chemical agents in the environment can result in changes to its structure. As discussed in Chapter 7, such lesions may become fixed, leading to mutation or cell death. However, bacterial cells (and indeed cells at all phylogenetic levels) have been found to be capable of reversing or repairing DNA damage. While repair processes are thought to have developed as mechanisms for reducing the level of spontaneous mutation (e.g. errors which occur during such normal cellular events as replication or recombination), they also serve to protect DNA from the consequences of damage caused by external agents. No cell is able to eliminate all damage, however, as perfect repair would have precluded the evolution that has led to that cell.

1 Classification of repairable lesions

DNA damage may be defined as any change to the structure of the molecule that alters its coding properties or its normal functioning in replication or transcription. Minor base modifications can cause an alteration of the DNA sequence while other types of damage may result in distortions of the double helix. Several categories of damage are recognized (Fig. 8.1).

1.1 Missing bases

Cleavage of the N-glycosylic bond that connects a purine or pyrimidine base to the deoxyribose sugar residue results in depurination or depyrimidination (i.e. loss of that purine or pyrimidine from the DNA) while leaving the sugar–phosphate backbone intact. The gap produced is known as an AP (apurinic/apyrimidinic) site (Fig. 8.2). Depurination is approximately 100 times more frequent than depyrimidination and is the most common spontaneous alteration to DNA, occurring at a rate of 0.5 events per cell per generation in *E. coli*. Base loss may also be induced by elevated temperature, low pH, UV light, ionizing radia-

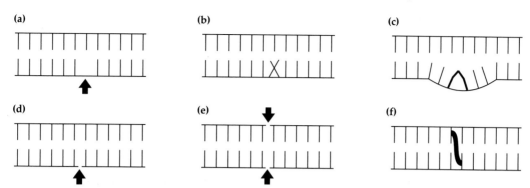

Fig. 8.1 Classification of repairable lesions. (a) Missing base; (b) incorrect base; (c) modified base (distorting the double helix); (d) single-strand break; (e) double-strand break; (f) interstrand cross-link.

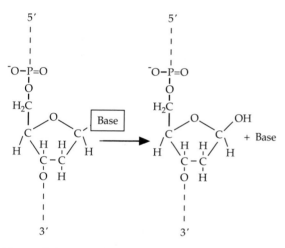

Fig. 8.2 Formation of an AP (apurinic/apyrimidinic) site.

tion and alkylating agents, all of which increase the instability of N-glycosylic bonds.

1.2 Incorrect bases

Mismatched (i.e. non-complementary) base pairs can arise as a result of: (i) errors in DNA replication which remain uncorrected after proofreading (Chapter 2); (ii) hybrid DNA formation during recombination between homologous but non-identical DNA sequences; or (iii) chemical modification of bases in duplex DNA (e.g. deamination of cytosine to uracil) (Chapter 7). DNA containing such mismatched regions is usually referred to as heteroduplex.

1.3 Modified bases

The bases in DNA are subject to a wide variety of modifications, some of which constitute damage while others represent normal processing and serve important biological functions. For example, enzymic methylation of bases is known to be important in distinguishing newly synthesized DNA from its template strand, in protecting DNA from endogenous restriction enzymes and possibly in gene expression (Chapter 6).

All oxygen and nitrogen atoms in DNA can be modified by chemical alkylation with the exception of the nitrogens in N-glycosylic bonds, the oxygens in phosphodiester bonds and the exocyclic amino groups. However, the most significant adducts are O^6-methylguanine, which can mispair with thymine, and 3-methyladenine, which is cytotoxic (Fig. 8.3). Ionizing radiation can break purine and pyrimidine rings and initiate a range of chemical changes, of which the formation of thymine radiolysis products are the most common. The best characterized base alterations, however, are those induced by UV light. Here the principal products are cyclobutane dimers and (6-4)-photoproducts. In the

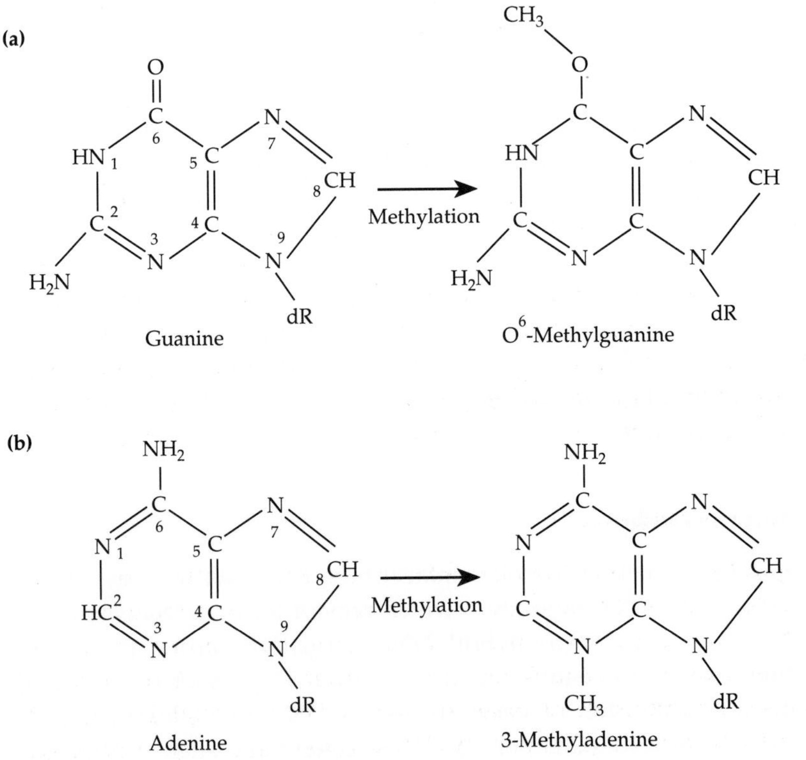

Fig. 8.3 Formation of O^6-methylguanine (a) and 3-methyladenine (b).

(a)

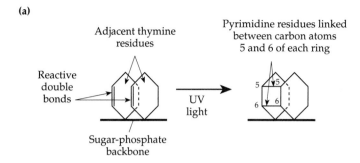

(b)

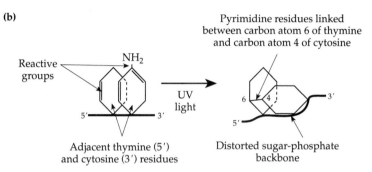

Fig. 8.4 Formation of a cyclobutane dimer and a (6-4)photoproduct, the two most important photoproducts induced by UV light. (a) Thymine-thymine (T-T) cyclobutane dimer. (b) Thymine-cytosine (6-4)photoproduct.

former, adjacent pyrimidines (most commonly a pair of thymine bases) are linked by a symmetrical reaction involving carbon atoms 5 and 6 of each ring, while in the latter linkage occurs between carbon 6 of the 5' pyrimidine and carbon 4 of the 3' pyrimidine (Fig. 8.4).

The major base alterations, including pyrimidine dimers, some alkylations and a variety of other bulky adducts which are formed during cellular metabolism (e.g. phenylalanine–base adducts) or induced by potent carcinogens (e.g. benzo(a)pyrene), are sometimes termed structural defects. They distort the helical structure of DNA and hence may interfere with replication through steric hindrance.

1.4 Single-strand breaks

Phosphodiester bonds can be broken by a variety of agents. Among the more common chemicals are peroxides, sulphydryl-containing compounds (e.g. cysteine) and metal ions (e.g. Fe^{2+} and Cu^{2+}). Both ionizing radiation and active oxygen species (e.g. hydroxyl radicals) can also be responsible for single-strand breaks. Although it is possible that the majority of single-strand breaks are caused by endogenous nucleases, scissions which occur as a consequence of normal metabolic events (such as during replication or transcription) are not considered to be damage.

1.5 Double-strand breaks

Two single-strand breaks on complementary strands which are staggered by a few base pairs as well as those which are directly opposite one another are classified as double-strand breaks. This type of damage is caused by ionizing radiation and is invariably accompanied by alterations to adjacent bases.

1.6 Interstrand cross-links

Covalent cross-linking of complementary strands of DNA prevents strand separation and hence blocks replication and transcription. It may also cause a local distortion of the double helix. Chemical cross-linking agents include nitrogen and sulphur mustards, mitomycin C, nitrous acid and various platinum derivatives, as well as certain furocoumarins (e.g. psoralen) via photocatalytic reactions (Fig. 8.5).

Although some types of lesion may be processed by several different sets of enzymes, a number of distinct repair pathways are recognized.

2 Direct repair

Some lesions are detected by enzymes capable of reversing the modification which caused them. Conceptually, therefore, such direct repair, which involves neither removal nor replacement of bases or nucleotides, may be regarded as the simplest class of DNA repair process. Currently two examples of this type of repair are well characterized.

2.1 Photoreactivation

The consequences of photoreactivation were first recognized in 1949 when it was observed that exposure to high-intensity white light enhanced the post-UV irradiation survival of a range of micro-organisms. This increased survival has subsequently been shown to require the presence of DNA photolyases (photoreactivating enzymes) which repair cyclobutane dimers *in situ* via a light-dependent process.

Studies on the photolyases from various organisms have revealed that all these enzymes contain two chromophores, one of which is invariably $FADH_2$ while the second may be either a folate or a deazaflavin. The former chromophore is largely responsible for the photolysis and the latter for light harvesting. While the precise wave-

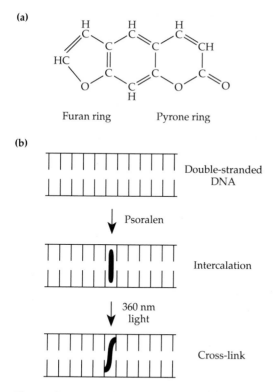

(a)

Furan ring   Pyrone ring

(b)

Double-stranded DNA

↓ Psoralen

Intercalation

↓ 360 nm light

Cross-link

Fig. 8.5 Formation of a psoralen cross-link. Psoralen intercalates with DNA and when subjected to light of wavelength 360 nm photoreacts forming a cross-link between adjacent bases on the complementary strands. (a) Molecular structure of psoralen. (b) Cross-linking.

lengths used in photoreactivation vary with the source of the enzyme and reflect the presence of different chromophores, all photolyases characterized to date utilize light in the near UV or visible regions of the spectrum (300–500 nm).

Photoreactivation has been observed in many organisms, but it is best characterized in *E. coli*. The *E. coli* photolyase, which is encoded by the *phr* gene, is normally present in low concentrations (10–20 molecules per cell). Genetic cloning and amplification techniques (Chapter 16), however, have permitted the purification of this protein to homogeneity in gram quantities. Its characteristics are compared with those of the *Streptomyces griseus* photolyase in Table 8.1. *In vitro* the purified *E. coli* enzyme catalyses light-dependent repair at a rate of 25 dimers per min per molecule, close to the rate observed *in vivo*.

The stages of photoreactivation in *E. coli* are shown schematically in Fig. 8.6. The *E. coli* photolyase binds equally well to UV-irradiated double-stranded DNA and single-stranded DNA, equal affinity being explained by the similarity in the backbone structure of the damaged

Table 8.1 Comparison of the photolyases of *E. coli* and
S. griseus.

Characteristic	*E. coli*	*S. griseus*
Molecular weight	53 994	50 594
No. of amino acids	471	455
Active structure	Monomer	Monomer
Chromophores		
Photolysis	$FADH_2$	$FADH_2$
Light harvesting	Folate	5-deazaflavin
λ_{max} (nm)		
Photoreactivation *in vivo*	365–400	436
Absorbance *in vitro*	380	443–445

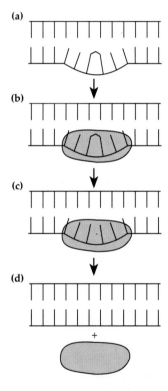

(a)

(b)

(c)

(d)

Fig. 8.6 Stages of photoreactivation. The sequence of enzyme activity is indicated, but the photolyase is not drawn to scale. (a) Cyclobutane dimer distorts the sugar-phosphate backbone of DNA. (b) The photolyase recognizes the distorsion and binds to the dimer. (c) The dimer is monomerized in the presence of light (345–400 nm). (d) The original DNA structure is restored and the photolyase released.

strand in the immediate vicinity of dimers in both types of DNA. Only cyclobutane dimers (thymine–thymine) are substrates for the photolyase. Enzymic and chemical probes have revealed that the *E. coli* photolyase interacts with a 6–7 bp region around the dimer, with intimate contact between the enzyme and DNA occurring only on the damaged strand. Although the enzyme can bind to damaged DNA in the dark, light in the 365–400 nm region is required for the cleavage of the C–C bonds in the cyclobutane dimer. Energy absorbed by the folate chromophore is transferred to the dimer, which dissociates with restoration of the original DNA structure. The enzyme is then released.

Although photoreactivation is probably the most primitive DNA repair process, it is not found universally, being absent, for example, in *Bacillus subtilis*, *Haemophilus influenzae*, *Streptomyces coelicolor* and *Deinococcus radiodurans*. It is, in contrast, present at very high specific activity in at least certain organisms which use light as their energy source, i.e. are photosynthetic, e.g. cyanobacteria. Being compulsorily submitted to the UV radiation present in standard daylight, these bacteria have had to develop a powerful photoreactivation capacity. In those species where it is present, it can be quite effective at low UV doses, reversing more than 80% of the potentially lethal damage. Furthermore, as monomerization of the dimers takes place *in situ*, photoreactivation is error-free (i.e. does not lead to mutation). Photolyases are synthesized constitutively and there is no evidence to suggest that this type of repair is induced by DNA damage.

2.2 Repair of O^6-methylguanine

O^6-methylguanine is the major mutagenic lesion produced in DNA by simple methylating agents. Direct repair of this lesion is accomplished in *E. coli* by an O^6-methylguanine-DNA methyltransferase. This enzyme, which is present at a very low level in undamaged cells, is induced as part of the adaptive response to alkylation (Section 8). During repair the methyl group is transferred directly from the alkylated guanine to a cysteine residue on the enzyme, thereby regenerating an unmodified base in the DNA. The methyltransferase, however, undergoes irreversible inactivation as a result because no mechanism appears to exist for demethylating the S-methylcysteine moiety produced. Methylated protein therefore accumulates as a dead-end product of the reaction instead of being regenerated.

The direct reversal of an O^6-methylguanine residue in double-stranded DNA by an active transferase is a rapid and error-free process, taking less than 1 second at 37°C. The transferase is sensitive to DNA conformation and demethylates O^6-methylguanine residues in single-stranded DNA much more slowly. Similarly, its reaction with the minor but potentially mutagenic lesion O^4-methylthymine proceeds less efficiently than with O^6-methylguanine. The enzyme is also able to demethylate one of the two stereoisomers of methylphosphotriesters (apparently innocuous lesions on the sugar–phosphate backbone). The acceptor site is a separate cysteine residue in a different functional domain from that involved in base demethylation (Fig. 8.7). This reaction is again terminal. Thus each enzyme molecule is able to repair one methylated base and one methylated phosphotriester.

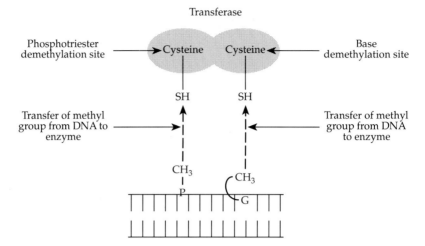

Fig. 8.7 Action of O^6-methylguanine-DNA methyltransferase on methylated DNA. Each acceptor site can function only once.

At first sight it may appear, therefore, that this type of direct repair, in which an entire protein molecule is consumed in order to achieve the correction of a single modified base and/or a single modified phosphotriester, may have evolved to provide a particularly efficient defence against a common highly mutagenic lesion. A more likely explanation is that this apparent suicide protein has a second molecular function. Recent studies in *E. coli* have identified the O^6-methylguanine-DNA methyltransferase as the product of the *ada* gene, which is known to play a key role in the regulation of the adaptative response to alkylation damage. This alternative function is discussed under inducible repair (Section 8).

Alkyltransferase proteins are also found in bacteria other than *E. coli*. However, a certain amount of variation occurs between species with regard to the substrate specificity and inducibility of these repair enzymes. For example, *B. subtilis* contains separate constitutive and inducible methyltransferase activities for O^6-methylguanine as well as an inducible methylphosphotriester alkyltransferase.

3 Base-excision repair

Excision repair processes play a significant role in the accurate (i.e. error-free) correction of damaged DNA in virtually all organisms. The damaged zone is cut out and replaced with newly synthesized DNA. This type of repair can only take place in double-stranded DNA as it is dependent upon the presence of an undamaged strand to use as a template for the resynthesis step. In base-excision repair an unusual or modified base is removed from the DNA, initially leaving the sugar–

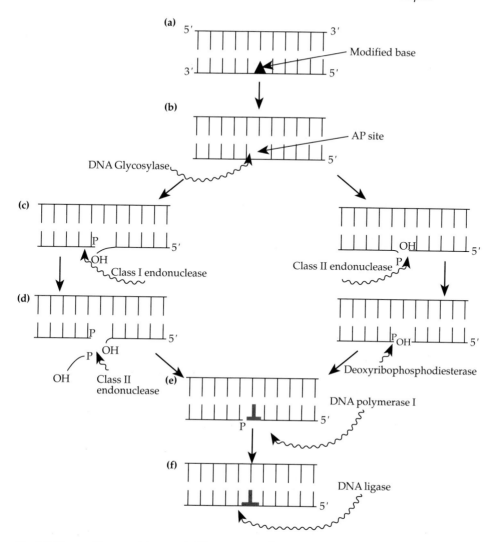

Fig. 8.8 Stages of base-excision repair. The sequence of enzyme activity is indicated but, for clarity, the enzymes themselves are not shown. (a) DNA containing a modified base. (b) The DNA glycosylase recognizes and removes the modified base, leaving an AP site. (c) An AP endonuclease cleaves the sugar-phosphate backbone: Class I enzymes cut 3' to the AP site, Class II enzymes cut 5' to the AP site. (d) Removal of the abasic deoxyribo-phosphate residue: Class II enzymes cut 5' to the AP site, deoxyribophosphodiesterase excises the 5' terminal deoxyribophosphate. (e) DNA polymerase I resynthesizes the missing DNA (heavy line). (f) DNA ligase seals the gap.

phosphate backbone intact. The resulting AP site is then further pro-cessed. The backbone is cleaved and the basic sugar removed, leaving a primer end from which resynthesis can be initiated. The stages of base-excision repair are shown schematically in Fig. 8.8.

A number of enzymes are involved in this process.

3.1 DNA glycosylases

DNA glycosylases catalyse the cleavage of the N-glycosylic bond between a deoxyribose residue and its base, thereby releasing the base. A number of DNA glycosylases have been identified in prokaryotic organisms. In *E. coli*, for example, the *ung*-encoded uracil-DNA glycosylase removes uracil which has been either misincorporated into DNA during replication or generated as a result of spontaneous or induced (e.g. by nitrous acid) deamination of cytosine. Similarly, a hypoxanthine-DNA glycosylase removes hypoxanthine produced by the deamination of adenine. Two 3-methyladenine-DNA glycosylases are known. The first, the product of the *tag* gene, removes only 3-methyladenine, while the second, encoded by *alkA*, excises 3-methyladenine and 3-methylguanine with equal efficiency as well as a range of other alkylation products at a slower rate. These two genes are differently regulated. The concentration of the AlkA protein, which is normally low, is increased some 20-fold as part of the adaptive response to alkylation damage (Section 8), while the Tag protein, which is constitutively expressed, accounts for 90–95% of the 3-methyladenine-DNA glycosylase activity in uninduced cells. As none of the known motifs for protein–DNA interaction, such as helix–turn–helix, zinc-fingers or leucine-zippers, seem to occur in DNA glycosylases, these enzymes presumably use a different mechanism to bind to damaged DNA.

3.2 AP endonucleases

AP endonucleases recognize AP sites generated by DNA glycosylases (as described above) or formed directly by spontaneous or induced depurination and depyrimidination. They hydrolyse phosphodiester bonds and are classified into two groups: Class I enzymes cleave the bond 3' to the AP site while Class II enzymes cleave the bond 5' to the AP site. The most extensively studied AP endonuclease is the product of the *xth* gene in *E. coli*. This enzyme, which is responsible for more than 85% of the organism's AP endonuclease activity, is somewhat misleadingly named exonuclease III because of the prior identification of a minor function (namely a $3' \rightarrow 5'$ exonuclease activity on double-stranded DNA). *E. coli*'s second AP endonuclease, the *nfo*-encoded endonuclease IV, is more typical of the AP endonucleases of other bacteria, and shows no nuclease activity on undamaged DNA. Compounds, such as paraquat, which generate superoxide radicals will induce endonuclease IV 10- to 20-fold. Like all major AP endonucleases, both exonuclease III and endonuclease IV are Class II-type enzymes and generate 3'-OH and 5'-P termini.

One group of AP endonucleases also have glycosylase activity. These enzymes release a modified base by cleavage of the N-glycosylic bond and then hydrolyse the phosphodiester bond 3′ of the resulting abasic sugar to create 3′-OH and 5′-P termini. A pyrimidine dimer-DNA glycosylase of this type has been identified in *Micrococcus luteus* and T4 phage-infected *E. coli*. Thymine and cytosine residues damaged by exposure to ionizing radiation or oxidative agents are similarly processed by endonuclease III in *E. coli*. In the latter case, in contrast to the pyrimidine dimer-DNA glycosylases, the two reactions are concerted so that no free AP sites are generated as intermediates.

3.3 Deoxyribophosphodiesterase

Incision on the 5′ side of an AP site generates a 5′ terminus with an abasic deoxyribophosphate residue. This remnant of a nucleotide has then to be excised prior to gap filling and ligation. Surprisingly, the 5′ → 3′ exonuclease activity of DNA polymerase I cannot act on this substrate and Class I AP endonucleases are unable to function once a 5′ cleavage has occurred. In *E. coli* this role is filled instead by a 5′ deoxyribophosphodiesterase, which converts the previously incised AP site to a small single-strand gap missing one nucleotide.

An incision on the 3′ side of an AP site generates a 3′ terminus with an abasic deoxyribophosphate residue. Subsequent repair in this case involves its removal by Class II AP endonucleases.

3.4 DNA polymerase I and DNA ligase

The gap created by the above base-excision repair enzymes is apparently filled in without nick translation as patch sizes of only two to four nucleotides are observed *in vivo*. In *E. coli* this single-stranded gap is resynthesized by *polA*-encoded DNA polymerase I and sealed by *lig*-encoded DNA ligase (Chapter 2).

4 Nucleotide-excision repair

In this type of excision repair, first described in 1964, lesions are cut out in a short oligonucleotide and the resulting gap is patched by resynthesis. Many kinds of damage can be removed from DNA, ranging from bulky adducts and pyrimidine dimers to relatively minor modifications, such as thymine glycols, apurinic sites and O^6-methylguanine. Base mismatches, however, are not corrected.

Like base-excision repair, nucleotide-excision repair is a multistage process (Fig. 8.9) involving a number of enzymes.

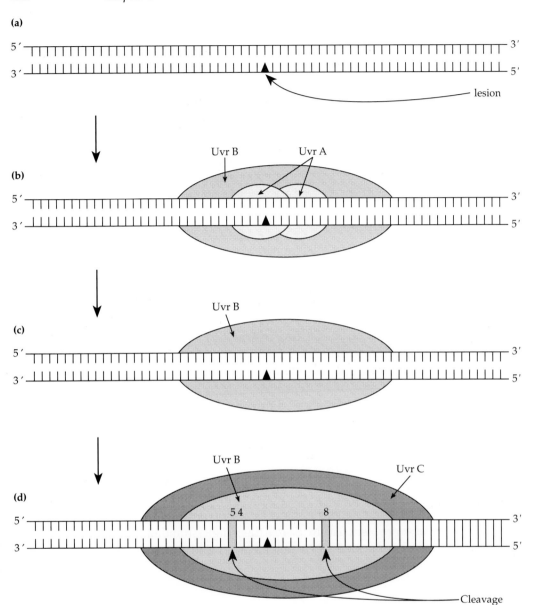

Fig. 8.9 (*Above and opposite.*) Stages of nucleotide-excision repair. The sequence of enzyme activity is indicated, but the enzymes themselves are not drawn to scale. (a) DNA containing a lesion. (b) A UvrA$_2$B$_1$ aggregate binds to the damaged site. (c) UvrA dissociates, leaving a stable UvrB-DNA pre-incision complex. Within this complex DNA is unwound (not shown). (d) UvrC binds, forming an incision complex. The excinuclease activity cleaves the sugar-phosphate backbone at the 8th bond 5′ and the 4th or 5th bond 3′ to the lesion. (e) (*see facing page*) DNA helicase II and DNA polymerase I displace the excinuclease subunits and DNA helicase II releases the DNA fragment containing the lesion. (f) DNA polymerase I resynthesizes the missing DNA (heavy line). (g) DNA ligase seals the gap.

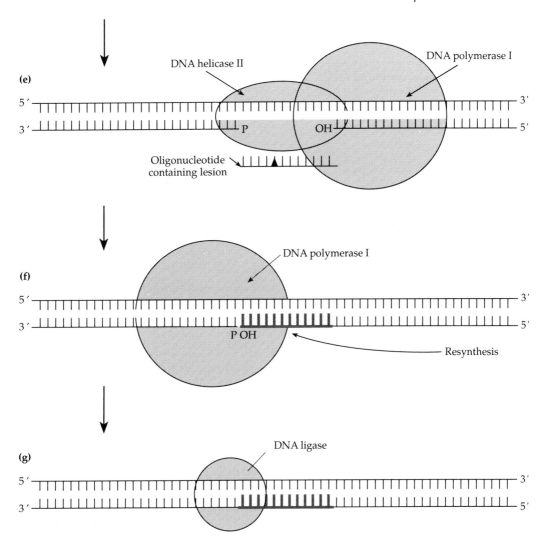

Fig. 8.9 (*Continued.*)

4.1 UvrABC excinuclease

UvrABC excinuclease is composed of three subunits, each of which has been well characterized in *E. coli* (Table 8.2). The first subunit, encoded by *uvrA*, has ATPase activity and will bind preferentially to damaged DNA via its zinc-finger domains. Although the second subunit, the *uvrB* gene product, has no independent DNA-binding capacity, it can form a complex with two molecules of the UvrA protein (namely UvrA$_2$B$_1$) which in turn will bind to DNA. A cryptic ATPase activity in the UvrB subunit is uncovered in this complex. The third sub-

Table 8.2 Comparison of the subunits of UvrABC in *E. coli*.

Characteristic	UvrA	UvrB	UvrC
Molecular weight	103 874	78 116	66 038
No. of amino acids	940	672	588
Active structure	Dimer	Monomer	Monomer
Metal content	2 zinc fingers	None	None
DNA binding ability	Yes	No	Yes
SOS regulation	Yes	Yes	No
No. of molecules per cell			
Constitutive level	25	250	10–20
Induced level	250	1000	–

unit, encoded by *uvrC*, will bind with equal affinity to both damaged and undamaged DNA. The first two subunits are under SOS control (Section 8).

There are three phases in the action of the UvrABC excinuclease on damaged DNA:

1 Substrate recognition. The relatively small size of the enzyme precludes the idea that it has a different active site for each of the many adducts which may form its substrate. All damage recognized by the enzyme must therefore cause similar helical distortions of the DNA and the enzyme in turn responds to this altered structure.

2 Formation of the incision complex. Interaction of the $UvrA_2B_1$ aggregate with a damaged site leads to the dissociation of the UvrA dimer and the formation of a stable UvrB–DNA preincision complex. Thus the formation of this complex occurs through a non-reversible process, in which the UvrA protein acts to direct the UvrB protein on to the DNA at the site of the lesion. The UvrA protein, however, may be recycled and serve in several rounds of UvrB loading. ATP hydrolysis is essential in the formation of the UvrB–DNA preincision complex within which DNA is unwound. This open configuration in turn creates a suitable binding site for the UvrC subunit, and the formation of the incision complex is completed by the attachment of the latter protein.

3 Incision of the sugar–phosphate backbone. The UvrABC excinuclease removes DNA damage by hydrolysing phosphodiester bonds on either side of modified nucleotides. The incision pattern shows minor variability, depending upon the size and type of adduct present, but in general cleavage occurs at the eighth phosphodiester bond 5′ and the fourth or fifth phosphodiester bond 3′ to the damage. A fragment (11–13 nucleotides in length) containing the lesion is liberated and a gap generated with 3′-OH and 5′-P termini. It is not yet known whether the excinuclease activity lies in either the UvrB subunit or the

UvrC subunit or whether it is created by domains from both of these proteins.

4.2 DNA helicase II, DNA polymerase I and DNA ligase

In *E. coli* both DNA helicase II (encoded by *uvrD*) and DNA polymerase I are thought to be involved in the turnover of the UvrABC excinuclease. DNA helicase II has been implicated in the removal of the UvrC subunit from the excised fragment. This oligomer can then be degraded by the nuclease activities of DNA polymerase I (Chapter 2) or by other cellular nucleases. DNA polymerase I (and possibly also DNA helicase II) binds to the 5′ incision site and releases the UvrABC excinuclease subunits located here. Repair synthesis is initiated and the gap filled using the intact complementary strand as template. DNA ligase then seals the newly synthesized material on to the existing DNA. In most cases the gap is repaired without any enlargement of its size (i.e. short-patch repair). The degradation and resynthesis of much larger areas (i.e. long-patch repair) is under SOS regulation and will be dealt with later (Section 8).

The *in vitro* rate of repair achieved by the six proteins described above (namely UvrA, UvrB, UvrC, DNA helicase II, DNA polymerase I and DNA ligase) agrees well with the calculated *in vivo* rate, indicating that these enzymes are all that are required to carry out nucleotide-excision repair in *E. coli*. Similar but less well-characterized systems operate in *M. luteus* and *D. radiodurans*.

The enzymes involved in nucleotide-excision repair may also interact with those which function in other repair pathways. For example, as DNA polymerase I plays an essential role in repairing the single-strand breaks which are introduced into damaged DNA at an early stage in nucleotide excision, it is not surprising that this same enzyme is involved in the repair of ionizing radiation-induced single-strand breaks. Repair of cross-links, which involves the concerted action of the nucleotide-excision and recombinational repair pathways, is con-sidered in a later section (Section 6).

5 Recombinational (or post-replication) repair

When DNA replication occurs in cells which have been exposed to UV irradiation or to certain types of chemical treatment, the newly synthe-sized DNA is observed to have a lower molecular weight than newly synthesized DNA in undamaged cells. This reduction in molecular weight is due to the presence of discontinuities about 1000 nucleotides in length ('daughter-strand gaps') in the nascent strands. Each time DNA polymerase III encounters a pyrimidine dimer or other bulky adduct in one of the template strands, replication ceases and reinitiates at a site beyond the lesion, thereby generating a gap opposite the adduct. By a series of recombination events, stretches of parental DNA from the sister duplex are transferred to fill in the daughter-strand gaps. The resulting discontinuities in the sister duplex are then eliminated by repair synthesis, while the original lesions (which, as a result of the genetic exchanges, may now be located on both parental and daughter strands) can be removed by other repair systems before the next round of replication. As the probability of mutation is small, recombinational repair is regarded as being an error-free system. Bacteria which use only excision repair and recombinational repair (e.g. *Proteus mirabilis, H. influenzae, Streptococcus pneumoniae* and *D. radiodurans*) are not mutated by UV irradiation.

Although the major stages of recombinational repair in *E. coli* have been known since 1968, recent purification and characterization of a major component, the RecA protein, has led to a better understanding of the underlying molecular mechanisms (Fig. 8.10). This protein binds to a daughter-strand gap and forms a long helical complex which promotes homologous pairing with the intact sister duplex. A reciprocal strand exchange then takes place. This event is facilitated by DNA polymerase I-directed repair synthesis from the free 3'-OH terminus of the damaged duplex and the gap is filled using the complementary strand of the intact duplex as template. RecA is able to promote such branch migration through a DNA strand containing pyrimidine dimers and other UV-induced photoproducts. The rate of branch migration is reduced by approximately 50-fold when a dimer is encountered, but it proceeds to completion even with heavily UV-irradiated templates. These reactions result in two paired but, because of topological constraints, not fully hydrogen-bonded molecules. After ligation by DNA

Fig. 8.10 (*Opposite.*) Stages of recombinational repair. The sequence of enzyme activity is indicated, but the enzymes themselves are not drawn to scale. For clarity only one lesion is shown and the parental DNA is drawn as strands 2 and 4 rather than the more usual configuration of strands 1 and 4 (see Fig. 8.11 for comparison). (a) DNA containing a lesion. (b) DNA polymerase III reinitiates DNA synthesis at a site beyond the lesion. (c) A RecA protein binds at the region of the daughter-strand gap. (d) The RecA protein promotes homologous pairing with the intact sister duplex. (e) Polymerase I initiates repair synthesis of DNA from the free 3'OH terminus end, using the complementing strand of the intact duplex as template, synthesizing the missing DNA (heavy dotted line), thus facilitating reciprocal strand exchange. (f) An intermediate with two Holliday junctions (sites for recombinational exchange, Chapter 9) is formed after ligation. (g) The duplexes are resolved, one containing the original lesion.

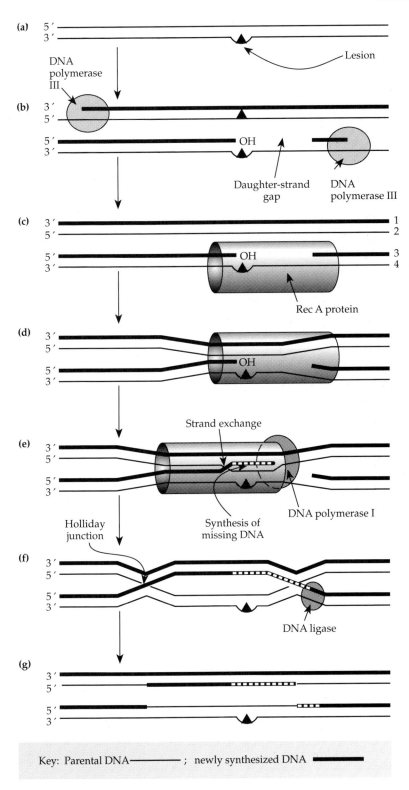

Key: Parental DNA————— ; newly synthesized DNA ■■■■■

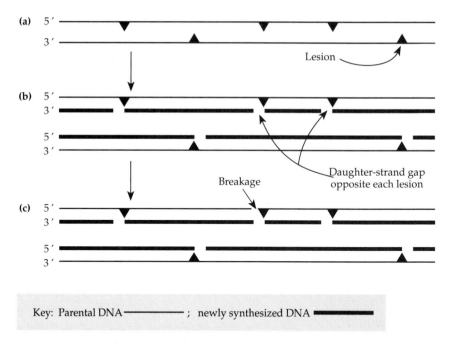

Key: Parental DNA ——————— ; newly synthesized DNA ━━━━━━━

Fig. 8.11 Formation of a double-strand break. (a) DNA containing lesions. (b) After replication newly synthesized DNA contains a daughter-strand gap opposite each lesion; (c) Breakage of the sugar-phosphate backbone at the 5′ side of a lesion converts the daughter-strand into a double-strand gap.

ligase, another intermediate with two Holliday junctions (Chapter 9) is generated, which, in turn, can be resolved to yield two intact duplexes.

Studies on the genetic control of recombinational repair have indicated that two independent pathways are involved in processing UV-induced damage. The first, known as the RecF-dependent pathway, is largely responsible for the repair of daughter-strand gaps and requires the products of at least eight genes (namely *recA*, *recF*, *recJ*, *recN*, *recO*, *recQ*, *ruv* and *uvrD*). However, apart from RecA and UvrD (which is probably responsible for unwinding DNA during recombination), the function of these proteins is still to be elucidated. The second pathway, known as the RecB-dependent pathway, plays little part in gap-filling *per se*, but its major role appears to be the repair of double-strand breaks that can arise from unrepaired daughter-strand gaps (Fig. 8.11). This process, which requires the RecBCD complex exonuclease V (also needed for conjugal recombination (Chapter 11)) as well as the RecA, UvrD and LexA proteins, may be related to an inducible RecN-dependent pathway known to be responsible for the repair of double-strand breaks caused by either ionizing radiation or mitomycin C. Although neither *recF* nor *recB* themselves are under SOS control, other members of these pathways are subject to induction (Section 8).

6 Cross-link repair

Lesions which covalently join the two strands of the DNA helix present a special challenge to repair systems as these cannot be corrected simply by using redundant information on a complementary strand. In *E. coli* such cross-links are repaired by the combined action of the nucleotide-excision and recombinational repair pathways and require the products of the *uvrA, uvrB, uvrC, uvrD, recA, lig* and *polA* genes as well as the presence of multiple copies of the genome. The current model of cross-link repair is shown schematically in Fig. 8.12.

In vitro experiments using psoralen cross-links at defined positions have revealed that the UvrABC complex incises on both sides of the cross-linked base on one strand only. Typically the ninth phosphodiester bond 5' and the third phosphodiester bond 3' to the base attached to the furan ring of psoralen are hydrolysed, while the strand attached to psoralen via the pyrone ring is not affected (Fig. 8.5). DNA polymerase I binds to the 3' incision site and its 5' → 3' exonuclease activity removes a lengthy stretch of nucleotides. The resulting enlarged gap 3' to the cross-link site provides a suitable substrate for RecA-mediated strand exchange from a sister duplex. Strand exchange proceeds past the cross-link site, physically displacing the oligonucleotide, which remains covalently joined to the intact strand. A three-stranded inter-mediate is thereby formed. The intact strand is then itself incised on either side of the adduct by the UvrABC complex. The concerted action of DNA polymerase I, DNA helicase II and DNA ligase results in the release of a double-stranded cross-link-containing fragment 11–12 nucleotides in length and concomitant repair synthesis to restore the integrity of the helix.

7 Mismatch repair

Repair processes capable of correcting base mismatches in double-stranded DNA were originally postulated to account for such phenom-ena as gene conversion (i.e. the non-reciprocal transfer of genetic information from one DNA molecule to another) and high negative interference (i.e. the clustering of genetic exchanges), which had been observed to occur during recombination in fungi. In bacteria the first evidence for mismatch repair came from studies on transformation in *S. pneumoniae* (Chapter 10). This type of repair has been best charac-terized in *E. coli*, however, where, in addition to the proof-reading step of DNA replication (Chapter 2), three mismatch repair systems are currently recognized.

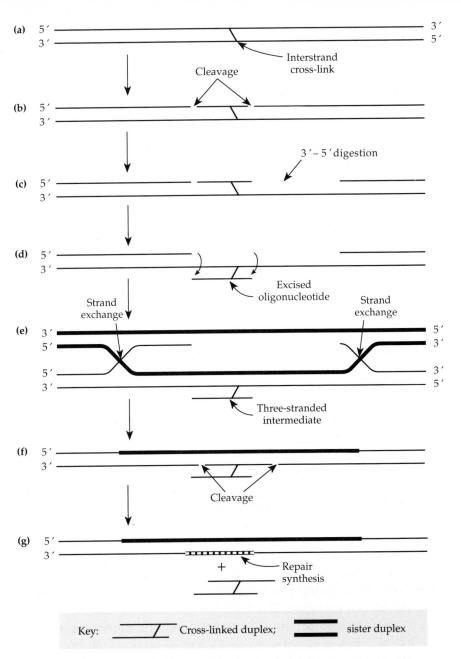

Fig. 8.12 Stages of cross-link repair. The sequence of enzyme activity is indicated but, for clarity, base pairs and individual enzymes are not shown. (a) DNA containing an interstrand cross-link. (b) A UvrABC complex cleaves the sugar-phosphate backbone of one strand on either side of the cross-linked base. (c) The $5' \rightarrow 3'$ exonuclease activity of DNA polymerase I enlarges the 3′ incision site. (d) RecA binds to the enlarged gap and displaces the excised oligonucleotide. However this remains covalently joined to the intact strand via the cross-link. (e) RecA mediates strand exchange from a sister duplex. (f) A UvrABC complex cleaves the sugar-phosphate backbone of the intact strand on either side of the cross-link. (g) The concerted action of PolI, DNA helicase II and DNA ligase brings about the release of the cross-linked oligonucleotide and restores the integrity of the helix (repair synthesis shown as heavy dotted line).

7.1 Methylation-directed long-patch repair

The consequences of random repair of base mismatches (e.g. A–C to A–T or G–C) are equivalent to the absence of repair (e.g. A–C to A–T and G–C after one round of replication). If newly synthesized strands of DNA containing potential replication errors could be distinguished from their template strands, however, then the correct direction of repair could be specified and any mismatches preferentially removed from the former strands. Methylation of adenine residues in GATC sequences (Chapter 6) is an obvious candidate for such strand discrimination. While parental chains are normally methylated on these residues in *E. coli*, nascent chains at the replication fork are transiently undermethylated. Unmethylated GATC sequences are the substrate of the *dam*-encoded DNA adenine methylase. The importance of this enzyme in error correction is confirmed by the mutator phenotypes displayed both by mutants deficient in this activity (which have lost their ability to discriminate between old and new DNA) and by methylase-overproducing strains (in which methylation takes place more rapidly than normal, thereby reducing the time available for mismatch repair to take place).

In vivo genetic studies and *in vitro* complementation assays have shown that the products of *mutH*, *mutL*, *mutS* and *uvrD* are also required for this methylation-directed repair process. MutS recognizes a base mismatch and mediates the formation of an α-shaped DNA loop within which the mismatch is enclosed. Although no enzyme activity has yet been attributed to MutL, this protein has been shown to bind specifically to MutS–heteroduplex complexes. MutH, activated by the presence of MutS and MutL, then cleaves the undermethylated strand immediately 5′ of the guanine residue of the nearest GATC sequence to the mismatch. The remaining steps are still to be elucidated but it seems likely that *uvrD*-encoded DNA helicase II displaces the nicked strand and the resulting gap is filled by repair synthesis. These events are shown schematically in Fig. 8.13.

A similar system capable of recognizing and eliminating base mismatches also exists in *S. pneumoniae* (Chapter 10). Two proteins, the products of the *hexA* and *hexB* genes, have so far been identified as being involved, and cells deficient in these activities, like their *mutH⁻*, *mutL⁻*, *mutS⁻* and *uvrD⁻* counterparts in *E. coli*, exhibit mutator phenotypes. No equivalent of the *dam* gene product is required, however, as strand discrimination in this organism relies upon the presence of transiently unligated single-strand breaks in newly synthesized DNA rather than its methylation state.

The spectrum of mismatch repair is virtually identical in *E. coli* and *S. pneumoniae*. Both systems act efficiently on transition mispairs, some

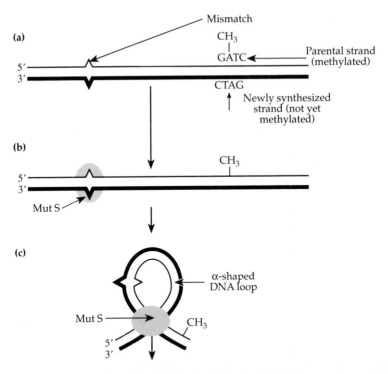

Fig. 8.13 (*Above and opposite.*) Stages of methylation-directed long-patch repair. The sequence of enzyme activity is indicated but the enzymes are not drawn to scale. (a) DNA containing a mismatch. (b) MutS recognizes and binds to the mismatch. (c) MutS mediates the formation of an α-shaped loop within which the mismatch is enclosed. (d) MutL binds to the MutS heteroduplex complex. MutH is activated by the presence of MutS and MutL. (e) MutH incises the undermethylated strand 5′ to the guanine residue of the nearest GATC sequence to the mismatch. (f) DNA helicase II displaces the nicked strand. (g) The resulting gap is filled by repair synthesis (heavy dotted line) which extends at least as far as the original mismatch.

transversion mismatches and short insertions and deletions. In general, mismatches that cause small deviations from the normal helical structure of DNA are substrates for the repair enzymes, while those that cause gross changes are not. Because strand breaks or unmethylated sites define the extent of repair as well as determining which strand is to be corrected, the natural spacing of these identifying signals means that, for each error corrected, extensive tracts (up to several kilobases in length) of nascent DNA are usually excised and resynthesized.

In addition to their role in the correction of replication errors, the products of the *mutH*, *mutL*, *mutS* and *uvrD* genes act on recombination intermediates formed during conjugation in *E. coli*. The incoming strand (which is apparently recognized by its possession of termini) is preferentially degraded if genetic exchanges result in heteroduplex DNA containing two or more mismatches. In *S. pneumoniae* the HexA and HexB proteins function in a similar way on heteroduplexes formed

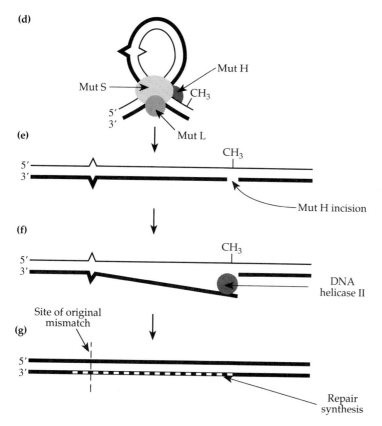

(d)

Mut S

Mut H

CH$_3$

5′
3′

Mut L

(e)

CH$_3$

5′
3′

Mut H incision

(f)

CH$_3$

5′
3′

DNA
helicase II

Site of original
mismatch

(g)

5′
3′

Repair
synthesis

Fig. 8.13 (*Continued.*)

during transformation. This type of mismatch repair, in which up to five kilobases of DNA are excised, may be regarded as a proof-reading system assuring high fidelity of homologous recombination. It may therefore play a fundamental role both in controlling genetic exchanges between members of multigene families and in defining the genetic boundaries which distinguish a species.

7.2 Very-short-patch mismatch repair

The very-short-patch type of mismatch repair is characterized by repair tracts that rarely exceed 10 nucleotides in length. The mismatch and site specificity (correcting internal G–T mismatches to G–C in CCAGG and closely related sequences) suggests that it may have evolved to rectify G–T mismatches that arise from the spontaneous deamination of 5-methylcytosine in G-5meC base pairs. In *E. coli* K strains the second C in CCA/TGG sequences is methylated by *dcm*-encoded DNA cytosine methylase (Chapter 6). Although uracil-DNA glycosylase can eliminate uracil, the deamination product of cytosine (Section 3.1), this activity cannot remove thymine residues which result from the deamination of 5-methylcytosine. Unlike methylation-directed long-patch repair, very-short-patch mismatch repair is not dependent upon the *mutH* and *uvrD* gene products, but the MutL and MutS proteins are required, as is the product of a small open reading frame which partially overlaps the *dcm* locus.

7.3 MutY-dependent mismatch repair

The pathway of MutY-dependent mismatch repair recognizes G–A mismatches where the adenine resides on a newly synthesized strand of DNA and, by short-patch excision and resynthesis, corrects them to G–C base pairs. The mispaired adenine is first removed by a G–A-specific adenine-DNA glycosylase encoded by *mutY*. Repair can then be completed by a Class II AP endonuclease, deoxyribophosphodiesterase, DNA polymerase I and DNA ligase (Section 3). This type of mismatch repair appears to be unrelated to methylation-directed long-patch repair as it is independent of the methylation state of the DNA, the presence of GATC sequences and the products of the *mutH*, *mutL*, *mutS* and *uvrD* genes. A similar G–A to G–C mismatch repair system, which functions in ATTCAT sequences, has been observed in *S. pneumoniae*.

G–A mismatches which occur in recombination intermediates are also processed in both *E. coli* and *S. pneumoniae*. While long-patch correction of such heteroduplexes results in reduced recombination frequencies, the action of these short-patch repair systems leads to the apparent clustering of genetic exchanges over a small region and hence in the recovery of higher recombinant yields than would be predicted from the physical distances in crosses involving closely linked markers. Although there is no direct evidence, it has been suggested that this effect may be of significance in genetic diversification.

8 Inducible repair

The occurrence of DNA damage leads to the increased expression of genes coding for certain repair enzymes. In *E. coli* three regulatory networks have been characterized which modulate the cellular concentrations of repair proteins.

8.1 The SOS response

The SOS response system, first described in the early 1970s, is the largest, most complex and best-understood damage-inducible repair network.

Exposure of cells to agents which damage DNA and inhibit DNA replication results in a number of diverse physiological reactions, collectively known as the SOS response. These phenomena, mostly studied in *E. coli*, include: (i) the induction of SOS mutagenesis (error-prone repair); (ii) the capacity to carry out long-patch excision repair; (iii) the increased capacity to repair double-strand breaks; (iv) the increased mutagenesis and plaque-forming ability of UV-irradiated bacteriophage lambda; (v) the induction of prophages, lambda ϕ80 (a λ-related one) and P22; (vi) the induction of colicin production; (vii) the alleviation of restriction; and (viii) the inhibition of cell division.

A negatively controlled regulon (i.e. all genes submitted to a common regulatory system, whether adjacent or dispersed on the chromosome, Chapter 15), consisting of at least 20 genes, is responsible for the SOS response. While many of the products of the SOS genes, including the regulators LexA and RecA, are produced at low levels in the repressed state, DNA damage greatly enhances gene expression (Fig. 8.14). In uninduced cells a repressor protein, the product of the *lexA* gene, binds to the operator site of each SOS gene, including *recA* and *lexA* itself. DNA damage is recognized by the RecA protein, which binds specifically to single-stranded DNA exposed by the initiation of repair or replication. This binding activates the RecA protein, enabling it to interact in turn with the LexA repressor. Catalytic cleavage and inactivation of the repressor, which occur at an insignificant rate in uninduced cells, are accelerated by the binding-activated RecA protein. As repressor molecules are cleaved, the SOS genes are expressed at increased levels and the physiological responses mediated by the products of these genes begin to be observed. Genes with operators which bind LexA relatively weakly are the first to be turned on fully, while those whose operators bind the repressor more tightly are expressed at their maximal level only when the intracellular pool of LexA has been further depleted. A return to the uninduced state is effected if the inducing signal (i.e. DNA damage) is eliminated. In the absence of activated RecA molecules, uncleaved LexA repressor accumulates and reduces the expression of the SOS genes to their respective constitutive levels.

Many of the components of the SOS response promote the survival of *E. coli* and its bacteriophages in potentially lethal environments. Filamentous growth, which results from the inhibition of cell division, allows more time for repair of DNA damage. As well as increasing the production of some of the enzymes which function in the nucleotide-excision, recombinational, cross-link and mismatch repair systems previously described in this chapter (Table 8.3), the SOS response brings additional error-correcting processes into operation. A long-patch nucleotide-excision repair pathway is induced which requires the

(a)

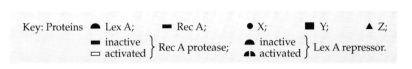

(b)

DNA damage

Key: Proteins ◖ Lex A;   ▬ Rec A;   ● X;   ■ Y;   ▲ Z;

inactive ⎫ Rec A protease;   inactive ⎫ Lex A repressor.
activated ⎭   activated ⎭

Fig. 8.14 Regulation of SOS response. (a) Absence of DNA damage. The LexA repressor binds to the operator site of each gene in the regulon including that of the *lexA* gene itself. The synthesis of some gene products (e.g. that of gene X) is completely repressed, while that of others (e.g. those of *recA*, *lexA*, genes Y and Z) are produced at low levels. (b) Presence of DNA damage. RecA binds to single-stranded regions of DNA exposed to the initiation of DNA repair or replication. Binding activates RecA enabling it to interact with the LexA repressor and enhance the cleavage of this protein. The alleviation of repression then allows the (increased) synthesis of the products of all the SOS genes. Re-establishment of *lexA* repression occurs on completion of DNA repair when the inducing signal (viz. single-stranded regions) is removed.

products of the *recF*, *uvrA*, *uvrB* and *uvrC* genes. Although at least 1500 nucleotides are removed and resynthesized for each lesion detected, this process is largely error-free. RecN-dependent repair of double-strand breaks is induced by ionizing radiation or mitomycin C damage. However, this pathway will only function in cells containing more than one copy of the genome. In addition to its action on the LexA repressor, activated RecA protein causes the cleavage of lambdoid repressors. During the SOS response, therefore, prophage induction occurs, enabling lambda to escape from a damaged cell.

SOS mutagenesis (previously called error-prone repair) can be regarded as a form of damage tolerance rather than a repair process. Although the exact molecular mechanisms have yet to be elucidated, current models suggest that during SOS induction the fidelity of DNA replication is relaxed, allowing incorporation of incorrect nucleotides opposite lesions such as AP sites and (6-4)-photoproducts in the template strand (Fig. 8.15). Translesion bypass depends on the forma-

Table 8.3 SOS-inducible DNA repair genes in *E. coli*.

Gene	Repair function(s)
lexA	SOS repressor
recA	SOS regulator; SOS mutagenesis; *recF*-dependent recombinational repair; *recB*-dependent repair of double-strand gaps; cross-link repair
recN	*recF*-dependent recombinational repair; repair of double-strand gaps
recQ	*recF*-dependent recombinational repair
ruv	*recF*-dependent recombinational repair
umuC	SOS mutagenesis
umuD	SOS mutagenesis
uvrA	Short-patch nucleotide-excision repair; long-patch nucleotide-excision repair; cross-link repair
uvrB	Short-patch nucleotide-excision repair; long-patch nucleotide-excision repair; cross-link repair
uvrD	Short-patch nucleotide-excision repair; *recF*-dependent recombinational repair; *recB*-dependent repair of double-strand gaps; cross-link repair; methylation-directed mismatch repair
dinA	SOS mutagenesis (?)
sulA	Inhibitor of cell division

tion of a multiprotein complex involving DNA polymerase III as well as the products of the SOS-inducible genes *umuC*, *umuD* and *recA*. This complex probably facilitates replication both by altering the template conformation to compensate for the distortion created by the lesion and by inhibiting the $3' \rightarrow 5'$ proof-reading exonuclease activity of DNA polymerase III. *dinA*-encoded DNA polymerase II, which is itself SOS-inducible and preferentially incorporates dAMP opposite a non-coding site, may also be involved in this bypass step. In order to participate in the multiprotein complex, UmuD must first be cleaved to its active form, UmuD′, by RecA. As this reaction is slower than that mediated by RecA on the LexA repressor, cells will preferentially induce the standard SOS response, particularly at low concentrations of RecA, i.e. when the repair process has just been triggered, whilst reserving this mutagenic pathway as a 'last-resort' function.

A number of other bacteria, including *Salmonella typhimurium*, *B. subtilis*, *P. mirabilis*, *H. influenzae*, *Bacteroides fragilis* and *Rhizobium meliloti*, exhibit physiological reactions similar to the SOS response of *E. coli* when challenged by DNA-damaging agents. The LecA repressor of *S. typhimurium* has virtually identical binding specificity to that found in *E. coli*, while *B. subtilis* and *P. mirabilis* contain proteins (named RecE and RecA, respectively) with regulatory functions analogous to that of RecA in *E. coli*.

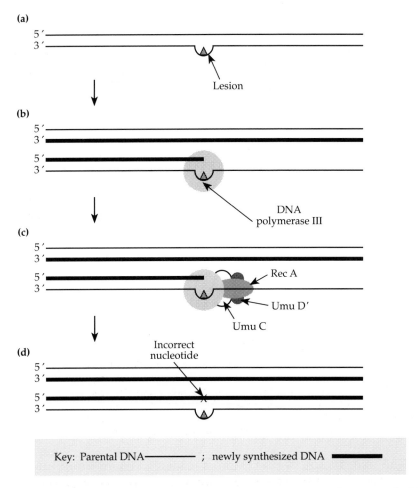

Fig. 8.15 Stages in SOS mutagenesis. The proteins involved are not drawn to scale. (a) DNA containing a lesion. (b) DNA PolIII stalls at lesion. (c) A multi-protein complex involving DNA PolIII, UmuC, UmuD' and RecA permits trans-lesion bypass. (d) Newly synthesized DNA contains a mutation opposite the lesion in the parental strand.

8.2 The adaptive response to alkylation damage

Treatment of *E. coli* with low concentrations of alkylating agents for extended time periods causes cellular resistance to the lethal and mutagenic effects of subsequent high doses of these agents. This adaptive response, first described in 1977, results from the induction of alkylation-specific repair enzymes and enables cells to cope with types of damage which are poorly processed by the SOS network. Methylation damage is the principal inducer and substrate of this system. Ethyl and propyl lesions are also recognized, although the inducing signal itself is reduced with increasing size of alkyl adduct.

(a)

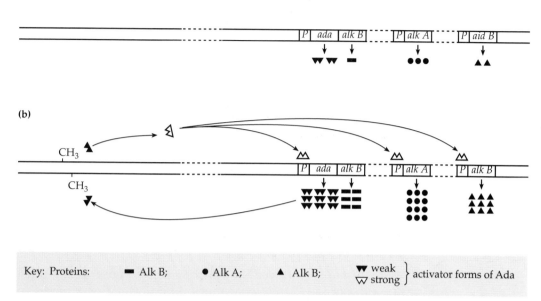

Key: Proteins: ▬ Alk B;   ● Alk A;   ▲ Alk B;   ▼▼ weak
W strong } activator forms of Ada

Fig. 8.16 Regulation of the adaptive response to alkylation damage. (a) Absence of alkylation damage. Low levels of the *ada, alkA, alkB* and *aidB* gene products are produced. (b) Presence of alkylation damage (usually methylation). The Ada protein is converted from a weak to a strong transcription activator when it de-alkylates a phosphotriester residue (see also Fig. 8.7). This activated protein then binds to each promoter in the regulon, enhancing transcription and thus bringing out the increased production of all four gene products. Re-establishment of the uninduced state occurs either by dilution or by proteolysis cleavage of the activated Ada protein.

The adaptive response network seems to be considerably smaller than the SOS regulon and is currently known to comprise four genes: *ada, alkA, alkB* (which forms an operon with *ada*) and *aidB*. The Ada protein, whose role in the direct repair of O^6-methylguanine has already been discussed (Section 2.2), acts as a positive regulator, increasing both its own expression and that of the other three genes in the presence of the appropriate DNA damage. As shown in Fig. 8.16, it is converted from a weak to a strong transcription activator when it dealkylates a phosphotriester residue and the relevant acceptor site becomes occupied. *alkA* is known to encode a DNA glycosylase (Section 3.1); however, the biochemical functions of *alkB* and *aidB* remain to be elucidated. A fifth gene, *aidC*, is also induced by alkylation damage. However, this differs from those involved in the adaptive response in being independent of *ada* control and being strongly induced by ethylating and propylating agents.

Although alkylated Ada protein is irreversibly converted into a strong transcriptional activator, increased resistance to alkylating agents gradually disappears when these agents are removed. The adaptive response may be shut down by simple dilution of the activator during growth or by its inactivation. An obvious candidate for the latter process is proteolytic cleavage. The Ada protein is readily cleaved by cellular proteases into two fragments, one of which can bind to the *ada* promoter but is unable to activate transcription. Shut-down could thus be effected by competition between Ada fragments and intact protein molecules for the promoter sequence.

Several other species besides *E. coli* have been tested for their ability to acquire enhanced resistance when challenged with low doses of alkylating agents. An adaptive response has been observed in *B. subtilis, B. thuringiensis, M. luteus* and *Streptomyces fradiae*, but does not seem to occur in *H. influenzae* or *S. typhimurium*.

8.3 The adaptive response to environmental stress

In a phenomenon reminiscent of the adaptive response to alkylation damage, exposure of *E. coli* to non-toxic levels of hydrogen peroxide increases subsequent ability to correct DNA damage caused by much higher concentrations of this agent. Adaptation to oxidative stress, however, is independent of both alkylation damage and the SOS response, and does not require the *ada*, *lexA* or *recA* gene products. At least 30 proteins in two regulons are induced in response to hydrogen peroxide challenge. OxyR functions as a positive regulator of nine genes, four of the encoded proteins have been identified (catalase/peroxidase, manganese superoxide dismutase, glutathione reductase and NAD(P)H-dependent alkylhydroperoxide reductase). However, neither the controlling element(s) nor the identity of the oxidative stress proteins belonging to the other regulon has yet been determined. In addition, *nfo*-encoded endonuclease IV, whose role as an AP endonuclease has already been discussed (Section 3.2), is induced by superoxide radicals and may belong to a third regulon induced by oxidative challenge.

A growing number of other stress-inducible genes are being identified; for example, error-free DNA repair systems are induced by elevated temperatures (the heat-shock genes) and by vitamin deprivation. It is possible that there is considerable overlap between such regulatory networks and that the induction of certain DNA repair pathways may be a relatively non-specific response to a range of chemical, nutritional and physical stresses.

9 Conclusions

The survival of a bacterial species is dependent upon the preservation of the informational integrity of the DNA in individual cells. In addition to the requirement for DNA polymerases capable of accurate duplication of the genome in each generation, a variety of repair mechanisms are needed to remove lesions ahead of the replication fork and to correct biosynthetic errors in newly replicated DNA.

During the last 30 years, much research has been directed towards an understanding of such DNA repair processes, with most emphasis being placed on *E. coli*. Early physiological investigations into the recovery of cells from DNA damage established the existence of a number of independent repair pathways and provided some insights into their methods of operation and control. Genetic studies then identified a large number of repair genes and permitted detailed analysis of the complex systems which regulate their expression. More recently, gene cloning techniques have provided sufficient quantities of repair proteins to enable their physical properties and biochemical activities to be characterized *in vitro*. Future developments are likely to include elucidation of the role of proteins in lesion targeting and analysis of the binding motifs involved in such DNA–protein interactions. For example, the presence of DNA photolyase is known to enhance pyrimidine dimer recognition by UvrABC excinuclease and a recent observation in *E. coli* that the transcribed strand of DNA is repaired at a much faster rate that the non-transcribed strand has similarly implicated RNA polymerase in damage location. Research into the actions of antimutagens (i.e. agents which reduce the yield of spontaneous and/or induced mutations) should help to clarify the molecular mechanisms by which repair processes are modulated and may also provide important information about the prevention of such diseases as cancer, which can be caused by genotoxic agents.

9: Homologous Genetic Recombination

1 General review

Recombination is the term used to describe the replacement of a sequence of DNA by another sequence, usually from a different but related genome. The size of the sequences involved can be variable but is usually long enough to allow at least one gene to be exchanged for its allele. Thus, when a bacterial cell possesses more than one chromosome, due either to a fast growth rate or to the introduction of another partial genome by any of the DNA transfer mechanisms (Chapters 10–12), homologous regions are likely to be present. This can promote a pairing between the DNA duplexes, known as synapsis, which could lead to the exchange of DNA, known as genetic recombination. This is most likely to occur between sites on separate DNA molecules, this recombination leading to hybrid, recombinant molecules, but it can occur between sites on the same genetic element, giving intramolecular recombination. The synapsis depends on extensive homology between the two DNA elements, and hence the reaction is known as general or homologous recombination.

An exchange of DNA material involves unwinding DNA duplexes (if relevant), pairing of complementary strands along homologous regions and cutting and reuniting nucleotide bonds, most of which reactions are common to the other DNA processes of replication and repair. Hence many of the enzymes involved in recombination are also involved in these other two processes (Chapters 2 and 8). It is of interest to note that the proteins specific to the recombination processes are not essential to the life of the cell.

Other types of recombination, which require little or no homology between the genetic elements involved and hence are known as non-homologous recombination, are independent of the above systems. Among these are the site-specific recombination processes utilized by some mobile sequences during their transposition process (Chapter 5) and for prophage integration (Chapter 4).

The first type of homologous recombination described in this chapter was thus named because it requires long homologous regions. Two mechanisms have been proposed. According to the mechanism known as copy choice, one of the two parental genomes must be replicating, during which process the newly forming strand suddenly switches to copying the other parental genome, thus creating a molecule having characteristics derived from both parents. Thus the recombinant would contain only newly synthesized DNA. This model was intended to account for situations in which only one of the expected reciprocal recombinants predominated, as can occur in bacteriophage genetics. Experiments using DNAs of which one originated from parents labelled with a heavy isotope proved that the recombinants acquired, in fact, original segments of DNA belonging to both parents. Hence the proposed mechanism was disproved.

The other mechanism, known as breakage-reunion, is currently largely accepted, in spite of elements that are still unclear. The overall genetic organization responsible for this process and the proposed models will be described in the case of *E. coli*, the best-known example. The models, if not always the enzymes, most probably also apply to other prokaryotes and to eukaryotes.

E. coli possesses several homologous recombination pathways, which require many genes distributed over the whole chromosome (Fig. 9.1). The functioning of at least some of these pathways is interconnected via either common constituents or common regulators. The preponderant pathway, the RecA pathway, requires proficient *recA* and *recB, C* and *D* genes. Mutants in *recA*, or in *recB/recC* exhibit 0.001% and 1% of wild-type level of recombination, respectively. The latter deficiency can be restored to 50% by mutations in either of two *scb* loci (thus named since they code for suppressors of RecBC), which put into action either the ScbA–RecE pathway or the ScbB protecting bypass (Fig. 9.2). Even if these pathways are rendered inoperable, very

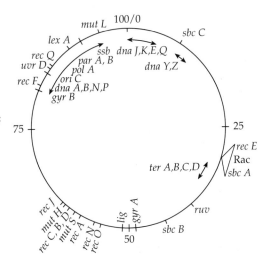

Fig. 9.1 Chromosomal location of recombination-involved genes in *E. coli*. *lexA*, repressor for recombination and repair systems; *mut*, mutator; Rac; deficient prophage; *rec*, recombination; *sbc*, suppression of RecBC⁻ phenotype; *ruv*, resistance to UV light; *uvr*, short-range excision repair. Loci indicated inside the circle are implicated in DNA synthesis and replication (see Chapter 2, Fig. 2.14 for explanation of these symbols).

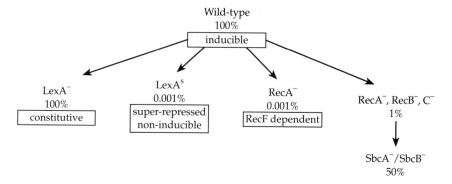

Fig. 9.2 Functional inter-relationships of the genes controlling the rates of recombination of the main recombination pathways in *E. coli*. Figures (representing % of wild-type capacity) represent recombination capacity as measured by the frequency of formation of recombinants after transfer of homologous DNA in the corresponding mutant.

low rates of recombination are still possible under the *recF, recJ* or *recO* genes.

2 The RecA pathway of *E. coli*: components involved and genetic organization

Two key proteins, RecA and the RecBCD complex, and particular sequences known as Chi sites are required for this process. Synthesis of the RecA protein is regulated by a complex system, the key protein of which is a repressor, LexA.

2.1 The LexA regulator

The LexA protein blocks its own synthesis and that of RecA (Fig. 8.14, p. 254), as part of a general repressor action on the SOS regulon concerned with repair of DNA damage (Chapter 8) and strand exchanges, including the recombination-involved genes *recN* (which belongs to the RecF pathway (Section 4.3)) and *recQ*. LexA mutants belong to two categories (Fig. 9.2), those which do not synthesize an active protein and are constitutive Rec$^+$ (such clones will continue to express the Rec$^+$ phenotype, as described in Chapter 8, once triggered to do so by the presence in the DNA of a lesion that activates the RecA protein) and the super-repressed ones, in which none of the functions negatively regulated by LexA can be induced and which are thus Rec$^-$ and SOS$^-$.

2.2 The RecA protein

The fact that *recA*$^-$ mutants form recombinants at a frequency 10^{-3} times lower than the wild type indicates that at least one other recombination pathway is present in these cells but that the *recA*-dependent one is the preponderant one. The RecA protein is a bifunctional molecule, having: (i) a DNA-dependent ATPase activity, used in the course of DNA unwinding, which promotes single-strand reannealing; and (ii) a specific protease activity, used to inactivate the LexA repressor (Chapter 8). Its synthesis is partially repressed by LexA, allowing for a background amount of RecA to be produced under conditions which do not require high levels of recombination, i.e. in the absence of DNA damage (Section 3.1). This amount is sufficient to initiate, when needed, the RecA-requiring processes (repair and recombination), which start through proteolysis of LexA, and to cope with the other cellular functions in which RecA is involved (e.g. the coupling of replication with cell division) and which are coregulated by LexA. RecA is synthesized as a proteolytic-inactive form. Conditions that require the system to function at high levels (Section 3.1 and Chapter 8) promote the protease activity, which results in degradation of the LexA repressor and thus further synthesis of RecA itself and of the other LexA-repressed genes.

Besides their decreased recombination capacity, RecA-deficient mutants show a very much increased degradation of their DNA, particularly when their DNA is damaged, giving rise to the 'reckless phenotype'. These cells show increased sensitivity to DNA-damaging agents (indeed, this is how they were initially isolated) and a lower growth rate. This pleiotropic phenotype results from the incapacity to inactivate the LexA repressor and thus to induce the LexA-regulated repair functions. Possible free DNA extremities resulting from DNA damage or from the normal functioning of the DNA replicating and error-correcting systems (Chapter 2), which are not repaired from lack of induced-repair processes, are targets for the RecBCD nuclease (Section 2.3), which acts as an exonuclease on these DNA extremities and degrades the DNA.

The RecA protein, originally known as protein X, functions as a polymer of 38-kD subunits. Several copies wind around a single-stranded DNA stretch – which may be present as such in the cell (delivered by a transfer system) or may originate from a duplex which it has previously unwound – promoting the pairing of the single-stranded DNA with a homologous region from the other (different) parental DNA to form a heterogeneous new duplex.

2.3 The RecBCD complex

The three genes *recB*, *recC* and *recD* are often referred to as *RecBCD* since they are located close to each other (Fig. 9.1) and all the encoded proteins are needed to form a complex enzyme, exonuclease V (ExoV).

It was hitherto known as the RecBC complex, before demonstration (1974 and 1984) that a third subunit, RecD, was included in the complex. While the mechanism of action of the RecB–RecC part or of the whole complex is fairly well understood (Section 3.2), the exact role of RecD is still unclear, although it might be responsible for an exonuclease activity. Deletion or insertion mutations in the *recD* gene do not alter recombination efficiency, indicating a sufficient level of ExoV activity.

ExoV is active in several ways, all being ATP-dependent, on different DNA substrates. It can act as an exonuclease on either double- or single-stranded DNA, producing mononucleotides. It shows endonuclease activity on single-stranded DNA or at Chi sites (Section 2.4) on duplexes, and can also display a DNA-unwinding, helicase, activity, the two latter characteristics being those mostly related to genetic recombination. Starting either from an extremity or in the middle of a molecule, the enzyme binds to a flush double-stranded extremity and, unwinding the duplex while moving towards the other extremity of the molecule, forms symmetrical single-stranded loops

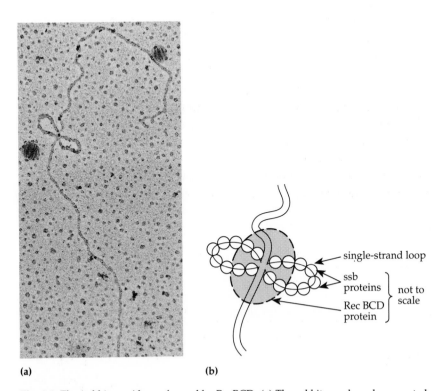

(a)   (b)

Fig. 9.3 The 'rabbit-ears' loops formed by RecBCD. (a) The rabbit-ears have been coated with ssb proteins for easier identification. (Potter & Dressler, 1988). (b) Diagrammatic representation.

known as rabbit-ears (Fig. 9.3). When encountering an active Chi site on one strand, it produces a nick some distance away (up to 10–20 kb) toward the 5′ side, the single-strand extremities thus generated usually being coated by single-strand-binding (ssb) proteins, and the loop on that strand thus being opened. The enzyme then moves forward, and the same process is repeated further. These single-strand regions might serve as potential substrate for RecA to bind and promote recombination with a double-strand homologous region.

2.4　Chi sites

The existence of particular sites capable of enhancing recombination by allowing higher activity of RecBCD was first observed on phage λ DNA. Similar sites used by the same RecBCD enzyme have since been well described on the *E. coli* chromosome. The latter, designated Chi after the Greek letter χ, the shape of which recalls that of the four-arm recombination intermediate, can exist as active, *chi*$^+$, or inactive, *chi*$^-$, forms, yielding Chi$^+$ or Chi$^-$ phenotypes, while the original χ notation has been retained for the λ ones, which are in a non-active, χ^0, form in the wild-type DNA.

The reality of such sites and the suggestion that they act as recombination–recognition sites for RecBCD were confirmed when several of them were shown to correspond to identical sequences of eight contiguous bases, 5′-GCTGGTGG-3′. Specific replacements of individual bases decrease recognition efficiency by RecBCD to varying degrees, in certain cases by huge factors (Table 9.1). In a standard *E. coli* chromosome, all fully or partly active eight-base Chi sequences, occurring more or less randomly on the chromosome, form a family of hotspots for recombination with a spectrum of activities. The whole chromosome possesses about 1000 fully active Chi sites, i.e. one every 5 kb. Since these sequences may occur anywhere and since the octamer

Table 9.1 Spectrum of activity of Chi site sequences.

		Octamer sequence									Efficiency (%)
		1	2	3	4	5	6	7	8		
Consensus Chi$^+$	5′	G	C	T	G	G	T	G	G	3′	100
Mutations		$\begin{cases} C \\ A \end{cases}$	T	A			C	A			0
					A						40
		A									10
			T								5

may represent sense triplets, they may often be encountered within ORFs.

2.5 Other constituents involved

2.5.1 *Single-strand DNA-binding proteins (ssb)*

Ssb proteins, which are important for maintenance of accessible stretches for DNA polymerase activity during replication (Chapter 2), can enhance recombination in several ways, due to their high affinity for single-stranded DNA. By an unknown mechanism, they stabilize the interaction between the RecA proteins and DNA. They also protect the intermediary single-stranded regions from degradation by exonuclease and, finally, by preventing secondary folding due to possible intrastrand complementary hydrogen-bonding, they maintain the whole length accessible to further recombinational events.

2.5.2 *Other proteins*

No demonstration of the *in vivo* need for any other DNA-unwinding enzyme has been made, although *in vitro* enhancement of recombination is observed when such an enzyme, acting in the manner of a gyrase, is provided.

Since the DNA exchange taking place during recombination most often leaves gaps in the strands' continuity, refilling synthesis of the corresponding sequences is required. This is performed by the PolI and ligase enzymes otherwise involved in standard DNA replication (Chapter 2).

3 Mechanism of recombination

3.1 Induction of the recombinational process

Conditions which stimulate recombination all correspond to the occurrence in the cell of an abnormal DNA structure. This may be either a lesion or the introduction of a fragment of exogenous double- or single-stranded DNA. All these situations result in increased synthesis of RecA, which, as seen above, is the consequence of the induced activation of the RecA-specific protease activity on LexA and the subsequent release of RecA-synthesis repression (Section 2.2). The actual chemical nature of the RecA activator is still subject to speculation (Chapter 8). A possible scheme, which fits in with both *in vivo* and *in vitro* observations, supposes that RecBCD (or another substitute DNase), through its activity on abnormal DNA regions,

results in production of single-stranded regions, or even of mono-
or oligonucleotides. Either of these could act as a trigger for RecA
activation.

3.2 Linear double-duplex recombination

In the great majority of cases, when recombination occurs the two
molecules of DNA involved are duplex, and hence most of the mechanisms that have been suggested appertain to the interaction of two
double-stranded components. Mechanisms were initially proposed for
eukaryotes (mostly fungi), in which two double-stranded linear DNA
molecules are involved, and give rise to two reciprocal recombinant
products. The model(s) also apply to prokaryote recombination involving two double-stranded DNA molecules, one of which is linear, but
reciprocal products are encountered only when two whole replicons,
e.g. plasmid and chromosome, are involved (Section 3.3).

Several models have been proposed over the years, but the model
which, at present, seems to fit the data best is that of Holliday (1964)
with the modifications of Hotchkiss (1974) and of Meselson and
Radding (1985) (Fig. 9.4), with the model of Whitehouse (1963) as the
closest contender. The process may be divided into three main stages:
(i) strand breakage; (ii) strand pairing and invasion; and (iii) crossover
formation.

When two homologous double-stranded DNAs are involved, initially
(Fig. 9.4i) either two single-strand breaks are made at Chi sites by the
RecBCD ExoV enzyme at opposite positions on strands of the same
polarity as the previously aligned double helices (Holliday model, Fig.
9.4a), or only one strand is cut (Fig. 9.4c).

In the Holliday model the single strands are unravelled and
stabilized by ssb proteins, allowing for pairing with the intact non-
sister strands of the homologous duplex, and are immediately ligated
to the extremities of their cognate opposite strands (Fig. 9.4a, iv). In
the Hotchkiss and Meselson/Radding models, the cut single strand
invades the other duplex by pairing with its complementary strands
and displacing the homologous one, forming a D-loop (Fig. 9.4b,c ii).
This causes a nick (i.e. the second nick) in the latter strand. Further
invasion, which may be associated with degradation of the displaced
strand the Meselson/Radding model, leads to the formation of heteroduplex regions, which need not be symmetrically located in the two
duplexes (Fig. 9.4iii).

In the Holliday model, ligation of the cut ends gives rise to a
cruciform, or χ, structure known as the Holliday junction (Fig. 9.4a, iv
and v). In both the Hotchkiss and the Meselson/Radding models,
previous DNA synthesis has to take place to fill the gaps left by the

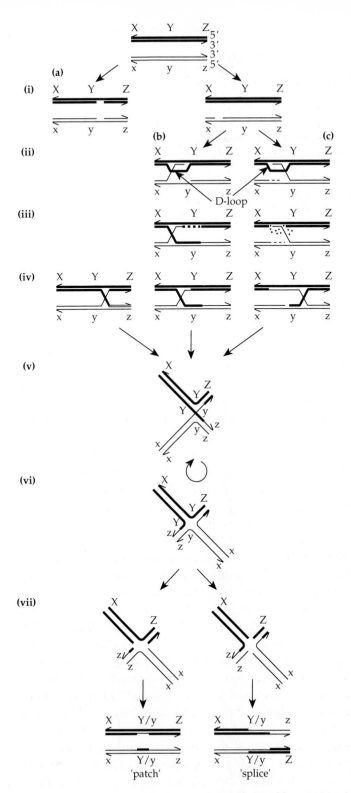

Fig. 9.4 Models of recombination. (a) Holliday. (b) Hotchkiss. (c) Meselson/Radding.

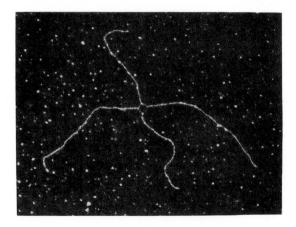

Fig. 9.5 A Holliday junction structure formed between two linear double-stranded DNA molecules (Potter & Dressler, 1988).

invading strand in the parental DNA (Fig. 9.4biii and biv, ciii, civ) before the Holliday junction can be formed (Fig. 9.4v). Actual observation of limited DNA synthesis accompanying recombination supports these models. These differ with respect to the type of end of the invasive strand, a 3'-OH in the Hotchkiss and a 5'-P in the Meselson/ Radding model, and in DNA synthesis occurring in both duplexes in the Hotchkiss but only in the strand opposing the invading one in the Meselson/Radding model. The cruciform structure (Fig. 9.4v) is central to all models. It has, in fact, been visualized in electron radiographs (Fig. 9.5), thus proving this part of the postulated models.

The initial search for the homologous regions on the two duplexes, the pairing and the further strand migration are driven by RecA proteins. Branch migration increases the length of the crossed-over section of DNA (Fig. 9.6).

All models suggest that the structural symmetry of the Holliday junction allows the helices to swivel (Fig. 9.4vi). The swivel, known as isomerization, produces with equal probabilities recombinants with either an insertion of a length of DNA from the other parent (the 'patch' recombinant) or a ligation of a reciprocal length of DNA as a terminal sequence (the 'splice' recombinant), depending on whether the Holliday junction is cut transversely or longitudinally (Fig. 9.4vii). Molecular models show that isomerization is feasible, but it has been suggested that, when complexed with proteins, the DNA may not be free to move. Experiments with laboratory-synthesized Holliday junctions attached to RecA proteins suggested that the RecA protein provided a framework holding the DNA in the two possible orientations required to give the two types of recombinants, which are actually

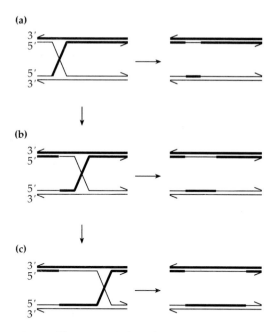

Fig. 9.6 Three stages in branch migration.

obtained in situations where reciprocal recombinants can be observed, i.e. eukaryotes (in particular, fungi).

The amount of DNA actually recombined will depend largely on the extent of any migration that may have occurred from the initial to the second crossover positions. Mismatch processes may then occur to erase structural defects, e.g. local mispairings resulting from differences in base sequences (Chapter 8).

Whereas the Holliday model and its modifications describe single-strand crossovers, the Whitehouse model, with more recent additions, involves an initial double-stranded crossover (Fig. 9.7). It is probably more appropiate to the conditions pertaining for fungi, i.e. a eukaryotic situation, than to those occurring in the prokaryotic world.

3.3 Circular double-duplex recombination

Circular double-duplex recombination is encountered between chromosomes and plasmids sharing homologous regions, i.e. F or F' plasmids, specialized transducing particles or artificially constructed recombinant plasmids carrying a fragment of chromosomal DNA. The general features described above apply, as has been visualized by the formation of 'figure-eight' structures, equivalent to the preceding χ structures (Fig. 9.8). Two different outcomes may occur from the initial Holliday

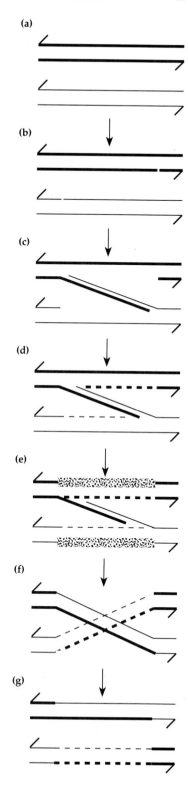

Fig. 9.7 The Whitehouse model of recombination. (a) Pairing of two homologous double-strands; (b) formation of a single-strand nick on one strand of each duplex; (c) pairing of the two complementary single-strand extremities thus created; (d) synthesis from the 5′ free extremity, of DNA complementary to the non-cut strand of each duplex; (e) degradation of the copied regions; (f,g) resealing of the free extremities with the nearest one in the same orientation, leading to reciprocal exchange.

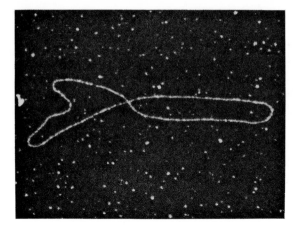

Fig. 9.8 A 'figure 8' structure formed between two circular DNA molecules (Potter & Dressler, 1988).

junction. Actual branch migration may, as seen previously, give rise to a second crossover between the same two strands a distance from the first one along the homologous region, leading to reciprocal exchange of genetic material between the two circular molecules. In contrast, nicks on the two intact strands may take place and lead, with possible need for local DNA synthesis, to the fusion of the two molecules and formation of a unique circular molecule. This mechanism is responsible for F integration into the *E. coli* chromosome, yielding an Hfr structure, the homologous regions being identical IS elements present on both molecules (Chapters 3 and 11). This has been called the single-crossover model of Campbell.

Similar events may occur between an F' plasmid and the chromosome and between a specialized transducing particle and the chromosome, leading to a merozygote chromosome which, in the latter case, is (possibly) defectively lysogenic (Chapter 12).

3.4 Recombination between single- (linear) and double-stranded DNAs

In bacterial natural transformation, the donor DNA is presented to the recipient chromosome in a linear, single-stranded form. The recombinational event is a standard breakage–reunion reaction, involving a three-strand intermediate, and producing, of course, only one recombinant type that can be visualized as viable. Mismatches resulting from formation of a heteroduplex will influence the final outcome after specific mismatch repair processes have taken place (Chapter 10).

3.5 General characteristics of recombination

As mentioned, the Chi sites induce a polarized orientation to the displacement of the RecBCD enzyme and thus of the site of the initial recombinogenic nick. Stimulation of Chi-promoted exchanges decreases with increasing distances, by a factor of 2 per approximately 2 kb.

Recombination frequencies, measured as the fraction of exchanges occurring between two markers, have led to the definition of recombination units. The classical phenomenological unit, defined in eukaryotes, the centimorgan (cM, named after the geneticist Morgan) corresponds to 1% exchanges between two markers on a pair of homologous chromosomes. This has been equalled to approximately 10^6-base-long segments (on the human genome), although actual probabilities are by no means constant. The same frequency of recombination observed after conjugation in *E. coli* requires only 2 kb, i.e. a frequency per unit length about 500 times higher.

A similar discrepancy is observed when the kinetics of recombination is considered. When an F' *lacZ* plasmid was transferred to a *lacZ⁻* recipient, recovery of β-galactosidase activity required 35 extra min if it implied recombination with a *lacZ⁻* allele brought by the F' plasmid, as compared with complementation after transfer of an F' *lacZ⁺* molecule. Analogous determinations in the eukaryote *Ustilago* suggested that $4-4\frac{1}{2}$ h were necessary.

Efficiencies may also vary with other undetermined factors, associated with, for example, the means of delivering the exogenous DNA into the recipient.

4 Other pathways

4.1 The SbcA–RecE pathway

Single or double mutants deficient in *recB* and/or *recC* show a decreased level of recombination (about 1% in the wild type), which can be restored to some 50% by mutation in a gene, *sbcA*, which then acts as a suppressor to the RecBC⁻ phenotype. The *sbcA* gene belongs to a chromosome-integrated defective prophage, known as Rac (for Rec activation) (Fig. 9.1), which also possesses an exonuclease-coding gene, *recE*, which is normally non-expressed. Mutations in *sbcA*, denoted *sbcA⁻*, probably eliminate a transcription terminator (Chapter 15) upstream of *recE*, thus allowing synthesis of the exonuclease (denoted ExoVIII). This enzyme is a 5' → 3' exonuclease active on double-stranded DNA, yielding as products both mononucleotides and long single-stranded stretches of the complementary chain with 3' protruding extremities. ExoVIII is very similar in its activity to the λ exonuclease active in recombination. In fact, partial suppression of the RecBCD⁻ phenotype is obtained by induction of the λ *exo* gene.

4.2 The sbcB–sbcC rescue system

The RecBCD⁻-induced decrease of recombination efficiency can also be partly overcome by mutation in another gene, *sbcB*. This gene codes for another exonuclease, ExoI, with a $3' \rightarrow 5'$ hydrolysis orientation, on single-stranded DNA. Absence of this enzyme does not, strictly speaking, participate in recombination enhancement, but does so indirectly by decreasing the probability that the intermediary single-stranded structures will be degraded.

Good viability of *recB⁻/recC⁻*, *sbcB⁻* clones is often associated with another mutation in an *sbcC* gene. Although little is known of this gene beside its localization (Fig. 9.1), it has been suggested that the SbcC product affects permeability or envelope structure.

4.3 The RecF pathway

The RecF pathway is a more poorly known process. It accounts for a limited recombination activity, since a deficiency in RecF requires, in order to be non-ambiguously observed, that it is studied in a multiple Rec-deficient background. RecF is known to be a single-strand-dependent ATPase which acts in conjunction with RecA. It might also have an endonuclease activity. Its functioning, like that of RecA, is initiated by that of RecB and its deficiency is partly suppressed by *sbcA* or *sbcB* mutations.

Two other genes, *recJ* and *recN*, are considered to be part of the RecF pathway. Although the genes have been cloned and sequenced, little information is available as to their function beside their implication in recombination.

4.4 Other genes detected as recombination-implicated

Several other genes have been detected by the appearance of a lower recombination level in an otherwise multiple Rec-deficient background when mutations occur in these loci. Thus genes *recO*, *recQ* and *ruv* (the latter two repressed by LexA) have been mapped and the corresponding proteins have been predicted from nucleotide sequence analysis (Chapter 16), but next to nothing is understood of their function.

Several approaches have been used to look for mutants showing increased recombination activity and have led to a different spectrum of loci. A first set of such Hyperrec mutants corresponds to mutations in genes involved with mismatch repair, *mutS, L, H* and *uvrD* (Chapter 8). Others may be phage-encoded mutations (e.g. P1), although the nature of the mutation is not known.

5 Recombination pathways in other organisms

The information available about recombination processes in other prokaryotes is very limited. No species has been detected in which recombination does not exist, as might be expected if recombination processes are, as in *E. coli*, largely shared by the repair processes.

recA genes seem universal and the RecA proteins' primary structure is generally conserved. Efficient strategies to demonstrate the presence of a *recA*-like gene have been to look for a DNA region, on the studied species, which would either complement an *E. coli recA*$^-$ mutant or hybridize with an *E. coli recA* probe (Chapter 16). Such genes have been found in various species, both Gram-negative and Gram-positive (Table 9.2). Further studies of these RecA-like systems have in general been performed in comparison with the *E. coli* one. Conservation of the amino acid sequence and of particular domains of the protein, and often of the nucleotide sequence, is frequent, even in species taxonomically distant from *E. coli*. Similarly, in most cases, the *E. coli* regulatory system involving LexA could control the synthesis of the heterologous gene when introduced into *E. coli*.

Availability of RecA$^-$ mutants, however, is limited, sometimes due to apparent non-viability of the potential clones.

Mechanisms and sometimes enzymes similar to RecA are considered to be present in at least some eukaryotes, in particular fungi, e.g. *Ustilago*.

RecBCD-like activity has been identified in several Gram-positive and Gram-negative species. This is the case in a dozen or so terrestrial and marine enterobacteria which have been thus analysed. A complete RecBCD complex is known to be present in *H. influenzae* and *S. typhimurium*. Reciprocity of interaction of the *E. coli* RecBCD enzyme on the heterologous Chi sites and at least of the *S. typhimurium* RecBCD enzyme on the *E. coli* Chi sites has been observed, suggesting a conservation of the Chi–RecBCD system during the evolution of enterobacteria, as a recombination-enhancing mechanism.

Chi sites are also involved in processes in eukaryotic genetics, e.g. in the tumour-producing *Agrobacterium tumefaciens* Ti plasmid there is a Chi site adjacent to the segment that will insert into the plant cell, and also in gene conversion in fungi.

In view of the biochemical and genetic relationship between recombination and repair, it has been speculated that recombination processes might in fact be a secondary consequence of the existence of

Table 9.2 Occurrence of RecA proteins in prokaryotes.

Species	recA-like gene cloning strategy	MW (kDa)	% Homology with E. coli		Regulation by E. coli LexA	Other properties
			Gene	Protein		
GRAM-NEGATIVE						
Enterobacteriaceae						
Escherichia coli	UV-sensitive mutant	38	100	100	+	
Proteus mirabilis	C	37–38		73	+	Protein crystallized
P. vulgaris	C	37–38		73	+	Regulates excretion of a nuclease
Serratia marcescens	C	37–38	85		LexA box present	
Shigella flexneri	C	37–38	99	100	+	
Erwinia carotovora	C	35	78	91	+	Regulates pectin lyase and carotovoricin
E. chrysanthemi	C	35				Not involved in pectin and cellulose lyases
Yersinia pestis	C					
Salmonella typhimurium	–	38				Protein purified
Pseudomonadaceae						
P. aeruginosa PAO	C	38	57	71	+	
P. putida	C		No			
P. syringae	C					
Rhizobiaceae						
Agrobacterium tumefaciens	C		High			Not involved in plant tumour establishment
Rhizobium meliloti	C		High			Not involved in induction of plant infection
R. japonicum	H		High			
R. leguminosarum	H		High			
Vibrionaceae						
V. cholerae	C	39	No	ICR	+	Involved in toxin operon duplication
V. anguillarum	C					
Aeromonas caviae	C	39.4	Low			

	C	MW	H	ICR	Comments
Neisseriaceae					
N. gonorrhoeae	C				Involved in antigen variation and virulence
Acinetobacter calcoaceticus	C				
Bacteroidaceae					
B. fragilis	C	39, 37	No	—	Produces two proteins
Methylomonadaceae					
Methylophilus methylotrophus	C	36	Low		
Cyanobacteria					
Gloeocapsa alpicola	C	39			
Synechococcus sp.	C	38		57	
Anabaena variabilis	C	38.4	—	58, ICR	
OTHER GRAM-NEGATIVE					
Thiobacillus ferrooxidans	C	38/40		66	Complements *S. typhimurium* RecA$^-$ mutant also
Haemophilus influenzae	C[a]	38		60–70, ICR	
Legionella pneumophila	C	37.5			
Three unclassified strains	C				
GRAM-POSITIVE					
Bacillus subtilis	C	42/45		ICR	
Actinomycetes					
Streptomyces fradiae					Mutants complemented by *E. coli recA*
BACTERIOPHAGES					
T4 (*E. coli*)	C	43.8		23, no ICR	Does not cleave *E. coli* LexA

C, complementation of *E. coli* RecA$^-$ mutant; H, southern hybridization with *E. coli recA* probe; ICR, immuno-cross-reaction with *E. coli* anti-RecA antibodies; MW, molecular weight.

efficient repair systems, on which evolution may have exerted a greater pressure than on recombination systems.

References

Holliday R. (1964) A mechanism for gene conversion in fungi. *Genetic Research*, 5, 282–304.

Whitehouse H.L.K. (1963) A theory of crossing-over by means of hybrid deoxyribonucleic acid. *Nature*, 199, 1034–40.

10: Genetic Transfer: Transformation

Three main mechanisms of transfer of genetic material, namely transformation, conjugation and transduction, are known to occur naturally among prokaryotes. In practice, the ability to transfer information seems widespread among prokaryotes although the information available does not make it possible to predict whether a new strain will perform genetic transfer or which of the mechanisms is most probable. Transformation may well be the most widespread, although the exact conditions for demonstrating it in the laboratory may not be known in all cases, whilst transduction (Chapter 12) requires the organism concerned to be inhabited by an appropriate bacteriophage. Conjugation, as currently known (Chapter 11), requires the donor/recipient organisms to possess recognition devices (e.g. a pilus in Gram-negative bacteria), but even this may not always be essential.

1 Definition of bacterial transformation

Bacterial transformation is the transfer of genetic material as free DNA to recipient cells, with no direct contribution from the intact donor cell. The DNA may consist of either chromosomal DNA (as linear molecules) or of plasmid DNA, which in laboratory conditions is a covalently closed circular structure.

Many artificial transformation systems have been developed, usually using plasmid transfer, but the mechanisms involved are quite different from those in natural systems. They will be described later (Chapter 12).

2 Discovery

In 1928, Griffith, who was studying the epidemiology of pneumonia, obtained two variants of the bacterium responsible for the disease, *Streptococcus pneumoniae* (then called *Pneumococcus* or *Diplococcus pneumoniae*). A smooth (S) capsulated variant led to the development of the disease when used to infect mice, and could therefore be defined as virulent, while a rough (R) non-capsulated variant could not induce the disease and was clearly avirulent. If a suspension of smooth cells was heat-killed prior to infecting mice their virulence was lost, but unexpectedly when a mixture of heat-inactivated smooth cells and living rough ones was used to infect mice many of the animals developed the disease. Furthermore, bacteria recovered from these mice were virulent when used to infect new mice. Griffith postulated the existence of a substance, liberated from the heat-killed smooth bacteria, which was capable of stably modifying the avirulent character into a virulent one. Griffith's experiment remained unnoticed, as has happened to many important discoveries, for some years.

It was then found that the transformation from R to S could be achieved using cell-free extracts in the test-tube without the mouse infection step, but it was only in 1944 that Avery, MacLeod and MacCarty isolated the molecule responsible for this transformation, which they called the transforming principle. They identified the purified transforming principle as DNA and hence as the carrier of genetic information, and gave a more complete description of bacterial transformation. Whereupon their findings were largely forgotten for several years. It was not until 1952, after Hershey and Chase's work on bacteriophage physiology (Chapter 4), that the key role of DNA in living organisms was widely recognized.

3 A standard transformation experiment

The transfer of DNA will be genetically visible only if it leads to modification of at least one of the characters of the recipient cells. Thus the experiment has to be performed using at least one mutant and the wild type of a given strain bearing distinctive characteristics (different alleles of the same gene), e.g. resistance to the antibiotic streptomycin, Str^r, in the mutant and streptomycin sensitivity, Str^s, in the wild type. It is possible to use two mutants to demonstrate transformation, for instance, one being sensitive to streptomycin and auxotrophic for leucine, Leu^-, while the other is resistant to streptomycin and prototrophic for leucine, Leu^+. One can use the former mutant, $Str^s Leu^-$ (say, of *Bacillus subtilis*), as the recipient population while chromosomal DNA is prepared from the $Str^r Leu^+$ mutant. The recipient population has to be used when it is in a particular state, called competence (Section 5.1), obtained usually by promoting synchrony and growing the cells in a specially devised medium, in order to show the highest efficiency of uptake of DNA. These cells are then incubated for periods up to an hour with an appropriate concentration of donor DNA (usually $1\,\mu g/10^8$ cells/ml) under optimum conditions for DNA uptake. For *B. subtilis* this would be at $37\,°C$. The donor DNA is extracted by standard procedures, and during this process is usually sheared by pipetting into fragments averaging 10^7 daltons (approximately 15 kb).

Table 10.1 Typical results obtained during transformation in *B. subtilis*. A strain of
B. subtilis bearing two markers, *leu*⁻ and *str*ˢ, is used as the recipient. After induction of
competence, it was mixed with DNA (1 μg/ml) extracted from a Leu⁺Strʳ mutant and
incubated for 60 min at 37 °C (A). Samples were then plated on appropriate media to
measure the number of transformants. A parallel suspension not treated with DNA was
used to measure spontaneous mutation frequencies (B). These must be deducted from
the figures obtained for the DNA-treated suspension, but they are usually relatively
negligible values.

Selected phenotype	Addition to the medium		Number of clones growing under selective condition			
			A		B	
	leu	str	Cells/ml	Frequencies	Cells/ml	Frequencies
All (numeration of the recipient cells)	+	−	1×10^8	−	1×10^8	−
Leu⁺	−	−	5×10^5	5×10^{-3}	2×10^1	2×10^{-7}
Strʳ	+	+	4×10^5	4×10^{-3}	<1	$<1 \times 10^{-8}$

Since transformants are recognized by their acquiring one or several
new characters (also called markers) from the donor DNA, they are
isolated by spreading known volumes of the mixture on appropriate
selective media (Chapter 7): Leu⁺ transformants and Strʳ ones will
be selected on media lacking leucine or containing streptomycin,
respectively, while only the rare spontaneous mutants in the recipient
population, which are checked by suitable controls, will also grow on
these. Results of a typical experiment are given in Table 10.1.

The transformation process does not depend on a predetermined
role as either donor or recipient for the two strains implicated (often
mutants of an original wild-type strain). Under laboratory conditions,
only the feasibility of selection for the transformed marker is a prerequisite for a successful experiment. In one involving the above
couples of alleles, only the allele distribution indicated between
recipient and donor DNAs allows easy selection of transformants. All
problems related to biases in selection (dominance, segregation, etc.
(Chapter 7)) must be taken into account.

4 Occurrence of natural transformation among prokaryotes

Although transformation was first discovered in the Gram-positive
Streptococcus pneumoniae, the second organism in which it was found
was the Gram-negative *Haemophilus influenzae* and then in the Gram-
positive *Bacillus subtilis*. These three strains have remained the archetypal
transformable bacteria from a study of which our knowledge of the
details of transformation has been derived. However, over the past

20 years an increasing number of naturally transformable bacteria, accepting either chromosomal or plasmid DNA or both, has been described, some of which are listed in Table 10.2. These clearly belong to very different genera, among which the known number of strains showing transforming ability varies, reflecting at least in part the efforts made to characterize this property.

A rapid assay for testing the ability of a strain to be transformed can be applied as follows: the putative recipient cells, bearing an appropriate marker, are plated as a lawn on the complementary selective medium. DNA from the donor carrying the complementary allele of the marker gene whose transfer is to be demonstrated is added as spots on the lawn. Transformation will be visualized as the appearance of micro-

Table 10.2 (*Below and opposite.*) Transformation capacities among prokaryotes.

Organisms			Transformation by:[a]	
Genera	Species	Characteristics	Chromosomal DNA	Plasmid DNA
Azotobacter	*vinelandii*	Gram-neg., rods	+	
Thermus	*thermophilus*	Gram-neg., rods	+	+
	flavus		+	
	caldophilus		+	
	aquaticus		+	
Xanthomonas	*phaseoli*	Gram-neg., rods	+	
Haemophilus	*influenzae*	Gram-neg., rods	+	+
	parainfluenzae		+	+
	haemolyticus		+	
	aegyptus		+	
	suis		+	
Neisseria	*gonorrhoeae*	Gram-neg., cocci	+	+
	meningiditis		+	
	sicca		+	
Moraxella	*nonliquefaciens*	Gram-neg., cocci	+	+
	bovis		+	
	kingi		+	
	osloensis		+	
	urethralis		+	
Acinetobacter	spp.	Gram-neg., cocci	+	
Cyanobacteria	*Synechococcus*	Gram-neg., rods, cocci	+	+
	Synechocystis		+	+
Agrobacterium[b]	*tumefaciens*	Gram-neg., rods	+	
	radiobacter		+	
	rubi		+	

continued

Table 10.2 (*Continued.*)

Organisms			Transformation by:[a]	
Genera	Species	Characteristics	Chromosomal DNA	Plasmid DNA
Rhizobium	*lupini*	Gram-neg.,	+	
	japonicum	rods	+	
	trifolii		+	
	meliloti		+	
Pseudomonas[b]	*stutzeri*	Gram-neg., rods	+	
Streptococcus	*pneumoniae*	Gram-pos.,	+	+
	viridans	cocci	+	
	sbe		+	
	chablis		+	
	sanguis		+	+
	lactis		+	+
Micrococcus	*lysodeikticus*	Gram-pos.,	+	+
	radiodurans	cocci	+	
Bacillus	*subtilis*	Gram-pos.,	+	+
	thuregiensis	rods	+	+
	cereus		+	
	licheniformis		+	
	amyloliquefaciens		+	
	stearothermophilus		+	
Staphylococcus[b]	*aureus*	Gram-pos., cocci	+	+

a, Absence of mention indicates absence of information.
b, Limited information or scarce mention in the literature.

colonies where the DNA was spotted. To ensure that the DNA is sterile a drop on the agar should show no growth. Only markers whose lag for phenotypic expression will not result in death of the transformed cells on the selective medium can be used for this test (Chapter 7). Auxotrophic markers are the most convenient, although they are not as easy to isolate from the wild type as antibiotic-resistant mutants. Resistance marker transfer to a putative recipient may be achieved by prolonged incubation (e.g. overnight) with the donor DNA in liquid medium, i.e. allowing expression and segregation of a possibly recessive marker, followed by plating on to the selective medium.

One may also wonder about the importance of the transformation process for the evolution and diversification of bacterial species in nature. There are constraints that limit the efficiency of the process (as we shall see below) and restrict its occurrence. However, in addition to

Griffith's original observation with *Streptococcus pneumoniae*, it has been observed to take place in several natural environments, in the soil among *Bacillus* and *Thermoactinomyces* strains, for instance, and in the respiratory tract of animals in the case of *Haemophilus influenzae*.

The reasons leading to the presence of free DNA in natural environments have not been well defined. While lysed cells obviously represent an important supply, this is probably not the only, and perhaps not the best, source. Indeed, it is well documented that many, if not all, bacterial species secrete a number of nucleases, more or less continuously during their life cycle. The average survival time of a free DNA molecule, most probably in a linear form and thus readily accessible to, and unprotected from, these degrading enzymes, may be anticipated to be rather short. However, in some strains, e.g. *B. subtilis*, there appears to be a correlation in the population life cycle between the liberation of DNA and competence, i.e. the ability to take up DNA (Section 5.2), and this may well be a frequent situation. Chromosomal DNA is actually liberated by populations of *B. subtilis* during the germination process, and also in the late stationary phase of growth, as a result of lysis of a fraction of the cells. This DNA is very efficient in transformation, since it is immediately taken up by an adjacent recipient, as if an active interaction between the two cells had taken place. This phenomenon has also been observed for plasmid transformation in *Streptococcus pyogenes*, in which the plasmid is excreted, as shown by its being subject to DNase action.

5 The successive steps of transformation

Elaborate studies have been performed mainly in *S. pneumoniae*, *H. influenzae* and *B. subtilis*. However, it seems to be a general rule that the initial steps of the process are different in Gram-positive and Gram-negative bacteria. In the following section we shall take the well-known cases as models but will mention the peculiarities of each group or individual strain.

Most of our knowledge concerns transformation by chromosomal DNA, so, except when stated, the following descriptions apply to the transfer of double-stranded, linear DNA molecules. A comparative summary is presented in Table 10.3. The efficiencies vary among the strains, although it should be borne in mind that optimal conditions may not have been determined in all cases.

5.1 The competence state

In most known cases, transformable strains are receptive to DNA only during a defined period of their cycle, visualized in terms of the life

Table 10.3 Characteristics of the transformation process in the well-known bacterial systems.

| | Transformation by: | | | | |
| | Chromosomal DNA | | Plasmid | Competence | Other |
Organisms	Bound	Transferred	DNA	phase	characteristics
Streptococcus					
pneumoniae	ds	ss	+	Inducible	Soluble competence
sanguis	ds	ss	+	Inducible	factor excreted
Bacillus subtilis	ds	ss	+	Inducible	Soluble competence factor excreted
Haemophilus					
influenzae	ds	ss	−	Inducible	11-bp specific
p-influenzae	ds	ss	−	Inducible	recognition sequence
Neisseria gonorrhoeae	ds	ss	?	Constitutive in piliated phase	Specific recognition sequence

ds, double strand; ss, single strand.

cycle of the population. This peculiar physiological state is called competence. Its appearance is usually the consequence of a nutritional shift-down (decrease of the concentration of one or several nutrients or of the available energy) and usually coincides with a decrease in the rate, or a blockage, of DNA synthesis. The mechanism is not well understood.

Empirical conditions have been developed for each strain and 'competence media' can be found in the specialized literature. The kinetics of emergence of competence varies greatly from one strain to another. Nearly 100% of the cells of a population of *S. pneumoniae* can become synchronously competent, although at least one strain is known in which the percentage of competent cells is significantly lower. Competence lasts for some 20 min, after which period the competent state is lost (Fig. 10.1a). Furthermore, competence may return in a cyclic fashion. On the other hand, no more than 20% of the cells of a population of *B. subtilis* are ever competent at the same time, but this level remains constant and maximal in the population for about 90 min (Fig. 10.1b). In several strains of cyanobacteria (photosynthetic bacteria), only a short period of competence is induced when growth of the suspension becomes limited due to suboptimal light conditions (Fig. 10.1c). Inhibition of the photosynthetic metabolism, by incubation in the dark or with an inhibitor of electron transfer during the uptake period, has even been reported to considerably increase the final yield of transformants in one strain, *Synechococcus* PCC7942. In contrast, suspensions of *Thermus thermophilus*, *Neisseria gonorrhoeae* and, in special conditions, *H. influenzae* show a constant degree of competence

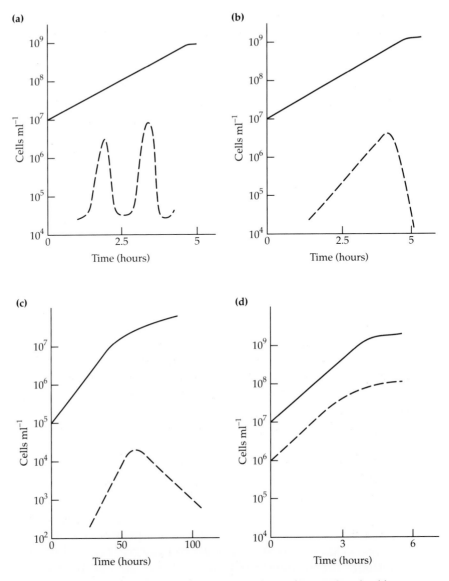

Fig. 10.1 Appearance and efficiency of competence during the growth cycle of four bacterial strains: *S. pneumoniae* (a), *B. subtilis* (b), *Synechococcus* PCC7942 (c) and *Th. thermophilus* (d). Samples of each suspension taken at varying times during the growth cycle, were treated with saturating concentrations of DNA. Total cell concentration (—) and transformants concentration (----) for a given marker were determined. The number of transformants formed for a given marker, is taken as representing competence efficiency.

throughout the growth cycle of the population, involving up to 100% of the cells in the last two strains (Fig. 10.1d). Under normal conditions, both *H. influenzae* and *Acinetobacter* become competent in response to metabolic shift-downs. A number of metabolites and inhibitors stimulate or prevent, respectively, the establishment of competence in various strains. Their mechanisms of action are usually not understood.

Important modifications of the cell envelopes are known to take place during the development of competence. In *B. subtilis* and *S. pneumoniae*, a protein complex appears at the surface of the cells, consisting of three to five proteins and including a labile competence factor, a specific endonuclease and DNA-binding polypeptides. The competence factor of *S. pneumoniae* has been purified and, when added to a non-competent culture of the same or a closely related strain, it can induce this physiological state (Fig. 10.2), probably by producing a discrete digestion of the cell wall, which exposes the DNA to the other proteins of the complex. An autolysin, synthesized and excreted at this period, may be responsible for this digestion. As a result, permeability is drastically increased, as seen by the leakage of cytoplasmic metabolites, or even proteins. The competence factor is in fact produced by non-competent cells and has the property of stimulating both its own production and the development of competence, which may be related to the achievement of higher levels of competence factor.

In *H. influenzae*, numerous small ($\approx 50\,\mathrm{nm}$) membranous structures, called transformasomes, protruding outside the competent cell and occurring at the same time as the appearance of new proteins in the cell envelopes, are considered to be a means of increasing contact between the exterior of the cell and its inside (particularly the chromosome) (Fig. 10.3). While also showing several modifications of their cell

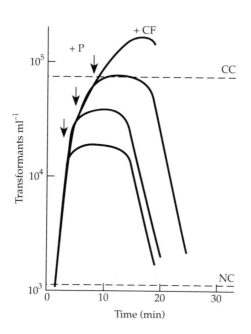

Fig. 10.2 Development of competence by competence factor CF. A suspension of non-competent *S. pneumoniae* non-treated (curve NC) or treated (curve CF) with competence factor was transformed with an appropriately marked DNA, and yields of transformants measured. Protease (+P) was added at times indicated by arrows. Development of competence in naturally-induced cells was used as control (curve CC).

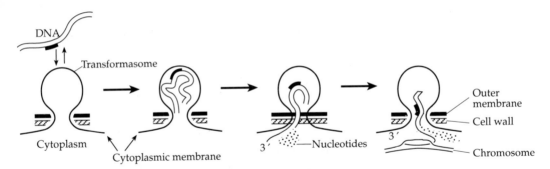

Fig. 10.3 Uptake and entry of transforming DNA in the Gram-negative bacterium *H. influenzae*. The heavy line on the DNA molecule indicates the specific recognition sequence.

surface, *H. influenzae* cells do not produce an external, labile, competence protein. Little information is available for other Gram-negative bacteria.

5.2 Binding and uptake of DNA

In general terms, there are three steps involved in the passage of transforming DNA from the outside of the cell to final integration into the host's chromosome, namely uptake of the donor DNA, transport of the adsorbed DNA to the chromosome and integration, which are common to both Gram-positive and Gram-negative organisms, although the detail of any step may differ in each case. The first step is itself divisible into a transient reversible state, in which the DNA is still accessible to DNase action or is removable by washing, from which it passes quickly to a state of stable adsorption, which is then irreversible.

5.2.1 *Gram-positive bacteria*

In these bacteria, DNA uptake is initiated by the protein complex described above. DNA molecules which come in contact with the cell surface will interact with various components of this envelope (reversible binding; Fig. 10.4). It is estimated that there are 50–80 DNA-binding sites (competence complexes) present on the surface of a single cell of *S. pneumoniae*. This association will become irreversible only when there is an interaction with a competence complex. The autolysin included in the complex exposes the DNA-binding polypeptides, which, being positively charged, decrease the general negative charge of the cell envelope, thus attracting DNA and probably orientating it in the proper position. However, even then the DNA molecules remain sensitive to degradation by any non-specific nucleases that might be present in the vicinity of the cell, a frequent situation. The mechanism leading to protection against this hydrolytic activity, although not well understood, is related to the effective transport of the DNA into the recipient cell.

When efficiently bound, the double-stranded DNA is processed by the endonuclease of the complex into segments of at most 10^7 daltons (15 000 bp, i.e. about 1/100 of the whole chromosome), which become bound to the cytoplasmic membrane. The double-stranded molecule then undergoes a second processing, through the activity of a specific exonuclease present in the cytoplasmic membrane, which degrades one of the strands. Simultaneously, the single strand is coated by proteins, the eclipse-complex proteins, and is protected against non-specific nucleases which could be present due to their normal pro-duction by the cells. Thus only single-stranded DNA finally enters the recipient and makes its way to be integrated into the chromosome. There is no strand specificity for this degradation. The operation requires energy, probably in the form of a transmembrane electro-chemical gradient and, in some cases, cations. In *S. pneumoniae*, adsorption, binding and uptake of a 10^7-dalton segment of DNA takes about 30 min.

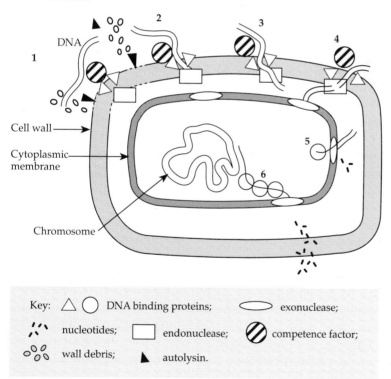

Key: △ ◯ DNA binding proteins;   ⬭ exonuclease;

⸝⸌⸍ nucleotides;   ☐ endonuclease;   ⊘ competence factor;

₀°₀ wall debris;   ▲ autolysin.

Fig. 10.4 Schematic representation of the steps leading to DNA entry during transformation of a Gram-positive bacterium. Stages 1 to 6 indicate a time sequence.

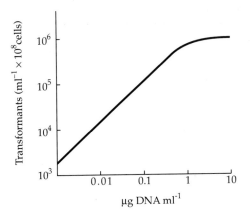

Fig. 10.5 Determination of the DNA concentration saturating a competent recipient population. Figures are indicative, since they vary with the strain considered.

The maximum amount of DNA that can be taken up by a cell is estimated by incubating a competent population with increasing amounts of DNA carrying an appropriate marker. The number of transformed cells is determined for each DNA concentration, until saturation is reached. The minimal saturating concentration that is necessary to allow maximal yield of transformation (Fig. 10.5) is usually of the order of $1\,\mu g/10^8$ bacteria, assuming that the DNA segments obtained after extraction are not shorter than about 1/100 of the total chromosomal length. More elaborate measurements, involving uptake of radioactively labelled DNA, have led to the conclusion that the transfer of one copy of a DNA segment carrying the marker studied is sufficient to transform the recipient for this marker.

This elaborate uptake process, well described only in *B. subtilis* and *S. pneumoniae*, explains the observation that, immediately after its entry into the recipient, DNA reextracted from these cells loses its transforming capacity (justifying the terms 'eclipse period' and 'eclipse complex'). Since double-stranded DNA is required for the initial binding to the cell-wall competence complexes, single-stranded DNA has a high probability of being degraded during the entry process.

The occurrence and the succession of these various steps were confirmed when mutants impaired in defined functions of the transformation process were isolated from *B. subtilis* and *S. pneumoniae*. The description of their deficiencies and complementation analyses have led to the definition of the functions impaired in the two strains. More extensive studies have led to the identification of a large number of genes involved in the expression of competence, i.e. binding, processing and uptake of DNA (Table 10.4). A complex network of regulatory systems triggers the synthesis of the corresponding proteins, the major initial signal being an unfavourable change in the growth conditions, resulting in a modification of the nutritional status of the cells.

Table 10.4 Mutations affecting DNA processing in transformation of several strains.

Organism	Mutation	Process affected
S. pneumoniae	*noz*	DNA uptake or degradation
	ent	DNA transport
	rec	Recombination
	trt	Sensitivity to trypsin
	ntr	Transformability
	comA	Competence protein synthesis
	comB	Competence protein synthesis
B. subtilis	*comA*	Early regulatory function, histidine-kinase activator[a]
	comB	Early regulatory function
	comP	Early regulatory function
	Classes I and II	DNA binding
	Class IV	DNA uptake
	Classes III and V	Early event in establishment of competence
H. influenzae	*rec*-2	DNA transport from transformasome

a, Pleiotropic activator (see Chapter 16).

5.2.2 *Gram-negative bacteria*

The Gram-negative bacteria, as exemplified by *H. influenzae* and *H. parainfluenzae*, exhibit a very different mode of entry of the transforming DNA, which is perhaps to be expected for two reasons: the presence of the external, lipopolysaccharide layer on the cell wall, and the observation that the original double-stranded DNA molecule is found inside the recipient. The small membranous vesicles projecting from the surface, or 'transformasomes', described above contain several polypeptides which are associated with competence. The transforming DNA in the form of intact duplexes, which can be in the form of circles, linear or closed hairpins, is taken into the vesicles (Fig. 10.3), where it very rapidly becomes protected from external DNases or cytoplasmic nucleases by binding with polypeptides, similarly to the system appertaining to Gram-positive organisms. In *H. parainfluenzae* the vesicles disappear from the surface and appear to move into the cells, but in *H. influenzae* the vesicles stay projecting from the surface. The DNA is then linearly translocated from the transformasome by the degradation of one strand (from the 5' extremity) and association with polypeptides. Unlike Gram-positive cell transformation, single-stranded DNA will transform but at about 50% efficiency of double-stranded DNA.

The two Gram-negative genera, *Haemophilus* and *Neisseria*, display a specificity towards the DNA that they will actually bind. The binding proteins of the complex need to be in contact with a specific sequence of the double-stranded DNA for the process to pass into the irreversible state. In *H. influenzae*, comparison of the base sequences of these regions has led to the identification of an 11-bp-long sequence, AAGTGCGGTCA, as the recognition site. Chemical modifications of the bases in this sequence decrease the efficiency of binding. When a fixed amount of a standard DNA segment of *H. influenzae*, identified by a transformable marker, is mixed with competent cells in the presence of variable amounts of non-genetically marked DNA from the same strain, the recognition sequences of all the DNA segments will compete for the binding complexes on the cells. The competition curves obtained have shown that approximately 600 of these specific sequences exist on the chromosome of *H. influenzae*. Similar experiments performed with foreign DNA, e.g. the *E. coli* chromosome, show very poor competition, indicating the presence of very few identical sequences. This phenomenon could be considered as a means of protecting the bacteria against invasion by heterologous DNA. The mechanism leading to the sequence-specific binding of DNA is not known.

One must assume that this adsorption and binding system is either less elaborate, or more efficient, than that in Gram-positive bacteria, since in *H. influenzae* 5 seconds are sufficient to protect a length of approximately 10^7 daltons of DNA against nucleases, and 1 min for its entry into the cell. The size limits of the DNA fragments involved have not been specified. Saturating DNA concentrations, whenever determined, are similar to those known in Gram-positive bacteria.

A similar mechanism of uptake is postulated in the case of *Neisseria*, which is known also to have a specific base sequence recognition site. No or very little information exists concerning the uptake processes of other Gram-negative transforming bacteria.

5.3 Transforming with plasmid DNA

Most naturally transforming strains can accept double-stranded circular DNA. In all studied cases, the binding and uptake processes seem to be like those for linear DNA transport, including a competence period. Thus competence mutants of *S. pneumoniae* and *B. subtilis* are deficient in transformation by both linear chromosomal and circular plasmid DNAs. There are indications that linear plasmid DNA may also be taken up by *S. pneumoniae*, which would require recircularizing or even a recombination between two incomplete linear molecules before stabilization in the host cell. The few cyanobacteria known to transform are capable of accepting both chromosomal and plasmid DNA according to the same physiological constraints, suggesting the existence of at least some common requirements for the two events.

However, plasmid DNA is a poor substrate in natural, as opposed to artificial (Chapter 12), transformation. Since the ratio of sizes of plasmid to chromosomal molecules is usually of the order of 1 to 100, a marker, present in one copy per molecule, will thus be 100 times more abundant per weight unit of DNA when located on the plasmid, an abundance that is not recovered in the final yield of transformants. Although the frequencies measured under DNA-saturating conditions may be of similar orders of magnitude, a calculation taking into account the relative numbers of copies of the markers transferred per µg of DNA leads to efficiencies of 10–100 times lower for plasmid DNA. In Gram-positive bacteria, this is thought to be a result of the incapacity of the plasmid DNA, which is transferred as a linear single strand, to recircularize and replicate in the recipient. Occasional reassembly of two complementary single strands transferred to the same cell is considered to be at the origin of the efficient recovery of plasmid markers. On the other hand, in Gram-negative bacteria, the plasmid would remain trapped in the transformasome and thus be prevented from replicating. Its circular structure might not be favourable for its release from the vesicle, using the mechanism adapted from linear molecules.

5.4 Integration of the donor genetic material

By using radioactively labelled DNA, it has been established, again in *B. subtilis* and *S. pneumoniae*, that the donor molecule is exchanged, as a single strand, with the equivalent region of the recipient chromosome, by homologous recombination (Chapter 9). A transient, hybrid, triple-stranded structure is formed by pairing of the two homologous regions. It should be noted that a single-stranded segment of at most 10^7 daltons can recognize the homologous double-stranded portion of a 2×10^9-dalton chromosome, in spite of the considerable complexity of folding of the chromosome. The efficiency of this pairing is close to 100% in *S. pneumoniae*. The endogenous double helix is then opened, allowing a tighter pairing with the invading DNA. Due to the action of specialized enzymes, i.e. nucleases belonging to the recombination system (Chapter 9), plus polymerase and ligase (Chapter 2), the exogenous segment, or at least part of it, is covalently bound in place of the original one (Fig. 10.6).

In laboratory experiments, where genetically marked DNA is used, differences in base sequence must exist between the two newly associated strands of DNA. *S. pneumoniae* has been shown to possess a correcting system, called mismatch repair, the *hex* system, whose action will attempt to erase these differences (Chapter 8): a specialized nuclease digests a single-stranded stretch of about 10^3 nucleotides in length, including the mismatched region, and the gap is then refilled by a polymerase–ligase system, which will copy the other strand, thus re-establishing a correct pairing (Fig. 10.6). A consequence of this pro-

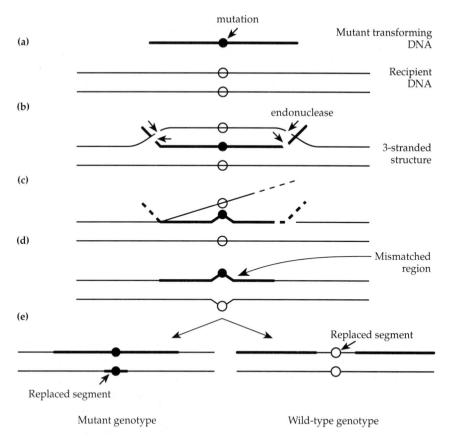

Fig. 10.6 Integration of transforming DNA in the homologous region of the recipient chromosome and the consequence of mismatch repair on maintenance of a transferred allele. (a) Pairing of homologous regions of the two DNA molecules; (b) local unwinding of duplex DNA; (c) exonuclease digestion of duplicate regions, on either molecule; (d) gap-filling by polymerase and nick-sealing by ligase; (e) repair of mismatch on either strand.

cess is the immediate appearance of identical genetic information on both strands. Depending on which strand has been excised, a mutation, say in the invading strand, will be either erased or stabilized. This process thus influences the probability of transmitting a given allele during a transformation event. A further complication is that the efficiency of mismatch repair is not constant over the whole chromosome, but shows high and low spots of activity, resulting in high or low probabilities of disappearance of the allele introduced by exogenous DNA. This explains the differences in efficiencies of recovery of transformants for different markers obtained during an experiment, which has resulted in markers being classified into four classes, namely very high, high, intermediate and low efficiency. The nuclease involved, the Hex enzyme, recognizes certain base sequences in or near the marker, and, depending even on one of the bases in the sequence, will exhibit differing affinities influencing its excising activity. There is good evidence to indicate that such a system also operates in *H. influenzae*. The mismatch repair process is reminiscent of other similar systems, particularly in *E. coli*, in which a recognition specificity is achieved depending on the methylation state of the DNA (Chapters 6 and 8).

5.5 Expression time

The time required for the expression of the newly acquired character may vary with the nature of the function, and also with the organism. While it is usually short for heterotrophic bacteria (of the order of a generation time), cyanobacteria transformed for antibiotic- or drug-resistance markers of either chromosomal or plasmid origin will be recovered only if the cells are challenged with a slowly increasing concentration of the corresponding chemical, even after a fairly long expression period in non-selective conditions. Appropriate conditions, including sufficient time for expression, should always be established to ensure maximal yields of transformation.

It should be recalled that bacteria growing under favourable conditions usually contain several (two to four) copies of their genome per cell (Chapter 2). In almost all cases, only one of these will be involved in a transformation event. A heterogeneous cell will thus be formed, and several generations may be necessary before segregation of the genetically different chromosomes is achieved. While this has no consequence when one is looking for transformation for markers of dominant phenotypes, it can lead to large losses of transformants in the case of a recessive character if the cells are challenged with the selective conditions too early.

6 Efficiencies of transformation and cotransformation

Each of the stages described above, competence efficiency, DNA entry and integration, can be limiting in the transformation process. The frequency of transformation of a given marker A can be written:

$$F_A = K \times U_A \times C \times Q \times T_A \tag{1}$$

where F_A is the transformation frequency for marker A; K is the reaction constant; U_A is the fraction of the DNA population bearing the marker A (U decreases when the DNA is broken into smaller pieces); C is the DNA concentration (under non-saturating conditions) (when C reaches the C_{max} (saturating) value, F_A becomes independent of C; however, with a large excess of DNA irregularities may appear); Q is the proportion of competent cells (it is extremely important to consider this parameter, since the competence is a transient state); and T_A is the probability of integration of marker A (this coefficient is related to the final step of transformation, i.e. recombination).

From this equation one can derive the frequency of cotransformation of one cell by two markers. If the two markers are distant enough for there to be no possibility that they are on the same fragment of DNA, the frequency will be that of two independent events of the same cell:

$$F_{A.B} = F_A \times F_B/Q \tag{2}$$

This frequency equals that calculated as the product of the frequencies of each individual transfer only when Q is close to 1, i.e. most of the recipient population is competent:

$$F_{A.B} = F_A \times F_B \tag{3}$$

This simplified form of the equation yields sufficiently precise information to be of general use in preliminary mapping.

If the markers are close enough to be frequently borne by the same DNA fragment, their cotransformation can be considered as a single event, and one obtains:

$$F_{A.B} = K \times U_{A.B} \times C \times Q \times T_{A.B} \tag{4}$$

where $T_{A.B}$ is an inverse function of the distance between A and B. Thus $F_{A.B}$ will tend towards F_A or F_B, which are usually of similar magnitude except for possible mismatch problems, when the distance between A and B decreases.

Comparisons of frequencies for individual versus simultaneous transfers of two markers considered in a given transformation will classify these markers as separated from each other by distances larger (when equation 3 applies) or smaller (if equation 4 fits) than the average length of the transforming DNA fragment. Figure 10.7 schematically illustrates this. Markers in the former case are said to be unlinked, or independent, while markers in the latter case are said to be linked (Section 7). Table 10.5 shows actual results obtained with *B. subtilis*. These data show that, in the case of the pair *dnaB-polA*, the experimental frequency of double transformants is significantly larger than that expected if these markers were transformed strictly independently. The distance between them is smaller than the average length of the transforming segment, and they would hence be described as linked. Means of estimating this distance are described below (Section 7).

Equation 2 can be used to evaluate the fraction of a population that is competent in any given physiological condition. As stated above, when this fraction, Q, is close to 1, the observed frequency of double transformants for two unlinked markers should equal the product of

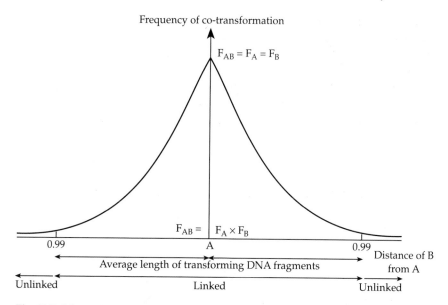

Frequency of co-transformation

$F_{AB} = F_A = F_B$

$F_{AB} = | F_A \times F_B$

0.99   A   0.99

Distance of B from A

Average length of transforming DNA fragments

Unlinked   Linked   Unlinked

Fig. 10.7 Schematic representation of the variation of the degree of linkage between two markers, A and B, with their distance on the chromosome. F_A, F_B and F_{AB}: frequency of recovery of transformants having inherited the donor allele(s) A, B, or A and B, respectively. Total length of chromosome taken as 1.

Table 10.5 Frequencies of double transformants in *B. subtilis*. Transformation of a $dnaB^-$ (synthesis of DNA), $polA^-$ (DNA polymerase A), $phoT^-$ (alkaline phosphatase) strain of *B. subtilis* by wild-type DNA was performed as described in Table 10.1. Concentrations of single and double transformants were determined (a). The calculated frequencies for double transformants (b) are obtained as the products of the experimental individual frequencies.

Transformants selected	Number of transformants $\times 10^{-3}$/ml	Experimental frequencies a	Calculated frequencies b	Ratio of a/b
DnaB$^+$	500	5×10^{-3}	—	—
PolA$^+$	600	6×10^{-3}	—	—
PhoT$^+$	400	4×10^{-3}	—	—
DnaB$^+$PolA$^+$	400	4×10^{-3}	3×10^{-5}	130
DnaB$^+$PhoT$^+$	5	5×10^{-5}	2×10^{-5}	2.5
PolA$^+$PhoT$^+$	2	2×10^{-5}	2.4×10^{-5}	1.2

the frequencies of transformants for each individual character. Any deviation from this equality, assuming all other parameters have been kept constant, is a measure of Q. Table 10.6 shows actual results obtained with *H. influenzae*. Competent cultures prepared in the standard way consisted of cells all of which were competent (State I), but an abnormal method of preparation (State II) yielded a reduced

Table 10.6 Evaluation of the fraction of competent cells in a population. Two suspensions of wild-type *H. influenzae*, in different physiological states, were transformed, under identical conditions, with DNA from a Str^r, Ery^r donor. The amounts of individual and double transformants were measured in each case, and frequencies were calculated as the ratio of transformants to total population.

Selected phenotypes	State I	State II
Str^r	8×10^{-3}	5×10^{-4}
Ery^r	5×10^{-3}	3×10^{-4}
$Str^r Ery^r$ 1	5×10^{-5}	5×10^{-6}
2	4×10^{-5}	1.5×10^{-7}

1 = experimental values; 2 = values calculated as the products of the individual frequencies.
From equation 2 (see text) applied to state II:
$F_{A,B} = F_A \times F_B/Q = 33 \times F_A \times F_B$;
Q (percentage of competent cells in the culture) = 3%.

percentage of competent cells. In experiment II, $F(Str^r, Ery^r) = 33$ $(F(Str^r) \times F(Ery^r))$. Application of equation 2 yields $Q = 3\%$, while a value of approximately 1 would be obtained in experiment I.

7 Elaboration of chromosomal maps

As for any transfer system, transformation is an important way of introducing modified genetic information into a cell, but it has also been used to perform chromosome mapping of some organisms. While the principle of the method is very similar to the general one used for any genetic system, the peculiarities of transformation should be kept in mind, since they introduce some differences in the interpretation of the data. The lengths of the DNA segments transferred usually do not exceed 1/100 of the whole chromosome and result from random breakage, and hence a single cell always has a very low probability of receiving a complete genome from the donor, even if it could adsorb 100 such segments. In standard experiments, the relative concentration of DNA per cell can be maintained at a level which saturates the possible uptake capacities of the recipient (10–100 segments per cell), when a high efficiency of transformation is required (the use of high concentrations of DNA to force multiple transfers is termed congression). But, when genetic mapping is the aim, the DNA concentration is reduced to low non-saturating levels of about 1 segment per cell in order to minimize multiple transfers which would give misleading results.

The elaboration of chromosomal maps thus operationally consists in measuring degrees of linkage, or cotransfer, between genes. The distance between two genes, visualized as markers, is proportional to the inverse of their degree of linkage, that is to the percentage of recombination between them, measured as the fraction of the selected transformants having undergone an exchange between the selected and the second marker. In all cases, however, the intrisinc characteristic of this transfer process, the cutting down of the transferred DNA to fragments that are small as compared with the size of the chromosome, would render the construction of whole chromosomal maps extremely laborious. Practically, transformation is thus an unsuitable tool for long-range mapping, but, on the other hand, it has proved very useful for fine mapping of regions of limited size (less than 1/100 of the chromosome). Depending on the strains or the conditions, several devices have been used to overcome the limits of this otherwise convenient system. Efficient methods have combined physicochemical techniques with genetics. Present-day methods, which rely on molecular analysis of DNA (Chapter 14), make less use of classical genetic approaches.

7.1 Mapping by direct estimation of transformation frequencies

One can compare the individual frequencies of transformation (measured as the ratio of transformed cells to the number of competent cells per unit volume), for each pair of markers considered, with the frequencies of their simultaneous transfer using non-saturating concentrations of DNA (Table 10.5). Cotransfer occurs if the two markers are frequently on the same DNA fragments, i.e. are linked, while unlinked markers, further apart on the chromosome, are never cotransferred except for the chance double adsorption per cell. Of course the terms linked and unlinked are relative since bacteria have only one chromosome, which furthermore is circular, and thus forms a single linkage group. This apparent contradiction with the above terminology reflects the fact that the chromosome is broken into a high number of artificially independent segments (visualized as linkage groups) during the transformation process. Within each of these segments, the degree of linkage between two markers varies with the inverse of their distance. Relative distances can then be determined within the segments and used to map the genes along the DNA molecules (Fig. 10.7). Figure 10.8 gives an example of mapping the genes *polA*, *dnaB* and *phoT* of *B. subtilis* from the results given in Table 10.5. The distance from *polA* to *dnaB* (linked markers) could be determined, but that to *phoT* could not be determined because the distance is too great for determination by this method. The results concerning the distance between *dnaB* and *phoT* are ambiguous, calculations yielding a figure close to 100, i.e. the average maximal length measurable by this procedure. Unless repeated assays or other results establish this

(a)

Pairs of markers	% of linkage	Relative distance (%)	Conclusion
dnaB, polA	73	27	markers linked
dnaB, phoT	≤1	≥99	limit of the method
polA, phoT	0.4	≈100	markers unlinked

(b)

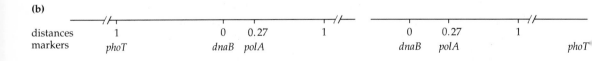

distances	1		0	0.27	1		0	0.27	1
markers	*phoT*		*dnaB*	*polA*			*dnaB*	*polA*	*phoT*

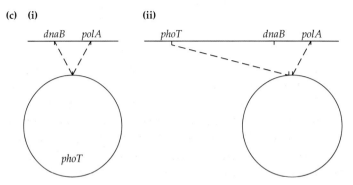

(c) (i)   (ii)

dnaB   *polA*   *phoT*   *dnaB*   *polA*

phoT

Fig. 10.8 Relative positions of the markers *dnaB*, *polA* and *phoT* on the chromosome of *B. subtilis*. Calculations are made from the data given in Table 10.5. (a) Degrees of linkage calculated as the ratio of the frequencies of double transformants to that of individual transformants. Relative distance as % of average length of a transforming segment being taken as 100 (= 100 − % linkage value). (b) Distances (in % of total chromosomal length) and possible relative positions of the markers (supposing the chromosome linear). (c) Circular map: (i) as deduced from results; (ii) as actually established.

distance as exact, these markers would also be considered, from these data, as unlinked. Thus two positions for *phoT* in relation to the other markers appear a priori possible, but it should be recalled that, since the chromosome is circular, this only indicates a unique non-determined position on the larger *polA–dnaB* portion, external to the transforming fragment, of the chromosome.

By adding the information thus obtained, and remembering that the chromosome is randomly broken, it is theoretically possible to build the whole map, although the limited size of the transferred fragments requires that a very large number of markers be available. Since this method would be extremely time-consuming, such analyses have been coupled with transductional transfer by the *B. subtilis* phage PBS1

(Chapters 4 and 12), which allows for transfer of segments that are 10% of the chromosome, and have provided evidence for the circularity of the chromosome of this strain.

This method of mapping, which compares the experimental frequency of double transformants with that calculated from those for the single transformation events, relies on accurate knowledge of the number of competent cells in the recipient culture (Section 6). This value is integrated in the calculation of the frequency of transformation. With many *S. pneumoniae* and *H. influenzae* strains, this number equals the total number of cells in the culture (Section 5.1), but this is not the case with *B. subtilis* or with some *S. pneumoniae* strains. As already stated (Section 6), this has a tremendous influence on the interpretation of the experimental data. A simplified example will demonstrate this: a standard transformation using *B. subtilis*, in which only some 10% of the cells in a competent culture are actually competent, would give frequency values for two markers, A and B, estimated at 1% as calculated on the basis of the whole population but at 10% if referred to the actually competent fraction of the population. A frequency of, say, 0.1% for double transformants AB, again calculated on the basis of the whole population, would at first sight be interpreted as showing linkage of the two markers, since $F_{AB} = 0.1\%$ is larger than $F_A \times F_B = 0.01\%$. However, if all calculations take into account the fraction of competent cells, then $F_{AB} = 1\%$, a value equal to the corresponding 1% recalculated for the $F_A \times F_B$ (10% × 10%) product, clearly indicating absence of linkage.

The first interpretation, which made no mention of the number of competent cells present, gave rise to the phenomenon known as spurious linkage. Because of this, the frequency method has limited acceptance and another method, the dilution-curve method, has been more often devised.

7.2 The dilution-curve method

The dilution-curve method is achieved by determining the variations in frequencies of transformation, calculated as ratios of transformants/total population, as a function of varying DNA concentrations below saturation. Dilution curves are made for the single and double transformants, as exemplified in Fig. 10.9 for transfer of tetracycline, erythromycin and optochin resistances in *S. pneumoniae*. The curve for the double Ery^{r}Tet^r is seen to follow the general form of those for the single transformants, whilst the double Tet^{r}Opt^r is seen to rapidly fall away as the DNA concentration is reduced. This indicates that the two latter markers are unlinked: double transformants, resulting from double hits on the recipient cell, would become rapidly less likely as

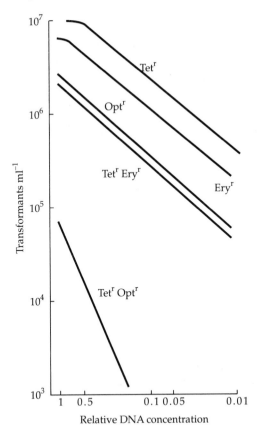

Fig. 10.9 Chromosomal mapping by the dilution method. A wild-type strain of *S. pneumoniae* was transformed with varying concentrations of DNA from an Ery^{r}Tet^{r}Opt^r strain. Single and double transformants were recorded.

the DNA concentration becomes less. In contrast Ery^r and Tet^r are linked since the curve for Ery^{r}Tet^r follows that of a one-hit system.

The dilution data have also been expressed as the ratio of double/single transformants, either AB/A or AB/B. When linkage occurs, by plotting the log of DNA concentration against the double log of AB/A (or AB/B), the curves can be straightened. The slope of the line so obtained can be taken as a measure of the distance between the linked markers.

7.3 Mapping by marker frequency

Mapping may be made by various forms of dosage of marker frequency, based on replication of synchronous cultures. The ratios of transforming activities for various markers in samples taken at varying times after initiation of replication are compared. One such system took

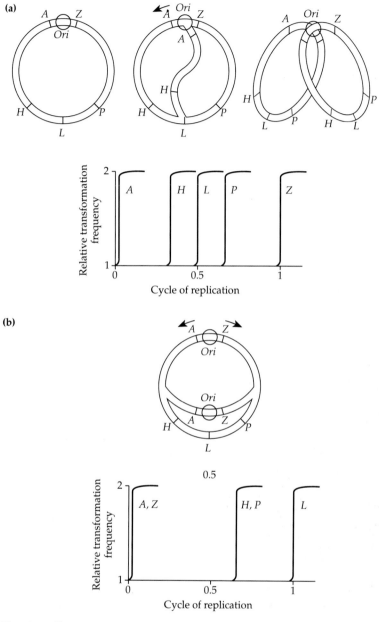

Fig. 10.10 Chromosomal mapping in *B. subtilis* synchronously germinating spores.
(a) Map deduced from hypothetical unidirectional replication from origin. (b) Actual
map taking into account bidirectional replication of the DNA from the origin, *ori*. The
expected relative frequencies of transformation for various markers, reflecting number of
copies of the corresponding genes per cell, are shown as a function of replication cycle.

advantage of the fact that DNA synthesis had been carried to completion in dormant spores of *B. subtilis* and that synchronous germination could be easily achieved. DNA isolated at increasing times from a synchronously germinating population differs from DNA isolated from dormant spores by the relative number of copies per cell, N, of each of the markers, depending on whether they have already been duplicated at the time of sampling. Taking into account the fact that initiation of replication starts at a fixed site on the chromosome (Chapter 2), the ratio N germination/N spore for each marker should theoretically change from 1 to 2 during the cycle of replication. By comparing transformation frequencies, which should reflect the marker ratio, one can determine the time of replication of any given marker during synchronous germination and hence its distance from the origin of replication (Fig. 10.10). The first map obtained (Fig. 10.10a) was, however, incorrect, due to the fact that bidirectional replication had not then been demonstrated, but Figure 10.10b allows for this.

All the above methods may give bad results in organisms in which a mismatch repair system (Section 5.4) is present. It is necessary then to allow for its interference either by adjusting for different integration efficiencies or using only markers of the same integration efficiency or by using a recipient deficient for this system.

7.4 Density shift

Another method depending on synchronous replication and used with success in both *B. subtilis* and *S. pneumoniae* is the density-shift method, wherein the newly replicating DNA is given a density label, e.g. ^{15}N or 5-bromouracil. The culture is sampled at intervals, the DNA of different densities separated in a CsCl gradient (Chapter 2) and the transforming activity for each marker assayed. A map in terms of the order of replication of the markers can be obtained, since no allowance is made for bidirectional replication.

7.5 The three-point test

As exemplified in Fig. 10.8, standard mapping methods occasionally end in ambiguous results as to the orientation of each couple of markers on the chromosome, when only two-marker cotransfer is considered. Furthermore, the level of resolution does not allow mapping of very close genes. A more precise method, used to achieve fine mapping, can be successfully used. It is known as the three-point test or three-point cross. This method can, in fact, be applied to any transfer system available as long as its main prerequisite, i.e. frequent cotransfer of the markers considered, is observed.

(a)

Markers				
citC	*polA*	*dnaB*	Number	%
+	+	+	22	33
+	+	−	14	21
+	−	+	25	39
+	−	−	5	7

(b) (i)   (ii)

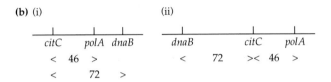

(c)

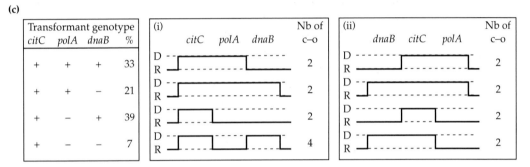

Transformant genotype			
citC	*polA*	*dnaB*	%
+	+	+	33
+	+	−	21
+	−	+	39
+	−	−	7

Fig. 10.11 Three point test mapping in *B. subtilis*. Mapping of the *dnaB* gene (involved in replication). DNA prepared from a *dnaB* mutant was used to transform a *polA* (DNA polymerase), *citC* (citric acid metabolism) recipient, under standard conditions, and single transformants selected. These were then tested, by replica plating on appropriate media and incubation conditions, so as to determine the complete genotypes of each isolated single transformant for each of the two other characters. (a) Results for *citC*+ transformants are presented. Recombination between the *citC*+ donor allele and the *dnaB*+ or *polA*− recipient alleles, calculated from (a), occurred in 72 (33 + 39) and 46 (39 + 7) % of cases, respectively. (b) The two possible maps (i) and (ii), deduced when recombination frequencies are taken as representative of relative distances. (c) Determination of the minimum number of crossovers (c–o) required to yield each type of transformant in either map option. (D, donor DNA; R, Recipient DNA). Comparison of the minimal number of crossovers with the relative % of each class of genotype (a) indicates that the less frequent class, *citC*+*polA*−*dnaB*− (7%) requires 4 crossovers in map option (i), which then appears most probable, with the relative distances as in (bi).

For any set of three genes of unknown relative position, but close enough to be transformed on the same segment most of the time, a transformation will be performed with appropriately marked strains. Cells transformed for any one marker are isolated, and their genotype for the other two characters is determined. The proportions of each of the four possible genotypes are then compared with their probabilities of appearance, with reference to the possible chromosomal maps. These probabilities are inversely related to the number of exchange events (crossovers) between the two DNA molecules necessary to obtain each genotype. Correspondence between the least frequent genotype and the largest number of necessary crossovers (i.e. four) unambiguously establishes the order of the three markers (Fig. 10.11). The actual distance between two markers is determined as seen before (Section 7.1).

7.6 Recent developments

Previous studies were carried out mostly on the two- or three-model transformable systems and required huge efforts to map a few hundredths of the whole genetic information contained on these chromosomes. Even less information was available concerning the chromosomal structure of other transformable species. The recent development of molecular biology technologies has generally confirmed the previously obtained genetic maps. These techniques, however, yield physical maps (Chapter 14), which need the availability of cloned genes, a difficult process, to transform the maps into genetic ones. Combining this methodology with transformation mapping, which only requires possession of mutants for the loci under study and not cloned genes, may be efficient for fine mapping purposes.

References

Avery O.T., McLeod C.M. & McCarty (1944) Studies on the chemical nature of the substance inducing transformation of pneumococcal types. I Induction of transformation by a deoxyribonucleic acid isolated from pneumococcus type III. *Journal of Experimental Medicine*, 79, 137–58.

Griffith F. (1928) Significance of pneumococcal types. *Journal of Hygiene*, 27, 113–59.

Hershey A.D. & Chase M. (1952) Independent functions of viral protein and nucleic acid in growth of bacteriophage. *Journal of General Physiology*, 36, 39–56.

11: Genetic Transfer: Conjugation

One of the reasons for the success of *E. coli* as a model bacterium stems from the discovery, made in the late 1940s, of its possession of a system for genetic exchange that may involve its entire chromosome. This is in contrast with the other methods of transfer (Chapters 10 and 12), which allow transfer and integration of limited portions of the bacterial chromosome. Establishment of the *E. coli* chromosomal map was thus greatly facilitated.

1 F-mediated conjugation in *E. coli*

1.1 Discovery of an oriented transfer

1.1.1 *Initial description of conjugation*

The discovery of the conjugation process in *E. coli* was one of the most fortuitous in recent microbiology. The possibility of exchanging genetic information in a way similar to eukaryotic sexual reproduction had been looked for in vain by many scientists. Lederberg and Tatum (1953) made another attempt, using a common laboratory strain, *E. coli* K12, of which they tested a few isolates. It so happened that this strain is one of the few bacteria able to undergo conjugation, and they picked up fertile isolates.

The experimental design used by these authors was both simple and elegant. They prepared polyauxotrophic derivatives of the original strains, and looked for simultaneous exchange of several characters. Even if they were expecting a rare event, its frequency could be expected to reach values higher than those of spontaneous multiple reversions. Positive results could thus be easily interpreted. Pairwise associations of their various strains indicated that some pairs were able to produce prototrophs at a frequency of 10^{-6} for all markers tested. The fertile associations allowed the definition of two types of strains: some were fertile whatever their partner, and were initially called F^+, for fertility; the others were fertile only when crossed with the F^+ ones and were called F^-. In such an $F^+ \times F^-$ cross, the latter cells were converted to the F^+ phenotype in the proportion of 70% after 1 h incubation.

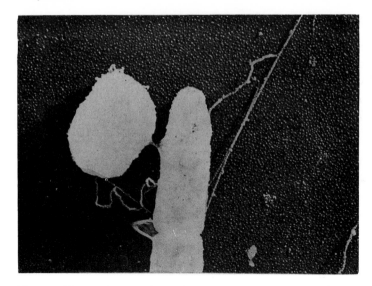

Fig. 11.1 Electron micrograph (shadowed) of conjugating *E. coli*. The elongated cell, undergoing division, is an *E. coli* K12 donor cell from strain HfrH; the plump bacterium is a recipient F⁻ from strain *E. coli* C. (Anderson, Wollman & Jacob 1959; courtesy of T.F. Anderson).

1.1.2 *Direct involvement and asymmetrical roles of the partners*

The formation of prototrophs was shown to require actual cell-to-cell contact (Fig. 11.1). No prototroph could be obtained if the strains were separated in compartments made by a microporous sintered-glass filter, which allowed passage of DNA or phages but not of whole cells, placed in a U-tube.

In order to test whether a process of cell fusion, as first proposed by Lederberg and Tatum, took place, Hayes used pairs of strains with particular sets of markers. All pairs carried the same arrangements of markers, covering several auxotrophies and resistance or sensitivity to streptomycin, but organized complementarily with respect to their F phenotype. In each of the corresponding crosses, clones obtained after selection for some of the available markers were tested for their complete genotype: all markers not used for initial selection were most frequently inherited under the allelic form present in the F⁻ partner. Thus the roles of the strains appeared to be different, there being a one-way transfer. The exchange phenomenon was reminiscent of sexual exchanges and was called conjugation. The F⁻ cells behaved as recipients or females, receiving DNA from the F⁺ donor, or male, cells. The resulting clones were called exconjugants (i.e. conjugants originating within the same species; transconjugants would refer to conjugants originating across different species).

1.1.3 *The F⁺/Hfr types of transfer*

The efficiencies of transfer obtained so far had been very low. Two donor strains endowed with new properties were then isolated independently from F^+ cultures by Cavalli-Sforza (1950) and by Hayes (1952). Crosses involving these strains and a polyauxotrophic F^- recipient yielded prototrophic exconjugants at frequencies which could reach

(a)

Recombinants	HfrCxF⁻ cross	HfrHxF⁻ cross	HfrH (λ⁺)xF⁻ cross
Thr⁺Strʳ	8.10^{-2}	$1.6.10^{-1}$	8.10^{-2}
Leu⁺Strʳ	9.10^{-2}	$1.4.10^{-1}$	5.10^{-2}
Gal⁺Strʳ	6.10^{-4}	6.10^{-2}	1.10^{-7}
Trp⁺Strʳ	1.10^{-3}	3.10^{-2}	3.10^{-7}
His⁺Strʳ	2.10^{-3}	1.10^{-2}	9.10^{-8}
Arg⁺Strʳ	$1.2.10^{-2}$	$1.5.10^{-3}$	1.10^{-8}
Met⁺Strʳ	4.10^{-2}	9.10^{-4}	5.10^{-8}

(b)

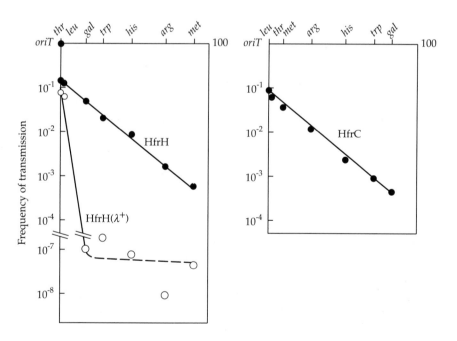

Fig. 11.2 Frequencies of appearance of different recombinants in HfrxF⁻ crosses. (a) Frequencies of appearance of recombinants in crosses involving HfrC, HfrH and HfrH(λ⁺) with the same recipient. Donor strains: HfrC and HfrH: wild type, Strˢ, λ⁻; HfrH(λ⁺): wild type, λ⁺, Strˢ; Recipient strain: F⁻: *thr⁻*, *leu⁻*, *trp⁻*, *his⁻*, *argG⁻*, *metB⁻*, λ⁻, Strʳ. (b) Chromosomal markers ordering according to gradient of frequencies of transmission.

10% of the mating pairs. These strains were designated Hfr, for high frequency of recombination, and were respectively called HfrC and HfrH.

In contrast with what had been observed in matings with F^+ males, the actual frequency of conjugants resulting from an Hfr $\times$ F^- cross varied widely with the markers used for selection. These could be arranged according to a decreasing gradient of transfer. The C and H Hfr strains differed with respect to the order of the markers when thus arranged (Fig. 11.2).

The conjugants also differed from those obtained after an F^+ $\times$ F^- cross in that the great majority retained the F^- phenotype.

1.1.4 *Discovery of F*

It was soon discovered that the fertility quality was not genetically linked to the other genetic properties of the donor cells. It could, for instance, be selectively lost by treatment with acridine orange. The term F factor was thus coined, to indicate a discrete genetic unit: the first plasmid had been discovered.

1.2 The mechanism of chromosome transfer

All the properties of *E. coli* fertility can be ascribed to those of the conjugative plasmid F. As already mentioned, this plasmid is derepressed for conjugative transfer (Chapter 3) and possesses insertion sequences (IS2, IS3 and $\gamma\delta$), of which several copies are present on the *E. coli* chromosome (Chapter 5). As a free plasmid, the F factor controls its own transfer by inducing the mobilization of its DNA (specific single-strand nick in the *oriT* region), the oriented transfer of the nicked strand and a concomitant replacement synthesis starting at the other extremity of this strand. Replication of a complementary strand also takes place along the transferred one in the recipient cell (Chapter 3).

1.2.1 *Chromosomal transfer by an F^+ population*

In an F^+ population, about 10^{-4} cells have spontaneously integrated the F plasmid into one of the chromosomal IS sites (Fig. 11.3). Upon mating, the integrated copies of F induce the same conjugative process as that of the free plasmid. The chromosome, now covalently linked to F, is thus also transferred into the F^- recipient. At the population level, assuming that the integration sites are randomly distributed along the chromosome, each chromosomal marker has the same probability of being transferred and recombined into the recipient cells. The low overall frequency of transfer of each individual character (10^{-5}), as was observed by Lederberg and Tatum of each individual character, is

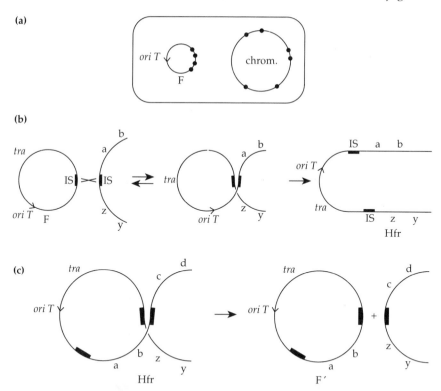

Fig. 11.3 The different states of the F plasmid and their relationship with the chromosome. (a) F$^+$ cells. F is an autonomous plasmid. The chromosome is not drawn to scale as compared to F. ● = ISs sequences. (b) Hfr cells. F integrates by rec-dependent recombination in one of the ISs present on the chromosome. The recombination process is reversible. (c) F′ cells are formed when an imprecise excision of the integrated F plasmid of an Hfr takes place. This event is not reversible.

accounted for by the low proportion of Hfr cells in the F$^+$ population. Although there is always an equilibrium Hfr ↔ F$^+$ of cells in a culture, by isolating a predominantly Hfr culture the transfer frequency can be increased (Section 1.2.2).

In spite of its low efficiency of transfer, this conjugation process has been utilized to perform chromosomal mapping.

1.2.2 *Mechanism of transfer by Hfr cells*

A particular Hfr strain in *E. coli* corresponds to a clone in which a particular insertion of F has been selected. Different Hfrs are known, corresponding to most of the IS locations recognized in the *E. coli* K12 chromosome (Fig. 11.4).

Jacob and Wollman (1959, 1961) proposed that transfer of the chromosome might be linear and oriented from a fixed origin. The

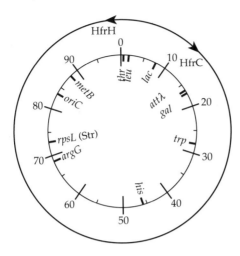

Fig. 11.4 Map of the *Escherichia coli* K12 chromosome, showing the F integration sites and orientations in the Hfr strains HfrH and HfrC, and the markers mentioned in the text.

gradients of frequencies of various exconjugants could be explained by two hypotheses: either a large piece of (or the whole) chromosome was transferred, but integration occurred according to a decreasing gradient on from the origin of transfer, or the donors transferred variable lengths of DNA starting from a fixed origin, integration then taking place at a constant frequency. To choose between these two possibilities, the authors made use of a phenomenon they had recently discovered: zygotic induction, that is, the induction of the vegetative development of a λ prophage present in an Hfr strain upon its transfer into a non-lysogenic recipient (Chapter 4). Vegetative development of the phage results in lysis of the recipient cells which have received the prophage DNA. The induction, resulting from the absence of repressor in the latter cells, does not necessitate the integration of the prophage DNA into the recipient chromosome. It was observed that certain categories of expected exconjugants were never recovered, corresponding probably to cells having received the prophage (Fig. 11.2). This result supported the hypothesis of a gradient of transfer.

The model then presented stated that DNA transfer starts at a specific point on the chromosome, which varies with the Hfr strains, and proceeds linearly into the F⁻ cell. The fragility of both the chromosome and the conjugative bridge, however, results in frequent interruptions of transfer after only a portion of chromosome has penetrated into the recipient cell. Thus most of the integrated F plasmids are able to efficiently mediate the transfer of only a short piece of the chromosome starting from their integration location.

To further test this model, the same authors devised an experiment to artificially interrupt the matings. Samples withdrawn from a mating suspension at various times were violently agitated with a blender, so as to separate the mating cells. Other devices designed to specifically

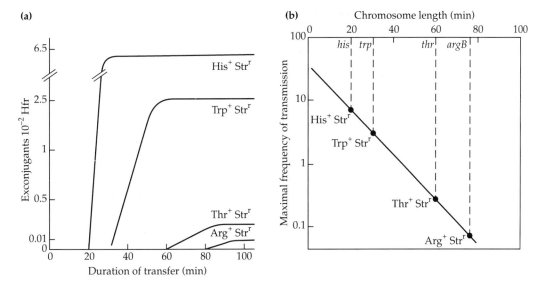

Fig. 11.5 Mating between a prototroph Str^s Hfr and a *his$^-$*, *trp$^-$*, *thr$^-$*, *argB$^-$*Str^r F$^-$ recipient. (a) Interrupted mating experiment. At the indicated times, samples were withdrawn, violently shaken to separate the mating pairs, and tested for the presence of the different types of exconjugants. (b) Gradient of transfer. Frequencies of marker transfers at time 100 of (a) were plotted, on a logarithmic scale, against distance on the chromosome.

affect the male cells were also used, such as rapid lysis of the Hfr cells by addition of a large quantity of phage (lysis from without (Chapter 4)) or interruption of transfer by nalidixic acid, an inhibitor of DNA replication. These interrupted mating experiments mimicked the natural gradient of transfer, confirming the previous hypothesis (Fig. 11.5a).

Provided a donor character has been transferred, it has a constant probability of being exchanged with the recipient allele by homologous recombination.

1.2.3 *Chromosome mapping*

The frequency of transfer decreases exponentially with the distance between the markers and the site of insertion of F into the male chromosome, i.e. the length of chromosome considered (Fig. 11.5b). This confirms the constant probability of spontaneous interruptions of transfer. A map of the *E. coli* chromosome, based on the frequencies of transfer of individual markers in each Hfr × F$^-$ cross considered, could thus be drawn.

The kinetics of transfer of individual markers obtained from interrupted mating experiments confirmed these results. It was assumed that transfer occurred at a constant rate, so that distances between markers were proportional to differences in times of entry of the markers (Fig. 11.5a). These experiments led to the determination of the minimum duration necessary for each marker of a given Hfr to enter the recipient. Transfer of the whole chromosome was estimated to require 90 min at 37 °C. The chromosome was thus said to be 90 min long, lengths being expressed in minutes. It was later decided, for simplicity's sake, to draw it on the basis of a 100-min scale.

When maps constructed for different Hfr strains were compared, it was observed that they were circular permutations of each other. This demonstrated that the *E. coli* K12 linkage map was circular (Fig. 11.4).

When a new character has to be localized, a simple procedure consists in looking for an Hfr strain which transfers the concerned marker with a rather high frequency (the marker is close to the origin of transfer of the Hfr chosen). A rough determination of its location can be gained from its relative frequency of transfer compared with that of other known markers after a standard mating. The marker is then mapped via an interrupted mating procedure. The precision of the technique is about 1 min, i.e. 1/100 of the whole map. A more precise localization can be achieved using three-point tests (Chapter 10) involving appropriate adjacent markers, usually performed after a transduction (Chapter 12).

The mechanism of transfer as described above implies that analysis of the frequencies of cotransmission of unselected markers among initially selected exconjugants yields information as to the relative positions of the characters on the chromosome. During a non-interrupted mating, proximal characters, located between the origin of transfer and the marker used as selection, have by construction been transferred.

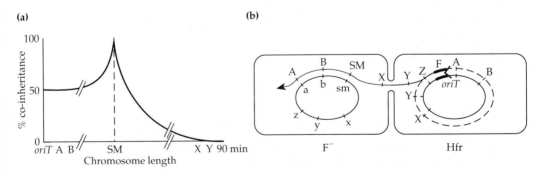

Fig. 11.6 Variation of the frequencies of co-inheritance of couples of markers (a) from the donor during a conjugational exchange (b). SM indicates the selected marker from the donor. The abscissa represents distances on the chromosome from the origin of transfer, *oriI*. The region from *oriI* to SM is proximal, the region from SM to 10 min is distal. This part of curve (a) is exponential.

Their probability of coinheritance with the selected marker is generally 50% (there is no gradient of integration), recombination being quite efficient (Fig. 11.6), although this rule does not hold for unselected markers that are very close to the selected one, for which the probability of coinheritance increases up to 100%. Reversely, the probability of coinheritance of distal characters, located downstream from the selected marker, decreases with their distance from the latter according to the decreasing exponential gradient of transfer.

1.2.4 *Formation of F' strains*

Another F-mediated event that can easily be foreseen, considering the presence of intervening insertion sequences, is the excision of F plasmids carrying a piece of chromosomal DNA (Fig. 11.3). The formation of such an F' strain from a donor F$^+$ strain results from two consecutive events: the integration of the plasmid into the chromosome (formation of an Hfr), followed by its imprecise excision due to the pairing of one of the ISs of the plasmid with an homologous IS located nearby on the chromosome. In an F$^+$ population these events are rare. No constraint exists as to the length of the chromosomal fragment carried away by the plasmid, but the displacement of long fragments (larger than one-half of the chromosome) usually leads to unstable structures.

An F' plasmid behaves like an F plasmid with regard to stability and conjugation. It is transferred with the same high frequency as the F plasmid. When transferred into a recombination-deficient (Rec$^-$) recipient strain, an F' plasmid provides a stable, partially diploid (merodiploid) state for the portion of chromosome borne by the F' plasmid. If transferred into a Rec$^+$ recipient, partial diploidy is also achieved, but recombination between the homologous regions can take place. This results in either an integration of the whole F' plasmid if only one crossover occurs (Fig. 11.7a) or the exchange of (or part of) the homologous chromosomal regions between the two DNA molecules (two crossovers) (Fig. 11.7b). Integration of the whole F' plasmid into the chromosome forms an Hfr characterized by the same origin and direction of transfer as the parental one from which the F' plasmid originated.

The original F' strain is haploid, but it keeps its complete genetic set of information even after conjugation, because of the replacement replication that occurs concomitant to the transfer.

1.3 Experimental achievement of a conjugation

As devised by Lederberg and Tatum, the first requirement for conjugation is the construction of pairs of strains with appropriate selective

(a)

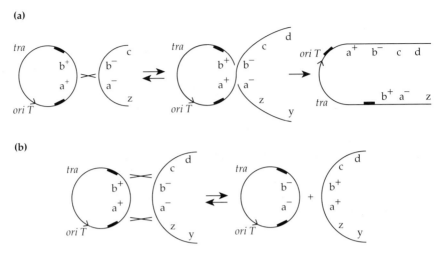

(b)

Fig. 11.7 Possible recombination events between an F′ plasmid and the chromosome in an F′ exconjugant. (a) One crossover. (b) Two crossovers.

markers. As already stressed, distribution of the allelic forms of each marker should, whenever possible, be such that direct selection of the exconjugants can be achieved.

Since a conjugation involves two whole cells, in contrast to transformation or transduction, the problem of selection of the exconjugants is complicated by the necessity to counterselect both parents from the mating mixture. An exconjugant must thus be genetically (phenotypically) defined by its inheritance of at least one character from each parent. Thus a suitable marker is allotted to the F⁻ recipient which can act as a selective marker when plating the conjugating pair, ensuring the demise of the Hfr. The anticipated association in the exconjugants of the appropriate allelic forms of each pair of parental markers will define the phenotype and thus the selective conditions.

Standard matings are performed by mixing exponentially growing suspensions of each parent. Although identical concentrations of each parent might seem to be ideal, it is often advisable to use a relatively lower concentration of the important parent, usually the donor, so as to ensure its maximal probability of actual participation in matings. The mixture is incubated, under appropriate conditions. From samples withdrawn at desired times, various recombinant classes are titrated by plating on appropriate selective media. The matings can also be performed on solid medium or on nitrocellulose filters. An advantage of this technique is to minimize cell movement and thus premature breakage of the conjugation bridge. Efficiencies are usually increased and the gradient is less steep.

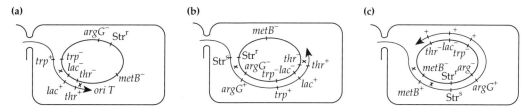

Fig. 11.8 Patterns of crossovers (X) leading to the formation of an exconjugant, depending on whether the donor selected character is transferred early (ex: *lac*+, *argG*−) or late (ex: metB+) as compared to the position of the donor counter-selection character (here Str^{s/r}). A protrophic Str^s HfrH was crossed with a poly-auxotrophic Str^r F−. (a) For Lac+Str^r exconjugants, the second crossover can occur anywhere between *lac*+ and Str^s. Most often Str^s will not be transferred, and all Lac+ clones will be Str^r. (b) For Arg+Str^r exconjugants, since *arg*G is close to Str, many recipient cells will received both *argG*+ and Str^s. Thus only those Arg+ exconjugants which will perform the second crossover between *argG*+ and Str^s (so as to inherit the Str^r allele) will be recovered. (c) For exconjugants having inherited a donor character located distally to the counter-selection character (ex: Met+Str^r), the two crossovers must occur distally to the Str^s gene. Thus at least 50% of the *metB*+ alleles transferred will not be recovered due to an initial crossover taking place proximally to Str^s, and integrating this allele in the final chromosome. x: Crossover.

Figure 11.5a gives actual values obtained in a mating involving a prototrophic HfrH donor and a polyauxotrophic, streptomycin-resistant recipient, mixed at 10^7 and 10^8 cells/ml, respectively, and incubated up to 90 min at 37 °C. Samples were withdrawn every 10 min, the matings were interrupted by drastic agitation and the samples were analysed. The recombinants are defined by two characters, one inherited from the donor (the prototroph marker) and one inherited from the recipient, here Str^r. The compulsory presence of this male counterselecting marker in the exconjugant requires a recombination event between the selected donor marker and the Str marker (Fig. 11.8). This counterselection marker was chosen because it is located distantly from the point of initiation of transfer of this Hfr, and thus the risks of disturbing the outcome of the process by the early introduction of an obligate crossover on the transferred donor fragment is avoided. Biases may be observed only for male markers transferred close to or distal from the Str region.

Frequencies of transmission are calculated as the ratio of concentrations of each class of recombinants to that of the less abundant parent cells, in this example the Hfrs. Figure 11.5a gives a graphic representation of the evolution of these frequencies as a function of time. Extrapolations of the curves to the abscissa axis yield the minimal times required for entry of the various characters. Assuming a constant rate of transfer, these values are proportional to chromosomal distances from *oriT* and allow construction of a chromosomal map (Fig. 11.5b, abscissa axis).

It is possible to consider the 90-min values for all the markers (Fig. 11.5a), which in fact represent the outcome of a non-interrupted mating. These values, when plotted as the ordinate, using a log scale, of a graph with chromosome distances measured, as just explained, as abscissa (Fig. 11.5b), yield the variation of frequencies of transmission as a function of length of chromosome from *oriT*. This represents the exponentially decreasing gradient of transfer, and its slope is a measure of the probability of spontaneous interruption of transfer per unit length.

A new marker, *X*, can be mapped by either of the two methods.

2 Conjugation in other Gram-negative bacteria

Polarized chromosomal transfer is obviously the easiest *in vivo* way to construct genomic maps. The F plasmid, however, has a narrow host-

Table 11.1 Bacterial species in which a chromosome map has been established by classical genetic methods. Physical maps, i.e. maps based on the distribution of restriction-enzyme cleavage sites, are also available for some of these strains (Chapter 14).

Species	Methods used
GRAM-NEGATIVE BACTERIA	
Escherichia coli	F-mediated conjugation and transduction
Klebsiella pneumoniae	Conjugation and transduction
Salmonella typhimurium	Conjugation and transduction
Proteus mirabilis	Conjugation mediated by several plasmids of different incompatibility groups (incP, incM, incC, incJ, incN, incV)
Proteus morgani	Conjugation mediated by RP4′ plasmids
Pseudomonas aeruginosa	Conjugation with two specific plasmids, and with incP derivatives (R6845 and R′)
Pseudomonas putida	Conjugation with plasmid incP::Tn501
Acinetobacter calcoaceticus	Conjugation mediated by RP4 and some endogenous conjugative plasmids
Rhizobium leguminarum and *R. meliloti*	Conjugation with incP plasmids
Caulobacter crescentus	Conjugation mediated by RP4
Neisseria gonorrhoeae	Transformation (still several unrelated linkage groups)
GRAM-POSITIVE BACTERIA	
Staphylococcus aureus	Protoplast fusion, computer-assisted analysis
Streptococcus pneumoniae	Density shift/transformation
Streptomyces coelicolor	Conjugation
Bacillus subtilis	Transformation and transduction

range, which has hindered or prevented its wide use. Apart from *E. coli*, it has proved efficient only in *Salmonella typhimurium* LT2 and in some *Klebsiella* strains. Two main strategies have thus been developed to achieve chromosomal transfer and mapping in other species. Either an endogenous plasmid endowed with properties similar to those of F proves to be available or an engineered one is prepared.

Conjugative plasmids able to mobilize the chromosome have been found in several instances. Mobilization is thought to happen by integration of the plasmid into the chromosome, although the existence of a *mob*-like sequence present in the chromosome has been postulated in some cases. None of the known systems shows the same high efficiency as F, and only one or a few origins of transfer are usually encountered. This is the case with the SP plasmids in *P. aeruginosa*.

The search for conjugative plasmids in each new species of interest may be a long and uncertain business. So, whenever possible, it is preferable to use a promiscuous conjugative plasmid engineered so as to make it able to insert efficiently into the new host chromosome, generating Hfr-like strains. Table 11.1 gives a list of bacterial species for which a chromosomal map has been devised by such methods. The incP promiscuous plasmids have been succesfully used in a majority of the Gram-negative species studied.

2.1 Chromosome mobilization promoted by integration of a promiscuous plasmid

Chromosome mobilization promoted by integration of a plasmid relies upon the construction of structures favouring either homologous recombination or site-specific integration. The strategies devised to use these plasmids for chromosome mapping depend on whether or not it is possible to isolate individual Hfr-like clones.

2.1.1 *Mapping in the absence of selection of individual Hfr strains*

Most methods aim at increasing the efficiency of integration of the plasmid into the chromosome with the help of the presence of a transposon (Table 11.2). Plasmids such as RK2 or RP4, belonging to the incP group, are generally used.

Another incP plasmid, R6845, a derivative of R68, possesses two tandem copies of the IS21 sequence present in all incP plasmids, the presence of which is responsible for its increased frequency of cointegrate formation with any other replicon, for instance a chromosome. The plasmid is inserted into the chromosome, with recombination occurring between the two direct copies of IS21. A strain harbouring an R6845 plasmid behaves as an F^+ male with regard to conjugation. All

Table 11.2 Utilization of some incP derivatives as fertility factors in several Gram-negative bacteria, for Cma or R' formation.

Bacterial species	Plasmids						
	RP4	R6845	PR4'[a]	RP4attλ.	PR4::Mu	RP4::mini-Mu	incP::Tn
E. coli	−	+	+	+	+	+	
Erwinia chrysanthemi					+	+	
Rhizobium sp.			+				
Acinetobacter calcoaceticus[b]	+	−					
Alcaligenes eutrophus						+	
Caulobacter crescentus	+						
Proteus mirabilis			+		+	+	
Pseudomonas aeruginosa PAO		+	−				RP4::Tn2521
Pseudomonas fluorescens						+	
Bordetella pertussis							RP4::Tn7
							RP1::Tn501
Rhodobacter sphaeroides		+					
Rhodobacter capsulatus		+					RP1::Tn501
Agrobacterium tumefaciens[c]		+					R6845::Tn5
Methylophilus methylotrophus ASI		+					

− = unsuccessful or very low frequencies. a, obtained either *in vivo* or *in vitro*. b, the presence of Tn3171 (related to Tn7) on the chromosome enhances the frequency of gene transfer. c, Tn5 was present in both the plasmid and the chromosome.

chromosomal markers have the same low probability (10^{-6}–10^{-7}) of being transmitted into a homologous recipient strain. This property is sometimes referred to as Cma (for chromosome mobilizing ability).

Phage Mu or the transposons derived from it, known as mini-Mu (Chapter 5), have been inserted in RP4, yielding plasmids showing a high efficiency of random insertion into the chromosome. This again allows transmission of donor markers with an equal low frequency. These plasmids can be used in numerous species, even outside the Enterobacteriaceae (e.g. *Alcaligenes eutropus, Pseudomonas fluorescens*).

Other transposons have proved useful in certain cases. Thus R68::Tn2521 induces the generation of a range of high-frequency donors transferring in both directions from numerous different origins.

Two strategies involving homologous recombination have also been used to promote higher frequencies of integration of the conjugative plasmid into the chromosome:

1 The same transposon is inserted in the plasmid and at random locations in the chromosome of a population of donor cells. Tn5 was thus inserted in a Kan[s] derivative of R6845 and in the chromosome of *Agrobacterium tumefaciens*.

2 Random chromosomal segments are inserted into the conjugative

(a)

Cross	Donor	Receptor	Counter-selection	Characters examined	Co-inheritance frequencies (%) A	Map distance 100 − A
(i)	$ura_{70}\,his_{19}$	$leu_6\,trp_{20}$	ura	Trp^+His^+	56	44
				Trp^+Leu^+	14, 18	86, 82
				Leu^+His^+	10	90
(ii)	$leu_6\,trp_{20}$	$ura_{70}\,his_{19}$	leu	His^+Trp^+	54	46
				His^+Ura^+	1, 2	99, 98
				Ura^+Trp^+	3	97
(iii)	$ura_{70}\,leu_7$	met ade	ura	Met^+Leu^+	18	82
				Ade^+Met^+	10, 28	90, 72
				Ade^+Leu^+	29	71

(b)

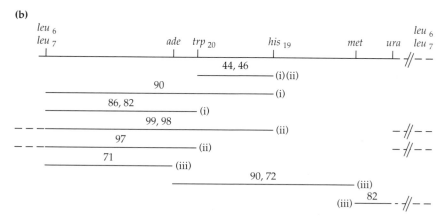

Fig. 11.9 Chromosome mapping in *Agrobacterium tumefaciens* by determination of co-inheritance frequencies. (a) For each cross, one auxotrophic donor marker was used as counter-selection by omission of the corresponding metabolite in the medium. Selection was made for one of its prototrophic characters and the complete genotype determined by replica-plating. (b) Partial map deduced from the co-inheritance values. Note that circularity of the chromosome must be postulated from these results.

plasmid. The derived molecules are then transferred by transformation into the strain to be used as donor.

In all these cases, the result is the formation of a mixed population of bacteria harbouring only the free plasmid and of bacteria possessing the free plasmid plus a copy integrated in the chromosome, either at random or in at least one of the various possible recombination sites. All markers from the donor have a quasi-equal probability of being transferred. Mapping is based on the analysis of the frequencies of cotransfer of groups of markers (coinheritance frequencies).

Figure 11.9 illustrates the application of this method for the mapping of six markers in *A. tumefaciens*. Computer analysis methods have been devised which make possible the simultaneous mapping of larger numbers of markers in a single cross. Up to seven have thus been simultaneously mapped in *Caulobacter crescentus*.

2.1.2 *Mapping after selection of Hfr-like strains*

Mapping after selection of Hfr-like strains is related to the previous strategies but requires the possibility of isolating a clone bearing a definite insertion of the plasmid. This can be achieved by using 'suicide' plasmids: integration of such non-maintainable plasmids into the chromosome of the cell is selected for by forcing recovery of a plasmid-borne antibiotic resistance gene. As in the case of the non-selected Hfr-like strains described above, recombination can be forced by inserting in the plasmid either a transposon also present as a single copy in the chromosome or a unique chromosomal fragment. These techniques may yield Hfr strains with plasmids inserted in either direction, in which case a secondary cloning is necessary.

Whatever the strategy chosen, the mapping procedure follows that described in the case of the *E. coli* Hfr × F⁻ matings. The timing of chromosome transfer might, however, be very different. Interrupted matings are often achieved via the utilization of drugs interfering with DNA replication, such as nalidixic acid or even rifampicin. The efficiencies of transmission may be markedly reduced as compared with those encountered in *E. coli*.

2.1.3 *Choosing the most appropriate method*

Not all methods apply to any given strain. Table 11.2 summarizes the most frequently used strategies in the case of several Gram-negative species. It is not understood why some transposons are able to promote chromosome mobilization in only certain species. Neither is it understood why some plasmids show different stabilities in closely related strains. Predictions aimed at avoiding or limiting unsuccessful tests when starting genetic studies on a completely new strain are thus risky.

2.2 Heterospecific matings

The *E. coli* promiscuous plasmid R6845 can promote chromosome transfer in *Methylophilus methylotrophus* AS1, a strain of industrial interest. The low frequency of recovery of recombinants and the lack of

Table 11.3 Complementation mapping of *Methylophilus methylophilus*. Two prime plasmids prepared in *M. methylotrophus* were transferred to a range of auxotrophic strains of *Pseudomonas aeruginosa*, and the frequencies of transfer of different characters by the same prime plasmid were determined.

Prime-plasmid Host strain	Marker selected		Other markers carried	
		Transfer frequency donor cell^{-1}		Transfer frequency donor cell^{-1}
pM0575 PAO462	*trpA*	$1.6.10^{-2}$	*trpB*	3.10^{-3}
			leu10	1.10^{-2}
			pur66	2.10^{-2}
			argG	2.10^{-2}
			trpF	$3.2.10^{-3}$
			argF	5.10^{-3}
pM0577 PAO40	*proC*	$3.8.10^{-2}$	*pyrB*	$1.2.10^{-3}$
			pur66	$1.5.10^{-3}$

mutants bearing suitably selective characters in this strain are important drawbacks for the utilization of this technique for chromosomal mapping. An alternative system, complementation mapping, has been established. Plasmid R6845 was used to generate a library of prime plasmids carrying random fragments of the *M. methylotrophus* AS1 genome. These plasmids were transferred into auxotrophic mutants of *Pseudomonas aeruginosa* PAO, and the genetic information carried by the chromosmal fragments from *M. methylotrophus* was identified by its ability to complement one or several of the *Pseudomonas* deficiencies. Fragmented maps were thus obtained. Comparison of these various linkage groups allowed the construction of a chromosomal map (Table 11.3).

Heterospecific matings have also been used to perform *in vivo* cloning of genes of interest from Gram-negative species into *E. coli*, used as the recipient strain. For instance, the genes coding for the pectate lyase from *Erwinia chrysanthemi* were identified by this technique. Matings were performed between a prototrophic *E. chrysanthemi* donor harbouring an RP4::mini-Mu plasmid and a polyauxotrophic *E. coli* recipient strain. Prototrophic clones from *E. coli* were recovered every time the R' plasmid carried a segment of the *E. chrysanthemi* DNA allowing complementation. The clones complementing the *E. coli* met and *ile* markers showed a pectate lyase activity, expressed in *E. coli* and detectable on polyglucuronate plates.

3 Conjugation in Gram-positive bacteria

3.1 Conjugation and map construction in *Streptomyces*

Recombination following conjugation between two different strains of *Streptomyces coelicolor* was first demonstrated in the late 1950s. This species – more precisely, strain A3(2) – has remained the main focus for genetic research in these organisms ever since. The term conjugation refers here to a process that markedly differs from that taking place in Gram-negative bacteria. Very little information is available about the actual mechanism of conjugation.

One difficulty in working with these filamentous organisms stems from the experimental devices that have to be used. Crosses are carried out on complex solid medium by mixing suspensions of spores from the two strains to be mated, conveniently marked. After germination of the spores, the hyphae of the two strains grow, and contact between the two mycelia can result in genetic exchange. Genetic analysis is performed on the uninucleate spores issued after further development. The spores are plated on appropriate selective media, so as to sort out and characterize the rare recombinants.

The whole experiment lasts several days, and does not lend itself to kinetic analysis.

3.1.1 *Conjugative plasmids of* Streptomyces coelicolor

The wild-type strain A3(2) harbours two plasmids, SCP1 and SCP2. These plasmids are occasionally lost during sporulation or regeneration of protoplasts, and isolates have been obtained which possess only one of them. Both plasmids are conjugative, but they share only one region of homology. SCP1 has turned out to be a giant linear plasmid of 350 kb (Chapters 2 and 3). Plasmid SCP2 is much smaller (31.5 kbp) and is present at an approximate copy number of one to two per cell.

In standard matings involving a plasmid-free and a plasmid-bearing strain, the plasmid is transferred to all the plasmid-free cells by the end of the conjugation process. If each new parent contains a marked variant of the same plasmid, the plasmids do not reassort randomly with respect to the parental chromosomes but tend to remain associated with their original host. This phenomenon has been called 'entry disadvantage'. Its mechanism has not been elucidated because of the technical difficulties mentioned above.

Plasmid SCP1 is responsible for most of the conjugational events observed (Table 11.4). Plasmid SCP2 shows approximately one-tenth of the conjugational efficiency of plasmid SCP1. Various derivatives of each plasmid are known. Stable integration of SCP1 into the chromo-

Table 11.4 Relative transmission frequencies (in % of the best pair, i.e. NF donor × UF recipient) observed in crosses involving *S. coelicolor* strains harbouring different fertility characteristics as a result of the presence of different conjugative plasmids.

Donor	Recipient		
	UF	IF	NF
NF	100	10	1
IF	0.01	0.01	10
UF	0.001[a]	–	–

UF (ultrafertility) = SCP1 absent from the cell; IF (intermediate fertility) = SCP1 autonomous; NF (normal fertility) = SCP1 integrated at nine o'clock. (Formally these structures are equivalent to the *E. coli* F⁻, F⁺ and Hfr structures, respectively.)
a, Residual fertility due to plasmid SCP2, abolished in SCP1⁻ SCP2⁻ strains.

some (NF strains) results in a high frequency of chromosomal transfer, while transfer from autonomous SCP1 plasmids (IF strains) yields lower fertility and absence of plasmid prevents transfer but allows the strains (UF) to behave as recipients. Autonomous prime SCP1 plasmids carrying segments of chromosomal DNA have been described. An SCP2 derivative promoting conjugation at a much higher frequency, SCP2*, is considered to be a mutant that has been derepressed for the transfer functions.

These are the only traits by which the conjugative process promoted by these plasmids resembles that of their equivalent from Gram-negative bacteria. The mode of genetic exchange they promote appears to be very different. For instance, the oriented transfer mediated by the integrated SCP1 form seems to be bidirectional, at least when the plasmid is inserted at the nine o'clock site of the chromosome.

Other transferable plasmids have been described for *S. coelicolor*, e.g. SLP1 and SLP4, which are integrated in the chromosome. However, they do not mediate genetic exchange in this strain, but they can be excised upon mating with *S. lividans*, and there SLP1 but not SLP4 can mediate mating with SLP1⁻ recipients.

3.1.2 *Conjugative plasmids of other* Streptomyces

Several small transferable plasmids have been discovered in other species of *Streptomyces*. Some, such as pIJ101 or SLP1, are much smaller than the smallest known conjugative plasmids from Gram-negative bacteria. This suggests that their mechanism of transfer is less elaborate. Their main properties are summarized in Table 11.5.

Examples of conjugal-like gene exchange are known in species of *Streptomyces* for which no sex factor has yet been described.

Table 11.5 Some transferable plasmids from *Streptomyces* species.

Plasmids	Host	Size (kbp)	Entry disadvantage	Remarks
SCP1	*S. coelicolor*	350	–	Can integrate into the chromosome
SCP2	*S. coelicolor*	31	+	No integration in the chromosome
SLP1	*S. coelicolor*	14.5		
SLP2	*S. lividans*			
SLP3	*S. lividans*		+	
SLP4	*S. lividans*			No genetic exchange
	S. coelicolor			No genetic exchange
	S. lividans			
pIJ110	*S. parvulus*	8.9	–	High copy number, high frequency
	broad host-range			of transfer
pJV1	*S. lividans*	10.8	+	High copy number
pIJ408	*S. lividans*		–	
SRP1	*S. rimosus*		+	
SRP2	*S. rimosus*			Prime plasmids described
pIJ101	*S. coelicolor*			
	S. lividans			
	S. glaucescens			

3.1.3 *Plasmids and lethal zygosis (pock formation)*

All plasmids known among the different *Streptomyces* species exhibit a property referred to as lethal zygosis. The initial description of this property mentioned that, when SCP2$^+$ cells were replica-plated on to a lawn of SCP2$^-$ suspension, narrow zones surrounding the patches of transferred SPC2$^+$ cells showed retarded growth and sporulation of the SCP2$^-$ cells. The minute region of retarded growth round the SCP2$^+$ cells was called a pock (Fig. 11.10). Zones of similar sizes were observed whether the SCP2$^+$ population was inoculated as individual spores or as patches from replica-platings, which ruled out the hypothesis of a diffusible inhibitor. The pock region contains only SCP2$^+$ cells of recipient genotype, while cells harbouring the genotype of the SCP2$^-$ parent are present everywhere else. The molecular basis of lethal zygosis is unknown. The phenomenon has proved very useful, since it has promoted the discovery of most of the known *Streptomyces* transferable plasmids.

3.1.4 *Chromosome mapping*

The first methodology used to construct the linkage map of *Streptomyces coelicolor*, as well as other species of the genus, consisted in analysis of the progeny of four-factor crosses. Both parents were double auxotrophs. The recombinants that had acquired all four possible associations of double prototrophic alleles inherited one from each

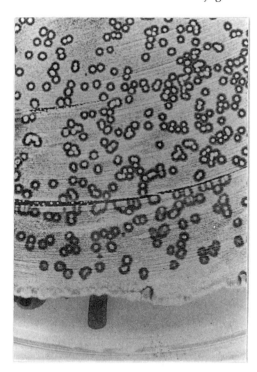

Fig. 11.10 Formation of a pock. Lethal zygosis yielding to the formation of pocks caused by plasmid SCP2 in *Streptomyces lividans* (magnification ×2.5). (Courtesy of J.L. Pernodet, Institute de Génétique et Microbiologie, Université Paris X1-Orsay)

parent, e.g. parents $A^+B^+C^-D^-$ and $A^-B^-C^+D^+$ giving rise to the four recombinant progeny classes A^+C^+, A^+D^+, B^+C^+ and B^+D^+, were selected. Their complete genotypic pattern was then characterized by first patching them on non-selective sporulation medium and then replica-plating the patches on the appropriate selective media. The compiled results are used to determine the most probable marker order, i.e. the order among the three possible ones that gives maximal yield of progeny with the minimal number of required crossovers in each class of selected recombinants. Figure 11.11 gives an example of such an analysis. Interpretation of the results for a large number of markers demonstrated the circularity of the chromosome. The precise map positions of close markers are difficult to assign, since their frequency of recombination is low, and thus very few clones are available.

A derivative of this method is the heteroclone analysis, which gives good results for the mapping of closely linked markers but is very time-consuming.

Protoplast fusion is also used instead of conjugation (Chapter 12). Analysis of the recombinants formed follows the same scheme but gives very limited mapping.

Chromosome maps have thus been established first in *S. coelicolor*, and subsequently in several other species of the genus *Streptomyces*, including some important antibiotic-producing strains. Many similarities

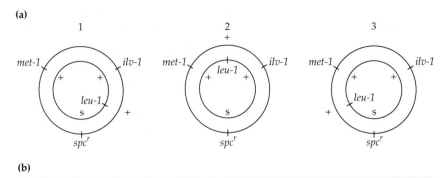

(b)

Class	Selected recombinants	Non-selected allele pairs Phenotype	Number	Number of crossovers required for chromosomal map		
				1	2	3
I	Ilv⁺ Leu⁺	Spcˢ Met⁺	138	2	2	2
		Spcˢ Met⁻	3	4	2	2
		Spcʳ Met⁺	9	2	4	2
		Spcʳ Met⁻	0	2	2	2
II	Met⁺ Leu⁺	Spcˢ Ilv⁺	140	2	2	2
		Spcˢ Ilv⁻	0	2	2	4
		Spcʳ Ilv⁺	9	2	4	2
		Spcʳ Ilv⁻	1	2	2	2
III	Ilv⁺ Spcʳ	Leu⁺ Met⁺	121	2	4	2
		Leu⁺ Met⁻	6	2	2	2
		Leu⁻ Met⁺	23	2	2	2
		Leu⁻ Met⁻	0	2	2	4
IV	Met⁺ Spcʳ	Leu⁺ Ilv⁺	81	2	4	2
		Leu⁺ Ilv⁻	45	2	2	2
		Leu⁻ Ilv⁺	14	2	2	2
		Leu⁻ Ilv⁻	10	4	2	2

Fig. 11.11 Four-marker mapping in *Streptomyces ambofaciens*. (Smokvina *et al.*, 1988). (a) Double-marked parental strains: Parent P1: *ilv1*, *met1*, *spcʳ*. Parent P2: *leu1*. (a) The three possible chromosome maps, assuming a circular chromosome. The chromosomes of parents P1 and P2 are shown on the outer and inner circles, respectively. (b) Phenotype of progeny. (c) Conclusion. In all four classes of selected recombinant progeny, the phenotypic arrangement given in the first line gives maximal yield with minimal number of crossovers in maps 1 and 3. Map 2 thus appears least probable. Further analyses have shown map 3 to be the correct one.

stand out from the comparison of these maps. The markers appear to be grouped along two diametrically positioned regions on a circular map.

3.2 Conjugation in other Gram-positive species

Although the number of conjugative plasmids described for Gram-positive bacteria is impressive and still increasing, no report of their utilization for chromosome mapping seems to be available. In lactic bacteria, the main reason for this situation may be that all the markers of industrial interest are localized on plasmids. Efforts have thus so far been directed towards plasmid mapping. The construction of conjugative plasmids akin to R' ones has, however, been reported recently in *Streptococcus lactis*.

Studies with conjugative plasmids and conjugative transposons are presently very active (Chapter 5). It is likely that much more information will emerge in the coming years.

References

Anderson T., Wollman E.L. & Jacob F. (1957) *Ann Institut Pasteur*, 93, 450.

Cavalli-Sforza L.L. (1950) La sessualita nei bacteri. Bolletino di Istituto Siëroter Milano 29, 281–9.

Hayes W. (1952) Recombination in *Bacteria coli* K12 — unidirectional transfer of genetic material. *Nature*, 169, 118–19.

Jacob F. & Wollman E.L. (1961) *Sexuality and the Genetics of Bacteria*. Academic Press, New York.

Lederburg J. & Tatum E.L.(1953) Sex in bacteria; genetic studies 1945–1952. *Science*, 118, 169–75.

Smokvina T., Francou F. & Luzzati M. (1988) Genetic analysis in *Streptomyces ambofaciens*. *Journal of Genetic Microbiology*, 134, 395–402.

Wollman E.L. & Jacob F. (1959) *La Sexualité des Bactéries*. Monographie de l'Institut Pasteur, Masson, Paris.

12: Other Transfer Systems: Natural and Artificial

1 Natural systems: transduction and capsduction

Two other natural transfer systems, namely transduction and capsduction, are known in prokaryotes, and both require the presence of a transfer agent. They both result in a similar pattern of transfer, but they differ in the nature of the vector and in their occurrence among prokaryotes. Transduction has been described in several species and is probably widely distributed, while only one case of capsduction is known.

Transduction refers to the transfer of bacterial DNA mediated by a bacteriophage. An initial infection leads to the association of bacterial DNA with phage particles resulting from a lytic cycle. The phage particles so produced, known as transducing particles, are then able to introduce the bacterial DNA they carry into other bacterial host cells on their next infection cycle. Two basic schemes of transduction exist, depending on whether it is performed by an integrative temperate phage or not. The model systems are those carried out by the *Salmonella* phage P22 and by the *E. coli* phages P1 and λ.

Capsduction is a transduction-like system except that it uses an unusual vector.

1.1 Generalized transduction performed by phages P22 and P1

1.1.1 *Formation of transducing particles*

Phage P1 shares many structural and physiological traits with the virulent phage T4 (Chapter 4). It is, however, a temperate phage which, in its prophage state, remains autonomous in the cytoplasm and reproduces as a plasmid (Chapter 3). Phage P22 is a temperate phage from *Salmonella typhimurium* LT2. Its temperate process resembles that of λ, while the morphology of its capsid and DNA and the features of its lytic reproduction are closer to those of T4. Both P22 and P1 perform transduction according to very similar mechanisms, but the genetic functions and molecular processes involved are better known for P22, which was the first case of transduction to be described. P1 transduction, on the other hand, has been used more often recently, because of its specificity for *E. coli*. During their infectious cycles, the genomes of P22 and P1 are reproduced as concatemers, which are then cut down to unit sizes and packaged into capsids by the headful mechanism common to most large phages (Chapter 4). This process is initiated by a system including a specific endonuclease which recognizes a particular site, *pac*, on the genome, which ensures proper sizing and packaging from the concatemer molecules. The size of the DNA units is larger than the actual genome, the excess DNA corresponding to a redundant region at the end.

Upon lytic infection, both phages induce the synthesis of a specific endonuclease whose action results in destruction of the bacterial genome. The activity of these enymes, contrary to the similar ones coded by phage T4, for instance, is low, so that the degree of degrada-tion of the bacterial chromosome at the end of the cycle of infection is limited. Fragments of size range varying from one to several phage genomes are frequently formed, and can be packaged in ready-to-fill

(a)

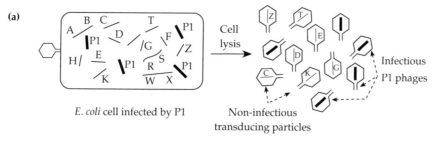

E. coli cell infected by P1   Non-infectious transducing particles

(b)

(i) Transducing particle   **(ii)** Infectious phage

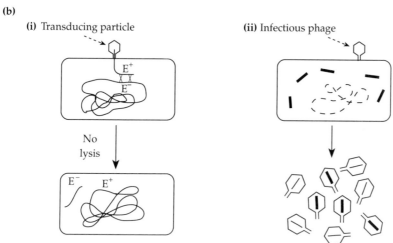

Fig. 12.1 Generalized transduction. (a) Formation of generalized transducing particles. (b) Secondary infection of a recipient cell by a transducing particle (i) or an infectious phage (ii).

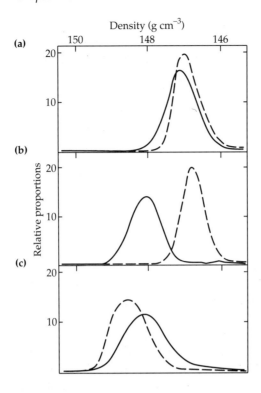

Fig. 12.2 Demonstration of the exclusive presence of host DNA in generalized transducing particles formed by Pl (adapted from Ikeda & Tomizawa, 1962). Cells were grown in either thymine (light medium, a) or 5-Br-Uracil (heavy medium, b and c). Infections with phage Pl were performed in light (a and b) or heavy (c) media. Lysates were tested for transducing (solid line) and infectious (dashed line) particles after centrifugation on a CsCl density gradient.

heads (Fig. 12.1a). Whether they are recognized as substrate and taken in charge by the packaging system or enter the capsids by other means is not clear. Similarly, it is not known whether a trimming of the bacterial DNA, similar to that controlling the size of encapsidated phage DNA, takes place. If the packaging system also functions for bacterial DNA, as seems probable, then more than one site equivalent to the P22 *pac* sequence must exist on the bacterial chromosome. This follows from the observation that the transducing particles formed contain fragments of DNA originating from different (random?) regions of the chromosome (Section 1.1.3).

The maturation process does not distinguish these pseudo-phage heads from normal ones, and assembles them with the phage tails. A standard phage lysate thus always consists of a mixture of normal phages and transducing particles, distinct from the phage only in their DNA content.

By specifically labelling either DNA with the thymine heavy analogue 5-bromouracil (Chapter 2), it has been shown that the transducing particles contain only bacterial DNA with no phage DNA at all, and reciprocally that the phage particles do not contain any bacterial DNA (Fig. 12.2). The differences in density observed between the DNA of the infectious and transducing particles in the light → light and heavy

→ heavy shifts (Fig. 12.2a,c) can be accounted for by the difference in the base ratio of the bacterial and phage DNAs. Single-burst experiments (Chapter 4) have shown that phage and transducing particles are formed simultaneously in the same cell.

The total proportion of transducing particles in a phage lysate can be evaluated as 0.3% for P1 and as 1–5% for P22. High-frequency transducing phage mutants have been isolated, but except for a general higher proportion of transducing paticles formed, no difference in the transducing process has been observed. The nature of the function mutated has not been elucidated.

1.1.2 *The transduction process*

Transduction *per se* requires that the bacterial DNA transported by the particles is transferred into recipient cells during a second cycle of infection (Fig. 12.1b). Since the donor bacterial DNA is coated by a genuine phage capsid, it can enter a recipient cell via the same process as that of a phage genome from a standard infectious particle. The recipient cells must possess the appropriate receptors, i.e. be sensitive to the phage. Since bacteria always possess a large number of receptors, if no precautions are taken the same cell can be infected by several particles. A cell receiving a transducing particle would most probably be simultaneously infected by one or several standard phages, due to the much larger proportion of the latter in the lysate. It would then be destroyed. To be efficient, transduction must be performed by applying multiplicities of infection (moi) small enough to ensure single adsorption event/cell (Chapter 4). The cells receiving only a transducing particle always survive, since these particles do not contain any phage genetic information.

Injection of the bacterial DNA fragment seems to take place with the same efficiency as that of phage DNA, suggesting that it does not involve a recognition specificity. When injected, the DNA fragment, contrary to phage genomes, remains linear and cannot be replicated. Being (largely) homologous to the equivalent region of the recipient chromosome, it can normally undergo recombination, resulting in an exchange between (usually only part of) the two regions (Fig. 12.3a). This yields stable transductants. Recombination takes place according to the classical homologous pairing and exchange process, involving two crossovers (Chapter 9). Both DNA strands can be exchanged; in *E. coli* and *Salmonella typhimurium*, the exchange is dependent on the RecA system, since transducing frequencies fall drastically in *recA* mutants.

It has been observed, however, that the frequency of recovery of genetically stable transduced bacteria amounts to approximately 2–5%

(a)

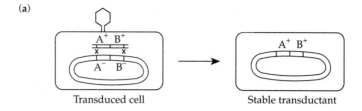

Transduced cell   Stable transductant

(b)

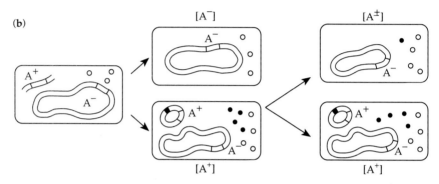

Fig. 12.3 Possible fates of transduced DNA in a recipient cell. (a) Formation of stable transductants. (b) Formation of abortive transductants, expressing both A^+ and A^- alleles of gene A, segregating A^- cells possessing a few copies of the A^+ enzyme. Proteins A^+ (●) and A^- (○). ⲧ, protein mediated DNA circularization.

of that expected from the proportion of transducing particles obtained in a phage lysate. In fact, more than 90% of the transducing fragments do not become involved in recombination, so the genetic information they bear is not stably transmitted to the recipient cells. Some of these molecules are degraded by cellular nuclease, but most of them remain intact in the cytoplasm, can express their genetic information and are persistently transmitted to one of the two daughter cells formed upon successive divisions of the initial recipient, a process known as abortive transduction (Fig. 12.3b). These DNA fragments are recovered as supercoiled structures, suggesting that they have been circularized. The DNA fragment itself cannot form a circle, since this would necessitate recombination between homologous (redundant) extremities, which do not exist in randomly cut bacterial chromosomes. Treatment of these circular structures with proteases releases linear molecules, indicating that they are formed by fixation of both DNA extremities to a protein, a structure sufficient to yield supercoils. It also provides protection of the DNA against bacterial exonucleases, and thus explains its maintenance through cell generations. In such clones issued from an initially transduced cell, only one progeny cell possesses the transferred fragment. Expression of the genes present on this fragment results in synthesis of each encoded protein. These become randomly distributed among the progeny cells, and thus diluted upon further divisions of the descendants. The whole process is thus equivalent to a linear transmission of the transferred genetic information.

These abortive transductions can be visualized every time a positive nutritional selection can be set up, and are easily distinguished from complete transductions. If a receptor cell bearing an auxotrophy is transduced with DNA originating from a prototrophic donor, prototrophic transductants can be selected on a medium devoid of the corresponding growth factor. Complete transductants will give rise to normal, homogeneous, colonies. Abortive transductants, in contrast, form small colonies. This apparently slower growth reflects residual divisions of cells which inherit copies of the proteins synthesized from the single DNA copy linearly transmitted in the colony.

It is generally accepted that the DNA fragments involved in abortive transduction usually do not undergo later recombination with the chromosomal homologous regions, even though they are stably maintained in the cell. The event that decides the fate of the transducing fragment is not known. The decision, however, seems to take place early and to be definitive.

1.1.3 *Genetic analysis using the transducing process*

Determination of frequencies of transduction for either a single marker or several markers simultaneously (cotransduction) can be performed as for transformation and conjugation (Chapters 10 and 11). The frequency of transduction is defined as the proportion of transduced cells (for one or several markers) in the host cell population infected by a transducing lysate.

Comparisons of transduction frequencies for individual markers have pointed to a difference between P1- and P22-mediated transductions. All characters are transmitted with similar frequencies during P1-mediated transfer, indicating an absence of selection during the packaging process. In contrast, however, P22 shows preferential sites for DNA pick-up on the bacterial chromosome, as depicted by the wide variations of efficiencies observed, although variations of integration efficiencies cannot be excluded.

Although the whole bacterial chromosome can take part in general transduction, each individual transduced cell receives only a limited portion of donor DNA, in a manner reminiscent of what takes place during transformation. Two markers can be transferred during a single transducing event provided their distance on the donor chromosome does not exceed the equivalent of the phage genome length. Longer fragments could not be inserted into phage heads. Frequencies of cotransduction decrease symmetrically with the distance of the markers from a reference one, from 100% to background values (unlinked markers). All intermediary values correspond to 'linked' markers. This rule applies whatever the marker used as reference. It reflects (and in fact was taken as the demonstration of) the random breakage and packaging of bacterial chromosomal fragments. The critical distance is slightly smaller than that of the phage DNA molecule, because cross-overs rarely take place at the extremities of the invading fragment. Biases, also depicted by the variations of individual transduction frequencies, are introduced by P22, which seems to need the presence of specific packaging initiation sites, and thus favours certain regions of the *Salmonella* chromosome. Fragments that are 1% of the *Salmonella* genome are carried by phage P22 (MW of its genome 26 MDa), while each P1 (genomic MW 60 MDa, equivalent to 2.3 min of *E. coli* DNA) transducing particle carries about 2% (2 min) of the *E. coli* chromosome.

All aspects of genetic analysis of transduction, including the consequences of mismatch repair and its use for fine mapping, are similar to those established for transformation (Chapter 10) and will not be described here. Contrary to the transformation situation, however, the length of the transducing fragments is fairly precisely defined, due to the necessity of forming correct phage heads, and is generally longer than a transforming molecule. Cotransduction frequencies determined by genetic analysis can thus be converted into actual distances on the bacterial chromosome (in min taken as arbitrary units in the case of *E. coli*). A formula has been established as follows:

cotransduction frequency $= (1 - d/L)^3$,

where d is the distance between the two characters tested and L is the length of the transducing fragment (i.e. length of the phage genome). Although the theory predicts total absence of cotransduction for markers too far apart, null values are not usually met with, due to the mutation frequencies of each marker studied. Contrary to the situation with transformation, chance double transductions due to a bacterium having received two DNA fragments by two independent transducing events are very unlikely, since low mois must be used (Section 1.1.2).

1.1.4 *Other generalized transducing systems*

The large *B. subtilis* phage PBS1 is the next-best-known generalized transducing system after those of P22 and P1, and its transducing process follows a similar pattern. The size of the phage (MW of its genome 150 MDa) allows it to carry fragments of about 8% of the *B.*

subtilis chromosome. It has been a very useful tool for the establishment of the host chromosomal map. Other naturally transducing phages are also known in *B. subtilis*.

Other transducing phages have been described in a number of bacteria, but seem rather rarely used for chromosome mapping, since more powerful systems (natural or artificial) are often available nowadays (Chapters 11 and 14).

The basic requirement that distinguishes a transducing phage from one incapable of transduction is the necessity to perform only partial degradation of the host DNA. It could be anticipated that nuclease-deficient phage mutants should be able to perform transduction. Indeed, a clone of phage T4 that is deficient in several of the enzymes implicated in the degradation of cytosine-containing DNA (T4 DNA contains OH-me-cytosine) has proved capable of transduction. Efficiencies of 10^{-4}/marker/phage particle, similar to those obtained with phage P1, are attained, with no bias on specific chromosomal regions. Here too the size of the phage is compatible with transferred fragments of approximately 7–8% of the *E. coli* chromosome.

1.2 Specialized transduction

The specialized transduction process relies on the induction of integrated prophages. Only limited and specific regions of bacterial chromosomes can be transferred, namely those adjacent to the insertion site of the prophage. The model system is phage λ of *E. coli*.

1.2.1 *Formation of transducing particles*

The insertion of a temperate phage genome into a bacterial chromosome, thus promoting the lysogenic state, is nearly always performed by a site-specific, non-homologous recombination involving a single crossover between the two circular DNA molecules (Chapter 4). The lysogenic structure can be reversed upon release of the blocking action of the phage repressor. The reverse crossover, in most cases, re-creates the exact initial molecules. Aberrant off-positioned crossovers (Fig. 12.4) in which pairing occurs with another nearby DNA sequence can happen at low frequencies, resulting in a free particle possessing most of the phage genome with a portion of the bacterial genome immediately adjacent to the site of integration. While a priori a large variety of such events can take place, only those yielding DNA molecules capable of recircularization and of a size appropriate for encapsidation (usually ±10% of the phage genome length) will be recovered as transducing particles. The latter constraint has two consequences for the structure of the molecule:

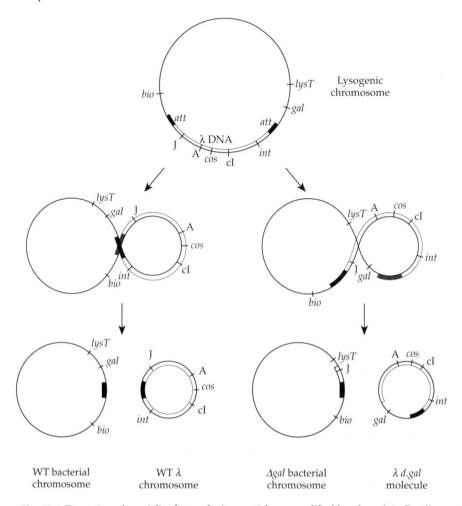

Fig. 12.4 Formation of specialized transducing particles exemplified by phage λ in *E. coli*.

1   Since the particle has gained a fragment of bacterial DNA covalently linked to the phage genome, this fragment must correspond to one of the two regions adjacent to the prophage insertion site, as the presence of DNA from both sides is not permissible because of size limitations, unless equivalent deletions exist in the phage genome. Transduction will thus be specialized (or localized) and limited to these particular chromosomal regions.

2   The phage genome has, in most cases, been partially deleted of the region located at the opposite extremity on the prophage map of the gained chromosomal fragment, and the missing phage fragment and the gained bacterial fragment must be of equivalent lengths (≈ 10%). The phage genome thus generated is defective. The nature and degree of deficiency depend on the size and location of the chromosomal

fragment carried. Complementation of the genes missing in the transducing molecules can usually occur, thanks to the frequent presence of
several genome copies per bacterium. These complementing phages
are called helpers.

A phage lysate from an individual bacterium can contain only one
type of transducing particle. The latter is designated by the letter *d* (for
defective) and the symbol of the transduced gene(s) appended to the
phage name. At the population level, the lysate is a mixture of normal
phages and a series of transducing, defective phages that are normally
encapsidated. The proportion of transducing particles ranges around
10^{-6}, except in particular phage mutants. The lysate is referred to as a
low-frequency transducing (LFT) lysate.

1.2.2 *The transduction process*

Secondary infections of sensitive bacteria with a specialized transducing lysate can lead to a variety of schemes (Fig. 12.5).

At low mois (<1 particle/bacterium), the cell receives only a defective phage, which can, however, circularize. Several mutually exclusive scenarios can then happen, reminiscent of those possible in F'
merodiploid cells.

1　Cell-mediated homologous recombination between equivalent
regions on the two DNA molecules can result in either of two structures: (i) a double crossover ends in the exchange of chromosomal
material, the phage DNA being lost (Fig. 12.5a); or (ii) a single crossover results in the formation of a lysogenic, partially diploid bacterial
chromosome (Fig. 12.5b). The presence of the two homologous regions
enhances the frequency of reverse recombinations, during which one
or the other copy of the duplicated region is carried off with the phage
genome and finally diluted away. Such transduced populations thus
continuously segregate the two types of haploid chromosomes.

2　A merodiploid structure can also be attained if a site-specific, phage-
induced integration occurs between the phage and chromosomal
attachment sites (Fig. 12.5c).

3　If the moi is high enough to produce multiple infections, the
bacterium, in most cases, recieves a normal phage together with the
transducing particle. Since the fate of most infections by temperate
phages is a lytic cycle, the phage helps the defective transducing
particle to reproduce. The mixed lysate formed differs from the initial
one in that the proportion of progeny transducing particles can reach
50–70% of the bursts. These lysates are referred to as high-frequency
transducing (HFT) lysates (Fig. 12.5d).

The helper phage can, on rare occasions, lysogenize the infected
cell. The stabilized phage genome can then promote insertion of the

(a)

(b)

(c)

(d)

(e)

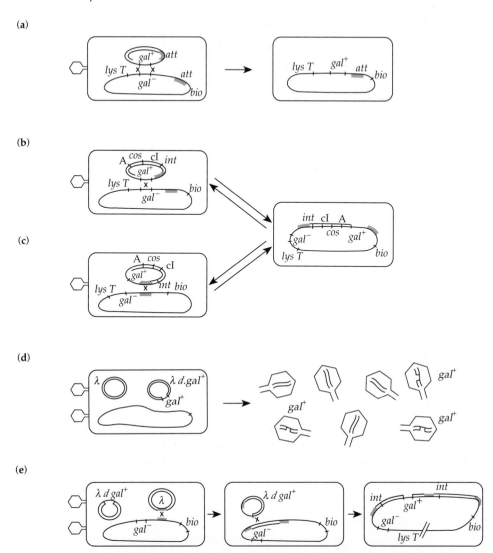

Fig. 12.5 The various posible recombination events during specialized λ transduction. (a) Integration of the host DNA fragment *via* a homologous double crossover. (b) Formation of a lysogenic, merodiploid recombinant by a homologous single crossover. (c) Site specific integration of the transducing particle at the *att* site. (d) Formation of a HFT lysate. (e) Formation of double lysogenic, merodiploid chromosomes. (See text for details).

defective transducing molecule by either homologous recombination or site-directed integration. Double-lysogenic (one prophage copy being defective) merodiploids are formed (Fig. 12.5e). Such structures revert frequently. If induced into the development of lytic cycles, they also produce HFT lysates.

1.2.3 *Specialized transduction by phage* λ

The *E. coli* phage λ integrates at a single site, *att*, on its host chromo-some (Chapter 4), situated between the *gal* and *bio* regions, and when induced can thus integrate either of these regions as defective phages. Such transducing particles are designated λ*dgal* and λ*dbio*. Other functionally unrelated genes localized in these regions can obviously also be carried.

Since λ integrates into the bacterial chromosome under only one orientation, the nature of the defects on the transducing particles can be deduced from the identity of the transduced genetic material (Fig. 12.6). λ*dgal* particles have lost genes coding for capsid structural pro-teins, while λ*dbio* ones are defective in recombination functions.

λ transduction in its normal pattern is of limited interest for present-day bacterial geneticists. It has, however, played an important role in the developments of several aspects of bacterial genetics. Fine mapping and functional analysis of the transduced region has con-tributed to the understanding of the *gal* region as organized in an operon (Chapter 15). The transducing particles then obtained con-stituted the first case of isolation of a specific fragment of bacterial DNA. Moreover, the system proved very important for the construc-tion of the λ prophage map and the understanding of λ physiology.

λ is an example of a family of related lambdoid phages. Hetero-logous recombinants have been constructed between them in which the attachment sites have been exchanged, and such modified phages have been used to transduce, in a λ genetic background, other chromo-somal regions (*trp* and *tonB* for phage 80, *gal–bio* for phage 434).

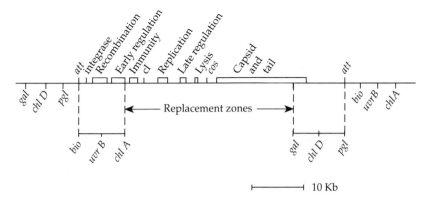

Fig. 12.6 λ transducing genomes. Maximal lengths of λ DNA possibly substituted by *E. coli* chromosomal DNA are indicated on the prophage map of λ. The genetic map of the chromosome regions transduced are also shown.

A temperate phage of *B. subtilis*, SPβ, carries out specialized transduction according to a mechanism very close to that of λ. Phage P22 can also act as a specialized transducer for a *pro* marker of the *Salmonella typhimurium* chromosome, being contiguous to the most frequent integration site of the prophage.

1.3 Capsduction

Capsduction is a transfer system that has, to date, been described only in the *Rhodobacter capsulatus* species, in which it is present in the majority of the strains tested. Its interest resides both in its peculiarities and as an example of the variety of systems one should keep in mind when studying new species.

1.3.1 *The transfer system*

The structure that carries the host DNA, called the gene transfer agent (GTA), is very similar to a transducing particle (Fig. 12.7), except that its size is smaller. Although heterogeneous within an individual cell production, the particles appear as icosahedric tailed phages. Their average size is 30 nm for the head diameter and 5–6 nm × 40 ± 10 nm for the tail, i.e. in the size range of small virulent phages.

The capsids are made of at least five major and three minor proteins. Immunoassays run with antibodies raised against the capsid components of the GTA produced by one strain indicated that GTAs produced by a dozen other strains are all recognized with similar high

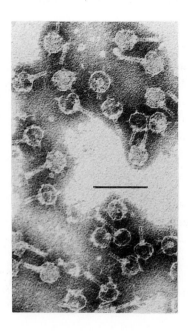

Fig. 12.7 Structure of GTA particles produced by *Rhodobacter capsulatus*. Bar = 100 nm (approx.) (Courtesy of B.L. Marrs, 1978).

efficiencies. The degree of conservation is comparable to that found among the capsid proteins of the T-even phages of *E. coli*. The same antibody preparation, however, failed to inactivate any other 'true' phage using the same host strains. All the GTAs thus belong to a single class, different from at least the known phages of this species.

The capsids contain linear double-stranded DNA of MW 3.6 MD (equivalent to three to four genes). Three criteria in particular have led to the conclusion that only host DNA, randomly chosen from the host chromosome, is encapsidated:

1   The sedimentation velocity in sucrose gradients and renaturation kinetics (Cot half values) are similar to those of host DNA.

2   The base sequence, at the population level, is variable, as shown by the absence of discrete, conserved fragments on restriction-enzyme digestion profiles. (A unique DNA molecule, because of its constant pattern of sites recognized by a given restriction enzyme, shows a well-defined, reproducible, profile of digestion by this enzyme, and can be visualized as discrete bands on an agarose gel electrophoresis; Chapters 3 and 16.)

3   No phage-like functions, such as cell-killing, plaque-forming or amplification of particular DNA, could be found. When transferred to a non-producing recipient strain, the particles do not confer to this strain the ability to reproduce similar particles. The particle has thus been named gene transfer agent (GTA), since this function seems to be the only one carried, and the process is named capsduction to recall the capsid morphology of GTA.

GTA particles are synthesized and released as one or two synchronized waves during the growth of GTA-producing cells (Fig. 12.8). The genetic information necessary for the synthesis of the capsid must be located on the host chromosome, since no conserved plasmid-like structure has been found among the various GTA-producing strains. Nothing is known of the mode of production and of the mechanism of extrusion of the particles. Standard yields could be measured only as gene transfer frequencies, since actual titres of GTA were too low for direct detection. Production reaches 10^5 GTA units/ml for a given marker, when the suspension enters the stationary phase of growth. Overproducing strains, showing yields $10–10^3$ times higher, do not differ in either GTA or host properties. The high level of GTA production, however, is accompanied by lysis of a fraction of the cell population, in a manner similar to that occurring after infection of *E. coli* cells by filamentous or RNA-containing phages.

1.3.2 *GTA-mediated genetic analysis*

Transfer of the DNA fragment is achieved by adsorption of the GTA to the cell surface and injection of the DNA into the recipient cell.

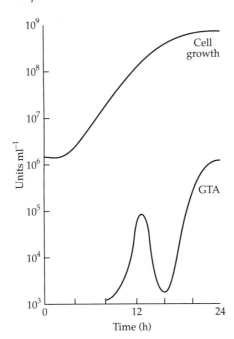

Fig. 12.8 Time-course of release of GTA activity by a growing culture of *Rhodobacter capsulatus*.

The incoming DNA is stabilized in the host genome by homologous recombination. Frequencies of transfer reach 10^{-5}–10^{-4} recombinant/marker/treated cell, a value slightly smaller than that usually obtained after generalized transduction. The GTA superproducers are one or two orders of magnitude more efficient.

All combinations of the two phenotypic traits, producer of and/or recipient for GTA, are encountered.

In spite of the small size of the DNA carried, the system has been useful for gene mapping in *R. capsulatus*. As studied for several markers, GTAs appear to mediate transfer of all regions of the chromosome equally. Cotransfer frequencies have been established as varying according to the formula:

$$F = 1 - (dL)^2$$

where d and L are defined as for transduction analysis (Section 1.1.3).

1.3.3 *Possible nature and origin of GTA*

The transduction-like properties of GTA strongly suggest that the particles originated from remnant genetic information of a phage. Except for capsid formation, all other, or at least all cell-deleterious, information seems to have been lost in all GTA-producing or possible

GTA-receptor strains. The small head volume, and thus the limited size of the encapsidated DNA also suggest a defective nature of the putative original bacteriophage. It is difficult to imagine that a DNA molecule of 1.5–2 kbp, even if particularly densely organized, could code for at least the five to eight proteins forming the capsid. Could some of these be bacterial host proteins routed away from their normal function and localizations?

The widespread distribution of GTA production capacity among *R. capsulatus* strains and the absence of detection of a related functional virus suggest that the putative original virus would have become defective only once and early in the evolution of the *R. capsulatus* species. The remaining genetic information would have been maintained and transmitted because it was stabilized on the bacterial chromosome. The high degree of conservation of GTA proteins, a surprising observation in regard to the absence of selfish function of the structure itself, could be explained by the genetic advantage it provides for the bacteria as a means of promoting genetic exchange. Alternatively, it can be hypothesized that the capsduction information would originally be bacterial, as (part of) a genetic exchange system.

2 Artificial DNA transfer systems

The development of microbiology and of the industrial use of prokaryotes has prompted the search for DNA transfer systems in bacteria other than the model ones. Except for representatives of species already well studied, which were so partly thanks to their possession of usable transfer systems, it soon became evident that transfer capacities among other species are either scarce or at least could not be demonstrated in most cases tested. Artificial means of transferring DNA have thus been set up. The development of molecular genetics technology has rendered these possibilities even more critical, and has simultaneously modified their required characteristics. It has become more useful to transfer plasmids, i.e. circular rather than linear DNA molecules, and the capacity to transfer large portions of chromosomes is no longer an important advantage.

Three principal processes have been devised in the last 20 years. They share one basic trait which distinguishes them from all natural systems: entry of the DNA is always a totally passive process, in which no cellular synthesis participates.

2.1 Protoplast fusion

Protoplasts can be induced to fuse and regenerate to normal cells, and undergo exchanges of genetic information between the incoming chromosomes. Initiated in the late 1950s simultaneously in prokaryotes and eukaryotes, the protoplast fusion methodology has since been largely developed for eukaryotes, in particular in fungal genetics. Attempts at developing this process in bacteria were restricted to a few strains belonging to different species, and proved successful only in some Gram-positive organisms. Studies with Gram-positive bacteria concerned mostly *Bacillus* and *Streptomyces* species. Fusions in *Bacillus* resulted in complex genetic reequilibrating processes, comprising either classical recombination and segregation of chromosomes, or maintenance of a diploid state accompanied by physiological extinction of one of the two chromosomes. In spite of the interest in regard to the understanding of the mechanisms underlying these particular characteristics, the system did not, however, answer the hope of establishing general, easy-to-handle means of performing genetic exchanges among species not amenable to other known systems; their study was not widely pursued.

Protoplast fusion was monitored with more success for *Streptomyces*, a prokaryote species showing morphological similarities with filamentous fungi. Although conjugational transfer exists in these cells (Chapter 11), protoplast fusion allows genetic exchanges to occur between sexually incompatible strains. The industrial importance of the streptomycetes encouraged the use of protoplast fusion in these organisms as a more convenient method than conjugation for strain improvement or, more generally, for promoting multimarker exchanges.

Protoplasts are obtained by various treatments, e.g. by growth of the mycelia in the presence of glycine, a procedure which leads to weakening of the cell wall, followed by lysozyme treatment. Protoplasts from two strains differing by one or several genetic markers are mixed, usually in the presence of an aggregating agent such as polyethyleneglycol (PEG), and then plated on appropriate media, allowing regeneration of the cell wall (in the presence of an osmoprotector). Selection can be made directly for all alleles looked for, or by a two-step process. Regeneration frequencies reach at best 50%. The proportion of fused recombinant cells among these amounts to several per cent in all *Streptomyces* strains studied. (Conjugation yields approximately 10^3-fold fewer recombinants except in NF $\times$ UF crosses (Chapter 11).) Two-step selections are thus easy enough to attain.

Protoplast fusion, as compared with standard conjugation, produces higher frequencies of recombination, depicted by a high proportion of multiple crossovers and an apparent lowering of marker linkage. A reason for this may come from the possible involvement of several genomes from each parent in successive rounds of recombination, leading to the formation of the final homogenote. Indeed, intermediate heterogenotes can remain stable for several generations.

Table 12.1 Mapping via a four-marker cross in *Streptomyces chrysomallus*. After fusion of the protoplasts obtained from two strains, (e.g. Arg^-Ser^- and Leu^-Phe^-, respectively), recombinants were selected as having inherited one prototrophic allele from each parent, (i.e. Arg^+Leu^+, Arg^+Phe^+, Ser^+Leu^+ and Ser^+Phe^+), and analysed for unselected alleles. The relative proportions of each category of recombinants were calculated (column I). This allowed the determination of the relative recombination frequencies in the observed intervals between markers among the whole population of recombinants examined, i.e. the number of clones showing a crossover in this interval (column III). For instance, the value for the Arg–Leu interval is $86.5 + 22 + 4.5 + 4.5 = 117.5$ (all Arg^+Leu^+ and Arg^-Leu^- recombinants). Comparison of these values led to two possible marker arrangements (Fig. 12.9). To discriminate between these two maps, the number of crossovers necessary for the formation of each class of recombinants in each possible map configuration was calculated (column II). The most probable map is that for which the least number of crossovers are necessary, i.e. for which the recombinant class requiring most crossovers is least frequent, here map a. The total number of these crossovers is representative of the corresponding distances. (After Keller *et al.*, 1985.)

Phenotypes				I Relative proportions of recombinants	II Number of crossovers for maps		III Total number of crossovers observed for each map interval	
Arg	Leu	Ser	Phe		a	b		
+	+	+	+	86.5	2	2	Arg–Leu	117.5
−	+	+	+	10.5	2	2	Arg–Ser	38.5
+	−	+	+	74.5	2	2	Arg–Phe	184.5
+	+	−	+	22.0	2	2	Ser–Leu	103
+	+	+	−	4.5	2	2	Ser–Phe	176
−	−	+	+	} 4.5	2	4	Leu–Phe	85
+	+	−	−					
−	+	+	−	} 1.5	4	2		
+	−	−	+					

Map positions are determined by a method equivalent to classical methods, consisting in a comparison of the relative number of crossovers observed within the total population of recombinants analysed for each couple of markers (or in each interval between two markers) (Table 12.1 and Fig. 12.9). However, the fact is that the particular traits of the recombination processes associated in protoplast fusion minimize its interest for chromosome mapping, although they render it efficient for the preparation of multiply marked strains.

Protoplasts can also be used as recipients for transformation by plasmid DNA. The aggregating agent PEG is generally added to the protoplast–DNA mixture so as to ensure better contact. Although the efficiency is usually not improved, as compared with natural transformation when available, some plasmids may be more efficiently transferred by this method. In contrast with classical transformation, linear molecules of DNA do not constitute a good substrate.

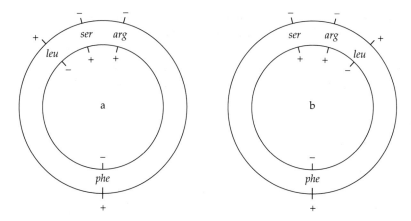

Fig. 12.9 The alternative positions of markers in *Streptomyces chrysomallus* predictable from analysis of relative recombination frequencies, as obtained from Table 12.1. Adapted from Keller *et al.* (1985).

2.2 Transformation

Curiously, the transformation system was originally devised for *E. coli* strains that are otherwise amenable to conjugation or transduction but not naturally transformable, since molecular biologists were starting to recognize the possible importance of strains of this species as intermediary or replacement hosts (Chapter 16). In doing so, they needed transfer systems that were more flexible than conjugation, implying vector molecules smaller than the conjugative plasmids.

The basic procedure consists in making the cell envelope non-specifically permeable. This is achieved by applying a combination of stresses which result in the creation of random pores in the envelopes. The most classical procedure associates temperature and ionic shocks. Transfers from 37 to 0°C and from low to high Ca^{2+} and/or Mg^{2+} concentrations and then back to normal temporarily allow entry of linear or circular DNA molecules. The treatment is operationally equivalent to the initial steps of spheroplast production (spheroplasts, the equivalent of the protoplasts of Gram-positive bacteria, correspond to cells in which the wall has been degraded with maintenance of the outer membrane). The yield of transformants results from a compromise between maintenance of a good survival of the cells and efficient permeability which tends to lead to cell lysis. Many variations of the initial protocol have been proposed, and are described at length in all technical manuals of molecular biology. Efficiencies in *E. coli* range between 10^2 and 10^6 transformants/µg DNA at saturating concentrations. Better yields are usually obtained with circular plasmids, probably because, as they remain circular during the transfer process, they are better protected against nucleases. Transformation with chromosomal DNA is not usual by this system.

This procedure has proved operative in many Gram-negative bacterial strains and is now largely used, particularly in genetical engineering processes. It does not, however, have universal applicability and it is difficult to anticipate its potency in a new strain.

2.3 Electroporation

The search for methods for making cells non-specifically permeable led to the use of electric shocks, which had previously been shown to increase membrane permeability. The rationale of the technique is the following: since all membranes are charged (negatively on the outside, positively on the inner side), application of an electric field, by displacing these charges, leads to an increase of the transmembrane potential. The amplitude of the potential modulation can be monitored through the characteristics (intensity, voltage, duration) of the electric field applied. It is also dependent on the diameter of the cells. As a consequence, the membrane can be reversibly reorganized, with regions becoming thinner and pores being formed. This does not explain, however, why the technique is efficient with Gram-negative bacteria, in which the non-charged outer membrane remains a barrier against penetration of macromolecules.

The diameters of the pores are large enough to allow transfer of macromolecules such as DNA. The entry is totally passive and non-specific, as in the case of artificial transformation. The process is, in fact, referred to as either electroporation or electrotransformation and the resulting cells are called transformants.

The damage created in the cytoplasmic membrane results in possible cell-killing. Best transformation efficiencies, i.e. the more favourable compromise between cell death and DNA entry, result from adjustments, specific for each organism, of treatment conditions (electric field strength, time of application). Values allowed by standard commercial devices are 5–12 kV/cm and 2–20 msec. Other factors of importance are the composition of the medium, the physiological state and concentration of the cells and the structure of the cell envelope (composition, thickness). Previous enzymic treatment of the cell wall may increase the yield of transformants. The conformation of the DNA should also be considered, circular molecules often producing better yields, probably because of their lower accessibility to endogenous nucleases. There is evidence that, at least for certain bacterial species, electroporation may discriminate between plasmid and chromosomal DNA. This may be related to the circular state of the plasmid molecule. Efficiencies usually range from 10^5 to 10^6 transformants/µg DNA, figures equivalent to those obtained by artificial transformation. Extreme values of $10^{10}/\mu g$ DNA have been reported for *E. coli*.

Electroporation appears to have widespread applicability. Of all available artificial systems, it presents most advantages. It has been successfully used for a large number of strains belonging to very different species or genera of eubacteria, both Gram-positive and Gram-negative. Its use for eukaryotic cells, both plant and animal, in which it was originally developed, has also become general.

References

Ikeda, H. & Tomizawa, J. (1965) Transducing fragments in generalized transduction by phage P1. I – Molecular origin of the fragments. *Journal of Molecular Biology*, 14, 85–109.

Keller U., Krengel U. & Haese A. (1985) Genetic analysis in *Streptomyces chysomallus*. *Journal of General Microbiology*, 131, 1181–91.

Marrs B.L. (1978) Genetics and Bacteriophage. In: Clayton R.K. & Sistrom W.R. (eds) *The Photosynthetic Bacteria*. Plenum Publishing Corporation, New York.

13: Genetics with Transposons

All the exchange mechanisms described in the preceding chapters involve recombination between homologous DNA molecules or regions, performed by Rec systems (Chapter 9). Other types of recombination are at work for transposition events (Chapter 5) as well as for integration of prophages (Chapter 4). Because these recombinations do not require DNA homology, they have been widely used to enlarge the scope of *in vivo* genetics. As early as 1977, Kleckner *et al.* had foreseen most of the properties of transposons that could be useful for geneticists. Since then, not only have all her predictions been realized, but the construction of modified transposons has now allowed even more elaborate uses. Any description of these applications at this present time would be incomplete, since new fields are still being opened as a result of the imagination of geneticists. However, some applications will be illustrated, using as examples the utilization of transposons for genetic studies of non-*E. coli* species. Although less developed, such approaches are often the most effective, if not the only, ones available with these organisms.

1 Transposon-induced mutagenesis

Using transposons to generate mutations has several advantages over classical mutagenic treatments:

1   Transposon-induced mutation frequencies are often higher than those obtained after most mutagenic treatments. Assuming a random insertion of the transposon used, the probability of mutation for a given gene (i.e. the ratio of the number of mutants in this gene to the total number of selected transposition events) is equal to the ratio of the size of this gene to the size of the genome, a figure which most often averages 10^{-3}.

2   Most transposon-induced mutations arise from complete blockage of translation (and often transcription) of the gene in which the transposon has inserted, conferring a non-leaky phenotype. Strong polar mutations (Chapter 7) are obtained if the gene is included in an operon. This, of course, must be modulated by the fact that some

transposable elements may enhance transcription activity of down-stream or upstream genes through mobile promoters.

3 In most cases there is a single insertion in each transposed cell, which avoids the risk of isolating double mutants.

4 The presence of the transposon closely associated to the mutated gene provides direct ways of selection, either genetically (using the transposon marker) or biochemically (using the transposon DNA as a probe).

Among transposable elements, the best candidates for transposon mutagenesis are those with random and stable insertions. No transposon can display these properties in all possible species. Thus Tn5, the most widely used, inserts in a non-random fashion in *Acinetobacter calcoaceticus* and shows multiple unstable insertions in some *Klebsiella* strains.

1.1 Suicide vectors

In order to achieve transposon mutagenesis, one must deliver the transposon into the cells in such a way that only those cells that have stably acquired it are selected. In other words, the vector used to introduce the element must disappear from the cell once transposition has taken place. Such vectors are referred to as suicide vectors (Fig. 13.1).

The most elaborate suicide vectors have been devised for *E. coli*. Thus derivatives of phage λ, bearing two amber mutations preventing lysogenization and lytic replication, respectively, have been constructed by addition of either Tn5 or Tn10. These phages can replicate in *E. coli* strains possessing strong suppressors, but act as suicide vectors in non-suppressive strains.

Several strategies have been devised for other species.

1.1.1 *Mobilization of non-replicating plasmids*

The first suicide vectors belonging to the category used in the mobilization of non-replicating plasmids were based on promiscuous conjugative plasmids modified so that they could not be maintained in the non-*E. coli* hosts to be studied, i.e. by replacing the origin of replication by a more strictly specific *E. coli* one. The resulting plasmid can conjugate into, but not replicate in, the host under study, thus behaving as a suicide vector for any inserted transposon. RP4 transfer functions have frequently been used for this purpose. For example, pRK2013 was constructed by insertion into plasmid colE1 of a fragment of RK2 (incP1) containing its transfer functions and only part of its replication apparatus. The plasmid was used in plant-infecting pseudomonads

(a)

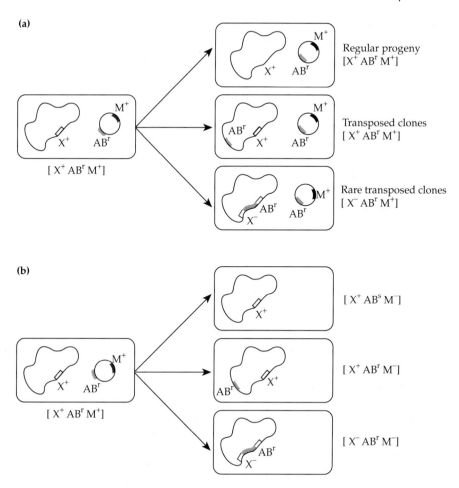

Regular progeny
[X⁺ ABʳ M⁺]

Transposed clones
[X⁺ ABʳ M⁺]

Rare transposed clones
[X⁻ ABʳ M⁺]

[X⁺ ABʳ M⁺]

(b)

[X⁺ ABˢ M⁻]

[X⁺ ABʳ M⁻]

[X⁻ ABʳ M⁻]

[X⁺ ABʳ M⁺]

Fig. 13.1 Functioning of a suicide vector. (a) Under permissive conditions or in a permissive strain, the vector replicates coordinately with cell division. The cells appear ABʳ, M⁺ (where ABʳ is a transposon-borne resistance to antibiotic AB, and M a vector-specific marker), whether or not the transposon moves to the chromosome. (b) Under non-permissive conditions or in a non-permissive strain, the vector cannot replicate, and the original copy is diluted out, as observed by loss of the M character, after a lag proportional to the number of copies of the plasmid initially present in the cell. The ABʳ cells recovered will be those in which transposition has occurred in any of its possible targets. A particular mutation (here in the chromosomal gene X, which will have become X⁻) looked for must then be directly or indirectly selected.

after addition of Tn903, in *Caulobacter crescentus* with Tn7 or Tn5-132, and in *Acinetobacter calcoaceticus* with Tn10. A derivative of R388 (incW group) bearing a colE1 replication region has been successfully used to deliver Tn5 into *Pseudomonas solanacearium*.

Alternatively, the *oriT* from a promiscuous conjugative or mobilizable plasmid can be inserted into an *E. coli*-specific plasmid carrying a transposon. The construction can be mobilized into any Gram-negative recipient, provided the transfer functions are available in the donor strain. An *E. coli* strain, S17-1, carrying the transfer functions of RP4 (in fact a copy of RP4 deprived of its resistance markers) inserted into the chromosome has been constructed. The strain can be used to transfer any plasmid possessing an incP1 *oriT* into (most?) Gram-negative bacteria.

1.1.2 *Unstable maintenance of plasmids*

In some species, incP1 plasmids can promote conjugation but are not stably replicated, and they behave as natural suicide vectors. This is the case in *Myxococcus xanthus*, *Pseudomonas cepacia* and some *Pseudomonas solanacearium* strains.

RP4 derivatives with thermosensitive replication functions have been used in several Gram-negative species, including *Pseudomonas aeruginosa*, *Legionella pneumophila* and *Erwinia chrysanthemi*.

Plasmid instability can be brought about by the presence of a mutation in the replication apparatus of the plasmid. Tranposon Tn917 carried by a thermosensitive plasmid from *Streptococcus faecalis* and introduced by transformation into *B. subtilis* proved to be mutagenic in the latter strain with loss of the vector. Tn917 is widely used with Gram-positive bacteria.

Derivatives of RP4 plasmids with the structure RP4::Mu::TnX appear to be unstable in many species, as a consequence (not under-stood) of the presence of Mu. Tn5 mutagenesis has thus been obtained in strains of *Agrobacterium*, *Rhizobium*, *Alcaligenes*, *Erwinia herbicola* and *Erwinia chrysanthemi* EC16. The vector is not efficient with other strains or species, such as *Pseudomonas*.

A P1::Tn5 phage can penetrate into but not replicate in *Myxococcus xanthus*, and thus behaves as a suicide vector in this species.

1.1.3 *Incompatibility between two plasmids*

Incompatibility between a plasmid carrying the transposon and a second one devoid of transposon (but conferring another resistance used as a selectable marker) can be sought by using two plasmids belonging to the same group. A strategy could be as follows. The first plasmid, either already present or introduced by any available transfer system, can afterwards be eliminated from the recipient cell by intro-duction of the second one and forcing selection on the marker borne by this plasmid. The copy-number control system, functioning collectively on both plasmids, will tend to eliminate the first one, which will be dispensable every time its marker, i.e. the transposon, moves into the host chromosome.

A variation of this procedure has been used in *Klebsiella* strains. A P1::Tn5 phage can be brought into a strain lysogenic for P1 (carrying its own selectable marker). Selection of the recipients can be made for acquisition of resistance to kanamycin, the marker borne by Tn5, together with maintenance of the initial P1 prophage. Only the super-infecting P1::Tn5 phage can thus be eliminated, Tn5 being stabilized by transposition into the chromosome.

1.2 Utility of transposon mutagenesis: some examples

Besides the advantages mentioned earlier, transposon mutagenesis is especially interesting in three experimentally difficult situations.

1.2.1 *When the mutation looked for does not confer a readily selectable phenotype*

The high frequency of mutants obtained by insertion of a transposon is especially advantageous when the mutation sought does not confer an easily selectable phenotype. The phenotype looked for is selected secondarily, by searching among all mutagenized clones (i.e. those that have acquired the marker borne by the transposon). Numerous examples are to be found, for instance in studies concerned with plant – bacteria relationships, in which selection is a long process. Mutants of a *Rhizobium* strain have been looked for which would be deficient in the ability to perform symbiosis with their legume host (Nod⁻) or to achieve nitrogen fixation (Nif⁻) in the symbiotic nodule. To screen for such mutants, sterilized seeds of the host plant must first be germinated for a few days, transferred individually to a com-bined nitrogen-free medium and each inoculated with a clone from a mutagenized bacterial population. The mutants can be picked up a few weeks later, after the phenotype has been expressed in the plants. Working with cowpea rhizobia, four Nod⁻ or Fix⁻ mutants could be isolated among 400 cells with random insertions of Tn5. At least 10^4 clones would probably have been necessary to reach the same result after classical mutagenesis.

1.2.2 *When the strain studied is not easily amenable to classical mutagenesis*

As mentioned in Chapter 7, classical mutagenic treatments may show low efficiencies in certain species. No metabolic mutants could be obtained in *Azotobacter beijerinckii* through classical mutagenesis, but both Nif⁻ and auxotrophic clones have been obtained by transposition with Tn76. Tn5 is unstable in this strain.

1.2.3 *When the strain possesses several similar activities*

The possession of several similar activities prevents straightforward isolation of mutants on a phenotypic screening. When a particular gene is to be mutated, this gene must first be cloned, if possible in a strain where it can be expressed (from its own promoter or in an expression

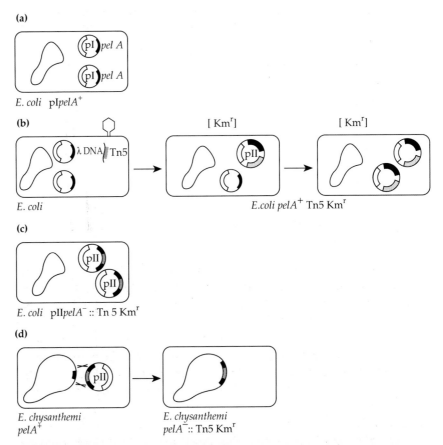

(a)

E. coli pIpelA⁺

(b)

E. coli   *E.coli pelA⁺ Tn5 Km^r*

(c)

E. coli   pIIpelA⁻ :: Tn 5 Km^r

(d)

E. chysanthemi   *E. chrysanthemi*
pelA⁺   *pelA⁻ :: Tn5 Km^r*

Fig. 13.2 Isolation of a *pelA* mutant in *Erwinia chrysanthemi*. (a) Cloning into the *E. coli* multicopy plasmid pBR322 of a segment of the *E. chrysanthemi* chromosome expressing a pectate lyase activity, *pelA*, in *E. coli* (plasmid I). (*E. chrysanthemi* possesses five major pectate lyases distinguishable by their electrofocusing migration. Each of them can be produced in *E. coli*). (b) Tn5 mutagenesis, in the pI-transformed *E. coli*, is obtained by infection with a suicide vector, λ::Tn5 Km^r (a mutant of phage λ unable to maintain itself in *E. coli*) followed by selection under a high concentration of kanamycin (400 µg/ml). Only the cells in which Tn5 has transposed into plasmid I, i.e. possess a high number of copies of the Kmr^r gene borne by Tn5, can resist such a concentration. The non-transposed plasmids are diluted out. (c) Assay for pectate lyase production in the transposed Km^r clones, and isolation of one with a Tn5 insertion in the *pelA* gene, that is lacking enzyme production (plasmid II). (d) Introduction of plasmid II, by transformation, into *E. chrysanthemi* and selection for Km^r clones. Since pBR322 does not replicate in this strain, such clones must result from an exchange, by homologous recombination, of the two *pelA* genes. Secondary transposition can be avoided by using a modified transposon (Section 3).

vector (Chapter 16)). Transposon mutagenesis is performed in this secondary host. The mutated gene is transferred back into the original wild-type cell and acquisition of the transposon-borne resistance is selected for after homologous recombination has taken place (Fig. 13.2).

This technique, often referred to as 'reverse genetics', is also used when transposon mutagenesis is not easily performed in a given strain, for instance through lack of a good suicide vector. Numerous transposon mutants have thus been generated in the Nif cluster (the cluster of genes that codes for all the functions required for nitrogen fixation) of *Klebsiella pneumoniae* after its cloning on *E. coli* plasmids.

2 Transposable elements and the development of genetics in new species

We have already described the use of transposons to insert a conjugative plasmid into a chromosome (Chapter 11). These elements can be useful in several other aspects of genetic analysis.

2.1 Supply of a replacement selectable marker.
Example: Isolation of *recA* mutants of *E. coli*

It is often useful to possess Rec⁻ mutants of strains being used for genetic studies. An efficient and very elegant solution has been provided in the case of *E. coli* K12. A strain has been found in which transposon Tn10 is inserted in the *srl* operon, located close to a known *recA* mutation (in fact, 6 kbp away from the *recA* gene). Transfer of the *recA* mutated allele to any Rec⁺ *E. coli* strain can be performed by homologous recombination of this portion of the chromosome after generalized transduction (Fig. 13.3). Transductants were selected for acquisition of Tn10-specified resistance to tetracycline. Most of these were *recA⁻*, their phenotype being easily detected by any of the relevant methods (lack of recombination, sensitivity to radiation, etc. (Chapters 8, 9 and 11)).

This method was adapted for the Gram-positive bacterium *B. subtilis*. Transposon Tn551, a Class II element encoding resistance to erythromycin (Em^r), is carried by a plasmid that is thermosensitive for its replication. A pool of clones containing random chromosome insertions of Tn551 is constituted during overnight growth at 43 °C, a non-permissive temperature. DNA extracted from this pool is used to transform different Em^s recipient strains, each containing an allele of interest of a chromosome gene. Em^r transformants showing co-inheritance of the donor allele of this gene are screened (Fig. 13.4). Recombinants carrying a transposon close to either the non-mutated or the mutated allele of the chosen gene can be constructed.

(a)

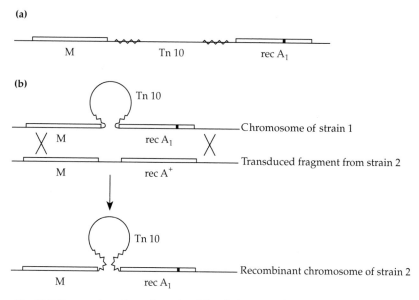

(b)

Fig. 13.3 Preparation of a $recA_1$ strain of *E. coli*. (a) Map of the P1 transducing DNA, from a lysate of Strain 1, carrying a Tn10 copy close to $recA_1$. (b) Transduction into Strain 2, $recA^+$, and selection for resistance to tetracycline. Among the Tc^r clones, a high ratio will result from recombinations including Tn10 and $recA_1$, yielding $recA_1$ strains.

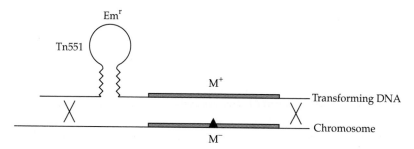

Fig. 13.4 Insertion of a transposon close to a chromosomal gene of interest in *Bacillus subtilis*. (i) Constitution of a pool of Em^r clones from a wild-type strain of *B. subtilis*, formed after Tn551 transposition. More than 10^4 clones ml^{-1} can be obtained. (ii) Extraction of genomic DNA from the pool of mutants. (iii) Transformation of a strain of *B. subtilis* mutated in the gene looked for, M$^-$, and selection for Em^r transformants. Among these, all the M$^+$ ones have acquired a copy of Tn551 close to the wild-type allele of marker M.

2.2 Handling of genes with no directly selectable phenotype. Example: Mapping genes involved in morphogenesis in *Myxococcus xanthus*

Myxococcus xanthus is able to undergo a complex morphogenesis process upon starvation. Mutants impaired in this process have been obtained by both classical and transposon mutagenesis, but the use

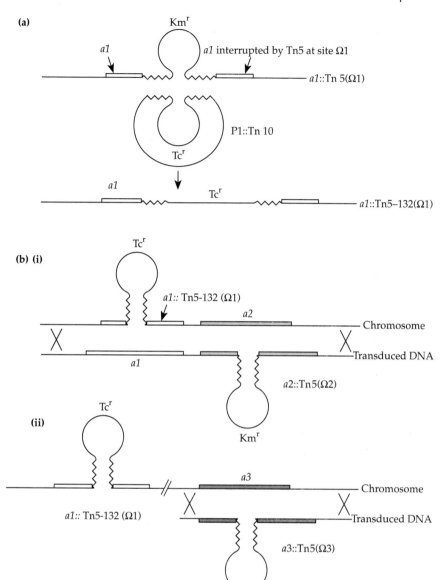

Fig. 13.5 Mapping of developmental genes in *Myxococcus xanthus*. Three clones, A1, A2 and A3, carry mutations due to the insertion of Tn5, denoted $\Omega 1$, $\Omega 2$ and $\Omega 3$, respectively, in three different genes, *a1*, *a2* and *a3*, responsible for developmental phenotypes. (a) Replacement of the Km^r marker of Tn5 by Tc^r from Tn5-132 in mutant A1. Isolation of a Tc^r Km^s recombinant, A1′, in which Tn5-132 is inserted at the $\Omega 1$ site. (b) Mapping by transduction. A transducing myxophage is grown on strains A2 or A3. The lysates are used to transduce A1′ (*a1*::Tn5-132). Km^r transductants are selected. Gene *a2* is linked to gene *a1*, the recombinants are Tc^s (i); gene *a3* is not linked to gene *a1*, the recombinants are Tc^r (ii).

of transposons has facilitated genetic studies of the mutations. For instance, it has been possible to map the corresponding genes using a generalized transducing phage, the recombinants containing the genes to be mapped being selected through a transposon-borne marker. Random transposition of the transposon brought about by the donor transducing DNA into the recipient DNA, which would introduce biases in the interpretation of the results, must be minimized. This can be achieved if a copy of the transposon is already present in the recipient, leading to expression of the transposition regulatory system previously to the transduction event. Using a derivative of Tn5, Tn5-132, obtained by replacing the internal part of Tn5 (Km^r) by that of Tn10 (Tc^r) (Fig. 13.5a), it was possible to exchange the two elements by homologous recombination without displacement of their insertion location. A collection of identical mutations associated with either resistance was generated, which can be used for transduction mapping similarly to regular resistance mutations (Fig. 13.5b). Intermarker distances calculated from recombination data are proportional to the physical distances on the chromosome.

2.3 Use of transposon insertion for targeting and cloning genes

Once a transposon is inserted into or near a gene of interest, this gene can easily be cloned, even though its phenotype is not directly detectable. Advantage was taken of the instability in *E. coli* of the conjugative transposon Tn916 − Tc^r to clone genes from Gram-positive bacteria. When a Tn916 element inserted in a shuttle vector is transferred into *E. coli*, its precise excision occurs at a very high frequency, whereas its insertion into the *E. coli* chromosome is much rarer. The cloning strategy used is as follows. A mutation is generated by Tn916 in the gene to be cloned from a Gram-positive bacterium, and selected as described in Section 1. The DNA containing the transposon and the mutated gene is cloned, together with other random chromosome fragments, in an *E. coli* plasmid and transferred into an *E. coli* strain to form a genomic library (Chapter 16). Those organisms having a Tc^r phenotype are selected and grown in the absence of tetracycline. Tc^s segregants can be isolated, and among them will be those in which the recombinant plasmid has lost the transposon but kept the DNA from the Gram-positive donor. The intact gene is thus cloned (Fig. 13.6).

2.4 Use of transposons for the generation of deletions

For many purposes the possession of deletion mutants is very useful. It is the best way to stabilize a mutation (e.g. to prevent reversions from occurring) when very large-scale cultures are needed. It is also the best

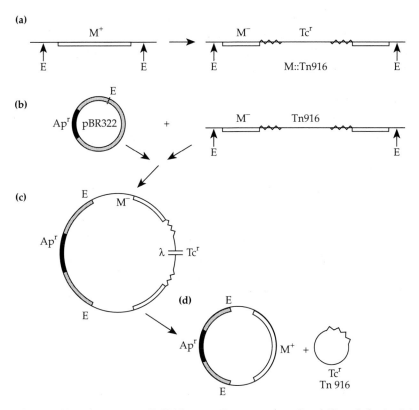

Fig. 13.6 Use of transposon Tn916 for targeting a gene from *B. subtilis* and cloning it into *E. coli*. (a) Isolation of a mutant, M⁻, by insertion of Tn916 in *S. aureus*. (E = site of restriction by the endonuclease E). (b) Cloning of the DNA of strain M::Tn916 cut by restriction enzyme E that has no site inside Tn916, into the multicopy plasmid pBR322 Apʳ. (c) Transformation of *E. coli* with the pool of pBR322 recombinant plasmids, and screening for an Apʳ, Tcʳ transformant. (d) Growth of this clone in the presence of only ampicillin. Tn916 excises and is lost. The resulting Apʳ, Tcˢ recombinant plasmid carries an intact M gene.

way to ascertain whether a given gene is dispensable for growth of the mutated bacterium (provided the cells do not maintain a heterozygous diploid state). Finally, it is a very efficient method for fine-structure analysis of clustered genes (Chapter 16). Some transposons, such as Tn10 or Mu, generate deletions with high frequencies (Chapter 5) and have been utilized for this purpose.

2.5 Determination of the physical location of genes

Knowing whether the genes involved in a given function are located on a plasmid or on the chromosome or even distributed over two or more replicons may not be an easy task. Transposon targeting of the genes greatly helps in providing the answers, as pictured in the following example. The functions involved in oncogenesis in plants infected by *A. tumefaciens* were thought to be borne by the Ti plasmid of the bacterium. Thirty-seven Tn5-insertion mutants of a Ti plasmid-containing strain showing altered virulence were isolated. The transposon present in each mutant and thus the gene in which it had transposed were physically located. The plasmid DNA and the chromosomal DNA from these clones were separated, using standard molecular techniques, which take advantage of the differences in size, structure and/or G + C content of these molecules. Each of them were then probed by hybridization with a radioactive Tn5 (Chapter 16). Twelve mutations appeared to be chromosomal and 25 plasmid-borne.

The presence of transposons (or of any easily detectable genetic elements) in a series of identified genes can also be utilized for chromosome mapping on a preestablished physical (i.e. restriction) map of that strain. The restriction fragments obtained after appropriate enzyme treatment of total DNA prepared from each mutant are separated by pulsed-field gel electrophoresis (Chapter 14) and probed with (part of) the transposon by hybridization.

2.6 Determination of the organization of genes in operons

The ability of most transposons to give strongly polar mutations has often been used to test the pattern of expression of related functions, i.e. the presence of operons, and their internal organization and direction of reading. The organization of the *nif* region of *K. pneumoniae* was determined by cloning it into an *E. coli* plasmid and comparing the consequences on the expression of the different cistrons of insertions of transposons Tn5, Tn7, Tn10 and Mu (Chapter 16). Similarly, the organization of the two operons of the NAH7 plasmid from *Pseudomonas* sp. responsible for naphthalene–salicylate oxidation have been determined using Tn5 insertions. That of the toluene-catabolism operons present on the TOL plasmid of *Pseudomonas putida* has been unravelled using Tn1000 and Tn5 mutagenesis of fragments of the plasmid cloned into plasmid pBR322 (Chapters 3 and 16) and introduced into *E. coli*.

3 Modification of transposons increases their range of utilization

The many fields in which transposons may be utilized, a few examples of which have been given, have been widened further, thanks to the availability of a great number of modified transposons. Replacement of its resistance marker has already been described in Tn5, yielding Tn5-132. Numerous other Tn5 derivatives bearing a large set of different antibiotic resistance markers are now available. These are useful in strains in which the kanamycin resistance gene is poorly expressed.

Thus Tn5 conferring the resistance to Hg from Tn501 has been constructed for studies in *Pseudomonas aeruginosa* strain PAO.

Other functions have also been introduced into transposons, of which the three following examples give an idea of the span of possibilities:

1   Addition of the *oriV* from plasmid Sel01 to Tn5 transformed the transposon into a plasmid replicable in *E. coli*. Isolation of the chromosomal segment bearing this entire transposon, called TnV, from the bulk of chromosomal DNA from a transposed cell was achieved by molecular methods (Chapter 16), followed by transfer into *E. coli*, into which this particular segment was maintained as a plasmid and the corresponding recipient cell selected by acquisition of the transposon marker.

2   Introduction of the *oriT* (*mob*) sequence from RK2 allows mobilization of chromosomes in which this modified transposon is inserted, provided a helper RK2 plasmid is present in the donor strain.

3   Particular sequences, such as a restriction site for a given restriction endonuclease (Chapter 6) can be introduced (by *in vitro* methods) near both extremities of a transposon. If this modified transposon integrates into a plasmid otherwise lacking this restriction site, it provides a unique site usable for gene cloning (Chapters 6 and 16): digestion by the endonuclease eliminates the internal part of the transposon (easily verified by loss of the transposon marker) and allows insertion of any other DNA fragment.

Another modification consists in the physical separation of the transposase gene from the rest of the element, which forms a 'minitransposon'. When the two parts are present in the same cell (on the same or different vectors), expression of the transposase (usually under the control of another, inducible, promoter (Chapter 15)) allows transposition of the minitransponson to its target(s) in the cell. Elimination of the transposase, by repression of expression of the encoding gene or by curing for the carrier plasmid, leaves a transposed element which cannot by itself transpose again and thus provides a stable marker and an irreversible mutation.

4   *In vivo* gene fusion

In vivo gene fusion is the most spectacular usage of modified transposons. It was devised by Casadeban in the late 1970s and has since been considerably improved. Mu*d*(Ap, *lacZ*) (Fig. 13.7) is a modified phage Mu in which the genes for lytic functions have been deleted and replaced by the *bla* gene from transposon Tn3 and the *lacZ* gene from *E. coli* K12. Two series of Mu*d*(Ap, *lacZ*) have been devised. In the first one (Fig. 13.7a), the translation signals (rbs) of the *lacZ* gene

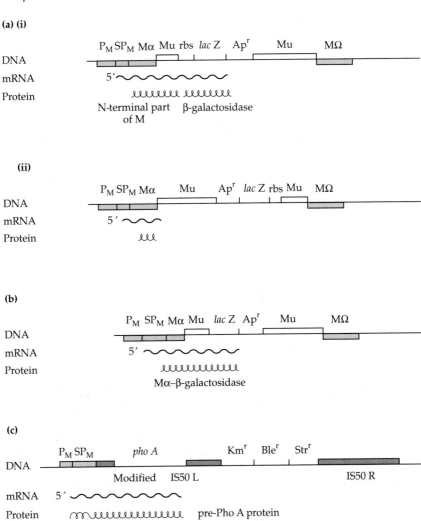

Fig. 13.7 *In vivo* gene fusion. (a) Insertion of the operon-fusion transposon *Mud* (Ap, *lacZ*) into gene M. In i, the orientation of the fusion allows expression of gene *lacZ*. In ii it does not (see text for details). (b) Insertion of the gene-fusion transposon *Mud* (Ap, *lacZ*) into gene M allows synthesis of a fused M-β-galactosidase protein showing β-galactosidase activity (see text for details). (c) Insertion of Tn*phoA* downstream of the signal peptide-coding region (SP$_M$) of gene M. The example illustrates the synthesis of a pre-PhoA protein which can cross the cytoplasmic membrane and be processed into its active form (see text for details).

have been maintained and its promoter has been deleted. When the transposon inserts distally from another promoter and in the correct orientation, the *lacZ* gene can be expressed. Observations of changes of its level of expression when environmental conditions are monitored allow the study of the regulation of expression of the gene (or the operon) in which the transposon is inserted. This device, termed operon fusion, has been extensively used to study the regulation of genes whose phenotype is difficult to monitor.

In the second series of modified Mu derivatives (Fig. 13.7b), the *lacZ* gene is further deprived of its translation signals and of the first nucleotides of its coding sequence. The gene can be expressed only if the transposon is inserted in the correct reading frame anywhere downstream from a complete expression signal. It will then promote the synthesis of a chimeric protein (gene fusion). In the presence of a helper Mu this transposon can be packaged into a Mu particle. Since such tools are extremely useful in deciphering gene regulation, other transposons, e.g. Tn5*lacZ*, have been engineered in a similar manner so as to bypass the narrow host-range of Mu.

Another fusion transposon, Tn*phoA* (Fig. 13.7c) has been devised to allow detection of genes coding for exported proteins. *phoA* is the *E. coli* gene coding for the alkaline phosphatase. This protein is periplasmic and only active as a dimer. The gene is translated as a longer precursor, of which the NH_2 terminal 'signal peptide' (SP) is cleaved during translation through the cytoplasmic membrane. If the polypeptide corresponding to the mature protein is synthesized from a *phoA* gene deleted of the region coding for the signal peptide, no alkaline phosphatase activity is found, presumably because of the inability of the polypeptide to form dimers in the physical conditions of the cytoplasm. Insertion of this deleted *phoA* gene in the correct reading frame downstream of the SP of a gene coding for another exported protein leads to restoration of an alkaline phosphatase activity. The region of gene *phoA* corresponding to the coding sequence of the mature protein has been inserted close to the inverted repeat of IS50L of Tn5. Transfer of this Tn*phoA* in a *phoA*-deficient strain and selection for PhoA activity in Km^r clones allow detection of such genes. This technique has thus revealed *E. coli* genes coding for exported proteins the synthesis of which is regulated by changes in osmotic pressure.

The use of transposons as a manipulative tool in genetic studies is still widely open to new applications. The interest of transposons has prompted many similar studies in other organisms, in particular in microorganisms, such as yeast, or higher eukaryotes, such as *Drosophila, Zea mays* or *Antirrhinum*.

References

Casadeban M.J. (1976) Transposition and fusion of the *lac* genes to selected promotors in *Escherichia coli* using bacteriophage gamma and mu. *Journal of Molecular Biology*, 104, 541–5.

Kleckner N., Roth J. & Boltstein D. (1977) Genetic engineering *in vivo* using translocable drug resistance elements: new methods in bacterial genetics. *Journal of Molecular Biology*, 116, 125–59.

PART 3
GENETICS AS A TOOL
IN UNDERSTANDING
GENETIC ORGANIZATION
AND FUNCTION

14: Construction of Genomic Maps

Understanding the architecture of the genetic information on the various units forming the genome of an organism has always been an aim of geneticists. In bacteria two main concerns back this ambition.

1 Observation of maps allows comparisons that are fruitful for taxonomic and evolutionary studies. As with the evolution of the fine structure of genes, the relative positions on chromosomes of independent or functionally related information constitute an important phylogenic landmark.

2 Possession of genomic maps is also a potent tool for geneticists. Linkage analysis used to locate newly discovered genes is facilitated when a larger number of reference markers (localized genes) is available. Conversely, isolation of mutants with unselectable phenotypes by their association with selectable neighbouring genes is possible only if linkage relationships have been established.

1 The use of conventional genetic methods

Construction of genomic maps based on determination of linkage values between markers is theoretically possible from the study of progeny after transfer of DNA from one strain to another with a limited number of genetic differences by any method (as long as the transferred DNA does not have the capacity to replicate by itself). Both the order of the genes on the DNA molecule and their relative distances can be determined by analysis of linkage values, that is, of proportions of the various types of progeny that retain the parental coupling arrangement for any pair of genes. Feasibility depends on the available tools. It is very easy in *E. coli* K12 and the related species *Salmonella typhimurium* LT2, because conjugation allows working at the scale of (almost) whole chromosomes in one transfer experiment (Chapter 11). Fine mapping, however, although possible, has seldom been performed by using exconjugants. This is performed after the introduction of relatively short pieces of DNA, as in transformation or transduction (Chapters 10 and 12), which facilitates analysis at scales down to the nucleotide. Construction of whole-chromosome maps with these two latter systems has never been performed, due to the enormous amount of work required to connect up short linkage groups, although the *B. subtilis* map was first elaborated by associating transformation or transduction with a particular physiological device: by taking advantage of the synchronization of replication in germinating spores, it was possible to follow the sequence of gene replication, i.e. the relative positions of genes on the chromosome map, transformation being used to assess gene dosage (Chapter 10).

A few maps have been obtained by these methods (Table 14.1). Due to lack of markers available or to limited efforts in this direction, some of them contain very sparse information. This, in particular, hindered generalization of the concept of circular bacterial chromosomes, as first observed in *E. coli* K12 and later in a few other strains (Chapters 10 and 11).

Map units were defined only in *E. coli* and *Salmonella*, but these are an arbitrary reference to the transfer system underlying map construction, i.e. minutes of transfer at 37 °C. Being arbitrary, these units have

Table 14.1 Chromosome maps obtained by conventional genetic methods.

Species (strain)	Methods used	Year first obtained	Observations
E. coli K12	Conjugation Transduction	1960s	About one-third to one-half of the genes mapped
Salmonella typhimurium	Conjugation Transduction	1960s	At least 50% identity of map organization with *E. coli* K12
B. subtilis	Transformation Transduction	1960s	Part of the maps obtained by order of gene replication in synchronized cells
Streptococcus pneumoniae faecalis	Transformation	1970s	Using density shift of synchronized cells and linkage by co-transformation Evidence for bidirectional replication
Haemophilus influenzae	Transformation	1970s	
Pseudomonas aeruginosa putida	Conjugation	. 1970s 1980s	Interstrain similarities of organization Determination of size Determination of circularity
Streptomyces coelicolor	Conjugation	1960s–1970s	Circular
Caulobacter crescentus	Conjugation Transduction	1977–1979	
Erwinia chrysanthemi	Conjugation		
Rhodobacter spheroides capsulatus	Conjugation Conj. + Capsd.	1984 1985	Not then proved circular Circular
Vibrio cholerae	Conjugation Transduction	1979; 1984	Two isolates; similar circular maps, except for an inverted region; 30 genes mapped, most located in one-quarter of the whole map.

now been normalized so that the whole chromosome is covered by 100 min. While tranformation data only yield relative linkage values along very short spans of DNA, similar data obtained from generalized transduction can be checked on the basis of the constant, and known, length of the transferred fragment. Nevertheless, in all these cases, the information relies on recombination values, and distortions introduced by non-homogeneity of local recombination capacities, such as the existence of recombination hotspots, e.g. chi sites (Chapter 9), cannot be avoided.

2 The use of pulsed-field gel electrophoresis and DNA hybridization

Since DNA molecular technologies have become widespread, restriction maps, giving the distribution of recognition sites for various site-specific endonucleases (Chapter 6), can be obtained. Conversion of these restriction maps (generally referred to as physical maps) to genetic maps has also become possible. The strategy consists in cutting the whole chromosome into fragments that can be handled, i.e. isolated and amplified on plasmid vectors, and which are easily amenable to physical, molecular and genetic analysis.

A crucial problem is the reconstruction of the whole map from these segments, i.e. the establishment of their order and orientation one to the other. Obviously, the longer, and thus the fewer, the fragments are, the easier the work is. As already seen (Chapters 10–12), this explains why conjugation, but not transformation or transduction, led to large-scale chromosome mapping. Modern molecular techniques allowing cloning of DNA fragments (Chapter 16) were extremely useful in that they opened genetic analysis to strains or species for which no natural genetic transfer was known. As first available, however, they were of limited use for chromosome mapping because maximal sizes of fragments thus analysed reached about 30 kb, i.e. approximately one to a few hundredths of an entire bacterial chromosome, as is the case for natural transformation or transduction processes. Construction of whole maps would thus have represented an enormous task. In fact, while plasmid and phage genomic maps could fairly easily be constructed, only one chromosome map, that of *E. coli* (Section 3.2), was obtained in this way, taking large advantage of the already available genetic data.

Another limit in using classical cloning devices is the difficulty in discriminating the fragments necessary to cover a whole chromosome, both because of their large number (approximately 500–1000) and due to the engendered overlapping of their sizes. Discrimination can indeed be achieved on the sole criterion of their size, by electrophoretic separation (Chapters 3 and 16). Molecules of up to a few thousand base pairs, which may display a superhelical coil conformation, are separated by the sieving action of the gel matrix, the sizes of the pores in the gel being about the same as the diameters of the DNA molecules. The migration velocity of the DNA fragments varies inversely with their size. Separation is possible but is strongly hindered since too many fragments will have very close sizes. DNA molecules larger than about 20 kb move through the gel as extended coils in a process known as reptation, the speed of which becomes independent of the size of the fragments, and thus also hinders separation.

All cloning analyses performed with these techniques were aimed not toward chromosome mapping but at fine analysis of particular regions, which could be identified and isolated from the bulk of fragments, thanks to other criteria, such as gene function or sequence homology.

The recent development of new techniques has allowed the manipulation and size estimation of large restriction fragments, and hence isolation of DNA molecules of sizes up to those of prokaryotic and small eukaryotic chromosomes, from which restriction maps can relatively easily be constructed. The much smaller number of fragments that need be handled and thus isolated in the initial stages greatly facilitates the process. Positioning markers, which is feasible by hybridization with cloned genes, transforms these restriction maps into genetic ones.

2.1 Construction of restriction (physical) maps

2.1.1 *Separation of DNA fragments by pulsed-field gel electrophoresis (PFGE)*

In 1982–1984 Schwartz and Cantor found a way of resolving DNA fragments ranging from 35 to 2000 kbp. The molecules were electrophoresed in an agarose gel subjected to electric fields alternately orientated at roughly right angles, a device which gave its name to the method. The rationale is that the longer the DNA molecules, the slower they reorientate at each change of direction of the electric field. Progress along the average direction of migration thus varies with the size of the molecules. The extreme interest of this method, especially for the analysis of eukaryotic chromosomes, prompted several teams to improve it in ways increasing the maximal size limit of the molecules and the resolution capacity of the gels, as well as the 'tidiness' of the resulting tracks. Figure 14.1 summarizes the most efficient devices used.

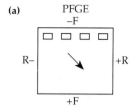

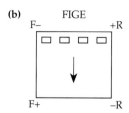

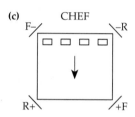

Fig. 14.1 Frequently used devices available for PFGE. (a) PFGE, Pulsed-field gel electrophoresis. (b) FIGE, Field-inversion gel electrophoresis. The forward pulse time is longer than the reverse one, thus generating a net forward migration. A good resolution is obtained for fragments up to 800 kbp long. (c) CHEF, Clamped homogeneous electric field. This system avoids distorsion and edge effects by use of fragmented electrodes − to +, direction of electric field application; arrow, direction of DNA migration.

The successful utilization of these techniques for chromosome mapping has two prerequisites, that the DNA can be cleaved reproducibly and that it is cleaved into only a few pieces. This implies that few random breaks due to shearing accumulate during extraction of the molecule and that restriction enzymes that have only a limited number of recognition sites on the whole chromosome, i.e. on average less than one in 10^5 bases, must be available.

The latter condition is met by several types of enzymes (Table 14.2). The choice regarding the appropriate enzymes used can be made on the basis of several criteria which reflect particularities of the base sequence of the chromosome to be studied, and thus probabilities of the presence of the recognition sequences of the endonucleases: (i) the observed rarity of certain short sequences, such as the stop codon TAG, more frequent in G+C-rich chromosomes, or of A/T and G/C clusters in G+C- and T+A-rich genomes, respectively; (ii) the specificity of codon usage among species or strains, i.e. specificity of occurrence

Table 14.2 Rare-cutting restriction enzymes.

*Not*I	GC/GGCCGC	8-nucleotide recognition site
*Sfi*I	GGCCNNNN/NGGCC	Hyphenated 8-nucleotide site
*Ase*I	ATT-AAT	⎫
*Dra*I	TTT/AAA	⎬ Rare cutters in G+C-rich genomes
*Ssp*I	AAT/ATT	⎭
*Sac*II	CCG/CGG	⎫
*Sma*I	CCC/GGG	⎪
*Rsr*II	CG/GNCCG	⎬ Rare cutters in T+A-rich genomes
*Nae*I	GCC/GGC	⎭
*Spe*I	A/CTAGT	⎱ TAG is a rare stop codon in prokaryotes (except
*Xba*I	T/CTAGA	⎰ in extremely A+T rich genomes)
*Avr*II	C/CTAGG	⎱ TAG is a rare stop codon and GC/CG or CC/GG
*Nhe*I	G/CTAGC	⎰ sequences are relatively rare in prokaryotes

/ indicates cleavage sites.

of a given codon among redundant ones, determines short-range base sequences; and (iii) biases in sequence arrangements compared with expected random distributions at the chromosome level (Chapter 6). These distortions lead to a limitation in the number of cleavage possibilities by restriction enzymes including such sequences in their recognition sites, even though these sites may consist of only six base pairs. A few enzymes with 8-bp recognition sequences are known, and on many genomes they are rare cutters.

Prevention of random breakage of the DNA is achieved by trapping the cells, before any treatment, into small plugs or beads of agarose. Proteins can circulate inside the agarose network, which allows restriction enzymes to penetrate and proteins and other small molecules originating from cell lysis to diffuse out. DNA, however, remains entrapped until an electric field forces it out. Cell lysis and restriction cleavage are performed inside the agarose, and the DNA fragments are electroeluted from the plug. The mixture is resolved by PFGE, and individual fragments can be isolated if desired.

2.1.2 *Construction of the restriction map*

Isolation and identification of each individual fragment generated by the restriction cleavage treatment is best performed if the total number of fragments does not exceed 20–25 (Fig. 14.2a). One to three sets of fragments are prepared by cleavage with different enzymes.

Two main strategies can be used to obtain the correct alignment of the fragments on the chromosome. In the first strategy, the fragments from one set are individually digested by the other enzyme(s) (Fig.

(a)

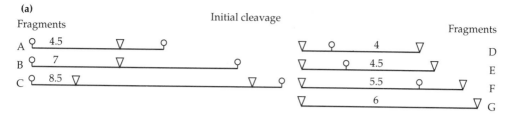

(b)

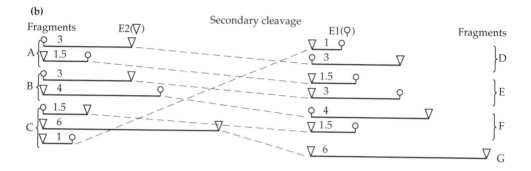

(c)

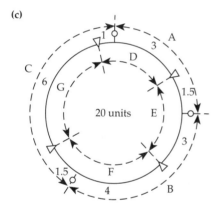

Key: ♀ Enzyme E1 restriction sites;   ▽ enzyme E2 restriction sites

Fig. 14.2 Construction of a restriction map. (a) Fragments obtained after initial restriction treatment of a whole molecule of DNA with either enzymes E1 or E2 (random physical breakage during purification has not been considered, for simplification). Figures represent arbitrary size units. The molecule has been supposed to include very few restriction sites, to make the representation easier. (b) Determination of the distribution and sizes of the fragments obtained by a secondary digestion of the above fragments. Fragments A, B and C obtained after restriction with enzyme E1 as in (a) were individually digested by enzyme E2, and reciprocally. Dotted lines indicate identical fragments in each of the restriction sets. (c) Comparisons of the sizes of the fragments and sub-fragments allows ordering of the restriction sites, and construction of the physical map. Here only a circular map is compatible with the restriction patterns.

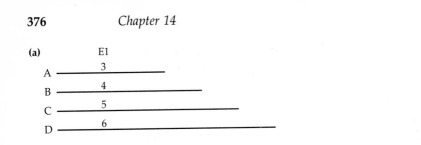

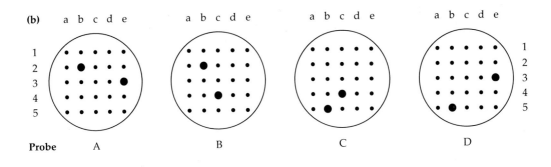

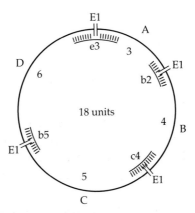

Fig. 14.3 Creation of a linking library. (a) Preparation of a set of fragments obtained by a rare-cutting enzyme E1 (numbers represent size of fragments in arbitrary units), and of a set of small fragments obtained by digestion by a frequent-cutting enzyme, En (average size of the fragments 5–10 kbp). The En fragments are amplified by *E. coli* cells after insertion into an appropriate cloning vector. A library is thus constituted. (b) The E1 fragments are labelled (with ³²P). Colonies obtained from a sample representing the En genomic library are transferred to replica filters, and the cells lysed *in situ*. Hybridization with each of the labelled E1 fragments allows the sorting of fragments as pairs hybridizing with the same clone from the library. (c) Alignment of the pairs yields the order of the E1 fragments on the map.

14.2b). Comparisons of the sizes of the resulting fragments allow the location to be made for the different cleavage sites on either individual fragment (Fig. 14.2b) and on the whole chromosome (Fig. 14.2c). The method is similar to that used for plasmid mapping (Chapter 3). A second strategy consists in determining which two fragments from a set of rare-cut fragments share part of their sequence with those from another set of fragments obtained after cleavage by a frequent-cutting enzyme (Fig. 14.3a) (forming a genomic library (Chapter 16)). This is performed by hybridizing the whole genomic library with each individual rare-cut fragment previously labelled, i.e. made into probes (Chapter 16), and sorting out the pair that gives the same hybridizing signal (Fig. 14.3b). The method is referred to as the generation of a linking library. Adjacent segments are called contigs, referring to their contiguous positions on the chromosome. In principle, only one map configuration can be deduced from the complete results (Fig. 14.3c).

This method has immensely accelerated and broadened accessibility to genome mapping, in particular because it does not require the existence of a convenient DNA transfer system. The first physical map published was that of *E. coli* in 1987. Table 14.3 gives a list of the first strains or species for which a physical map has been constructed, more having been regularly published since. It is interesting in two aspects: (i) the large variety of species, most of which are devoid of any known transfer system; and (ii) the very short lapse of time over which they have appeared, as compared with the time-scale associated with conventional genetic analysis (Table 14.1).

2.1.3 *Utilization of yeast artificial chromosomes (YAC)*

The YAC system, devised in 1987, allows the cloning of DNA fragments ranging from 100 to 1000 kbp. It takes advantage of the possibility of introducing and maintaining artificial chromosomes in the yeast *Saccharomyces cerevisiae*, provided the molecule is linear, contains an active yeast origin of replication (*ARS*) and a centromeric sequence (*CEN*), and is terminated at both ends by a telomere region (*TEL*). The molecule must also contain yeast-specific selection markers (e.g. URA^+, TRP^+, HIS^+, *SUP4*, a suppressor for a nuclear ochre ADE^- mutation) and sequences from the artificial plasmid pBR322, *ori* and Ap^r (Chapters 3 and 16), which allow replication and selection in *E. coli* K12 (Fig. 14.4a). Linearization (restriction at the two *Bam*h1 (B) sites) liberates the *TEL* extremities, an operation that can be controlled by loss of the His^+ marker (acquisition of a His^- phenotype) (Fig. 14.4b). A unique cleavage site (UCS) situated in the *SUP4* gene can be used to insert the fragment of DNA to be analysed (Fig. 14.4c). Disruption of the YAC *SUP4* gene, yielding an Ade^- phenotype, allows the identification of the clones bearing an insert. The whole construction is transferred into a TRP^-, URA^-, Ochre ADE^-, HIS^- auxotrophic yeast cell (usually using yeast spheroplasts), selection of transformed cells being performed through the acquisition of the prototrophic Trp^+

Table 14.3 (*Below and opposite.*) The first physical and genomic maps constructed by PFGE methods.

Species	Size (Mbp)	Genes mapped	Observations	Year
E. coli K12	4.75	≃450	Only genes sequenced have been mapped	1987–90
Borrelia burgdorferi	1.0 ± 0.1	0	The only known linear chromosome; also has 3 linear (0.16–0.5 Mbp) and 2 circular (0.27 & 0.3 Mbp) plasmids	1987–89
Caulobacter crescentus	3.5	43		1988
Brucella (several strains)	0.5–16	0	Minor polymorphism of the restriction patterns for 2 enzymes	1988
Pseudomonas aeruginosa	5.86	63		1988–90
Haemophilus				
influenzae Rd	1.96	24	Has a 45-kb duplication	1988–89
influenzae b	2.1	23	Limited restriction pattern homology within b isolates	1990
parainfluenzae				1989
Mycoplasma				
mycoides	1.2	6		1988
mobile	0.78	0	No information available about	1988
pneumoniae	0.8 ± 0.3	4	possible genome organization	1989–90
5 others	0.9–1.3	0	homologies	1989–90
Rhodobacter spheroides		23	First species known with two chromosomes (I, II). Ratio I/II = 1/1	1989
	I 3.09			
	II 0.91	8		
Anabaena PCC7120	6.37	30	No conserved arrangement for equivalent metabolic genes in *E. coli* or photosynthetic genes in chloroplast genomes 3 large (0.1–0.5 Mbp) cryptic plasmids	1989
B. cereus	5.7	14		1989
Clostridium perfringens	3.6	24		1989
Chlamydiae	1.45	0		1989
Rickettsiella	1.7–2.1	0		1989
Porochlamydia	1.3–2.6	0		1989
Lactococcus lactis	1.75	0	80% identity of restriction site locations in related strains; 20–40% in distant strains	1989
Streptococcus				
salivarius	2.5	0		1989
mutans	2.16	5		1990

continued

Table 14.3 (*Continued*).

Species	Size (Mbp)	Genes mapped	Observations	Year
Myxococcus xanthus	9.45	18 + 100	The largest prokaryote map constructed; mapped markers are 18 cloned genes and 100 Tn5-132 insertion mutants	1989–90
Campylobacter jejuni	1.72	0		1990
Methanococcus voltae	1.9	53	Circular map. All genes were mapped by hybridization	1990

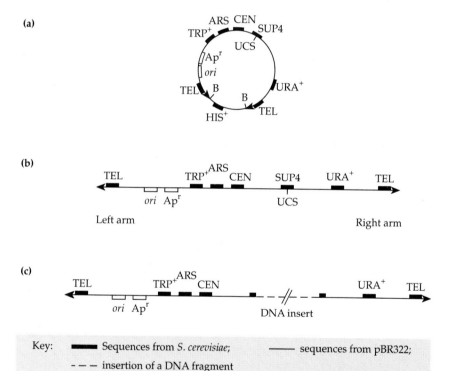

Fig. 14.4 Construction of a yeast artificial chromosome (YAC). (a) Circular structure of the YAC vector, as present when replicated in *E. coli* cells. Sequences from *S. cerevisiae* are: ARS, active yeast replication sequence; CEN, centromere; TEL, telomere; SUP4, suppressor for a nuclear ochre ADE⁻ mutation; URA⁺, TRP⁺, HIS⁺, prototrophic alleles implicated in the biosynthesis of uracil, tryptophan and histidine, respectively. Sequences from pBR322 are: *ori*, *E. coli*-specific origin of replication; Ap^r, resistance to ampicillin. UCS, unique cloning site; B, *Bamh*1 sites. (b) Linearization of the molecule by digestion with *BamH*1. Liberation of the TEL extremities and loss of the HIS⁺ allele. (c) Insertion of a DNA fragment into UCS, with concomitant inactivation of the SUP4 allele.

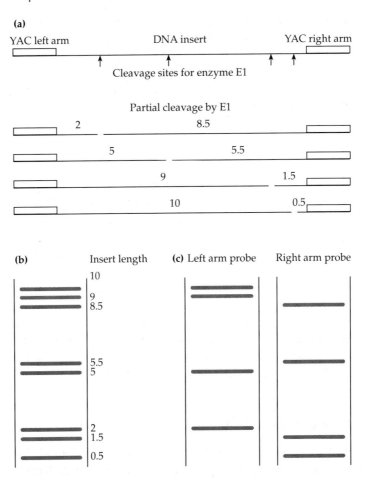

Fig. 14.5 Rapid analysis of the restriction pattern of a YAC insert. (a) Partial digestion (only 1 out of 4 sites are cleaved) of the YAC with enzyme E1, known not to cut the YAC vector itself. (b) Separation of the fragments by PFGE. Insert lengths are in arbitrary units. (c) Identification of the positions of the fragments by Southern blotting using the left and right arms of the linear YAC as probes.

Ura$^+$ phenotype. Such YAC constructions are stably propagated in the yeast cells provided they are at least 80 kbp long.

The haploid *S. cerevisiae* cells reproduce both their own 8 chromosomes and the ninth, smaller, recombinant, artificial chromosome. The latter is recovered from cells of individual clones lysed after embedding in an agarose plug. PFGE directly performed from the plug allows elution of the chromosomes. The YAC exits faster than the larger native ones and can be easily recovered. The possibility of identifying the left and right arms bracketing the exogenous insert provides a rapid and elegant means of constructing the restriction map of the insert (Fig. 14.5).

The resulting genome libraries (i.e. collections of clones in which the inserts statistically represent the whole genome to be studied) comprise a small number of very large DNA fragments. This technique, extremely valuable for mapping eukaryotic genomes, is also very useful for prokaryotic ones. An example of application of this technique will be described below (Section 2.4).

2.2 Conversion of physical maps into genetic maps

Restriction maps can be converted into genetic ones by locating genes with reference to the various restriction sites determined by hybridization with probes of cloned genes, a technique known as 'Southern blotting' (Chapter 16).

The limiting factor of this conversion lies in the number of cloned genes available. These, however, can originate either from the host itself or from heterologous but somewhat related species possessing the same function, in which sequence conservation (i.e. base sequence homology) can be expected to be sufficient to allow efficient annealing.

Precision of mapping is directly dependent on the degree of resolution of the restriction map, and thus on the number of different enzymes with which each initial fragment has been tested.

2.3 The limitations of the method

The reliability of this method depends on the isolation and identification of the complete set of fragments generated by the restriction treatments. It seems likely that a few enzymes will be found that produce appropriate digestion profiles. In contrast, difficulties resulting in distortions, omissions or additions in the maps may occur from several causes:

1 Non-ambiguous resolution of the digestion fragments may be hindered if the fragments are too numerous or if two (or more) have very close lengths, and thus similar migration velocities.

2 The possibility cannot be excluded that one or several recognition sites for an endonuclease may be protected in the host but no longer protected when the cloned subfragments are amplified (usually in appropriate *E. coli* strains) after ligation into a cloning vector.

3 Very small fragments may be generated, which would be too small to be detected (detection, performed by fluorescence of the intercalating dye ethidium bromide (Chapter 3) is proportional to the amount of DNA, and thus to the size of the fragments) or so small that they would have eluted from the gel before it was examined.

4 The presence of repetitive sequences (multifamily genes or copies of

mobile elements) may lead to false alignments biased by erroneous apparent identity of the corresponding regions.

Only agreement between several sets of data will help overcome such drawbacks.

2.4 An example: construction of the *Myxococcus xanthus* map

No easy *in vivo* chromosomal mapping can be performed in *M. xanthus*, and its chromosome is among the largest known in bacteria. PFGE mapping was thus a very convenient approach.

Two restriction enzymes were utilized, *Ase*I and *Spe*I, which generate 17 and 22 fragments, respectively, from the *M. xanthus* chromosome (Fig. 14.6). Most of the fragments could be placed on a circular structure, yielding a macrorestriction map, and an estimation of the size of the whole chromosome was given as 9450 ± 100 kbp.

Southern blotting of the fragments performed with cloned genes allowed some conversion of the physical map into a genetic one. The latter was enriched by using a very elegant technique. A large set of mutants available in *M. xanthus* was generated by insertion of the

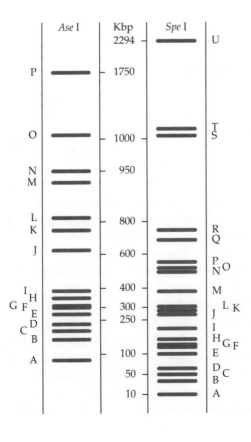

Fig. 14.6 An example of the utilization of PFGE to analyse a bacterial chromosome. CHEF mapping of the *Myxococcus xanthus* chromosome. Identification of the restriction fragments obtained after digestion with *Ase*I and *Spe*I.

transposon Tn5 derivative Tn5-132 (Chapter 13), which possesses a unique *Ase*I site. Comparisons of the CHEF profiles of wild-type and Tn5-132-mutant chromosomes digested by *Ase*I indicated the rough positions of Tn5-132, and thus of the various genes affected in these mutants.

A more complete mapping was achieved by resorting to the YAC technique. This was the first attempt to map a bacterial chromosome using a yeast as the intermediary host (Kuspa *et al.*, 1989). 409 YAC containing-fragments $\simeq 50$ kb long, obtained after partial digestion of the *M. xanthus* genome with the endonuclease *Eco*RI, were screened. By comparing the restriction patterns of these clones, overlapping frag-ments were identified and the chromosome resolved into 78 contigs. Hybridization of 18 known cloned genes with these 78 YACs led to a more accurate localization of these genes than the map already obtained with the 17 and 22 fragments previously examined. It was found that several genes hybridized with two YACS bearing different contigs, a result which led to a revised map covered by only 60 contigs.

Further completion of the *M. xanthus* map has been achieved by using the YACs as probes for hybridization with the CHEF fragments obtained after *Ase*I and *Spe*I cleavage.

3 Conclusion

3.1 Comparison of the various mapping methods

The PFGE methods have considerably broadened the span of species in which mapping and thus genetic analysis have become possible. The rapid rate of appearance of maps of very varied species, however, does not mean that these methods are easy to apply. Adjustment of the technique to a new strain generally requires a lot of work, and, when adapted, the task is heavy. This does not mean, either, that *in vivo* genetics is no longer useful. Most information concerning regulation of gene expression must be looked for in the original host, a study which is highly facilitated by the availability of mutants and transfer systems (natural or artificial). Also, when an *in vivo* mapping technique is possible, it is still generally faster and less expensive. Thus conjugation with Hfrs is still the best way to localize a gene on the *E. coli* chromo-some, and transduction is the easiest way to perform fine mapping. The situation may reverse for species for which a physical map has been constructed, since the identification of the functions of new characters is often concomitant with their cloning. The case of *E. coli* K12 (Section 3.2) is typical of such a transition situation. Access to the ordered genomic library allows a rough localization of the gene by hybridization to a set of clones covering the whole genome. The clones

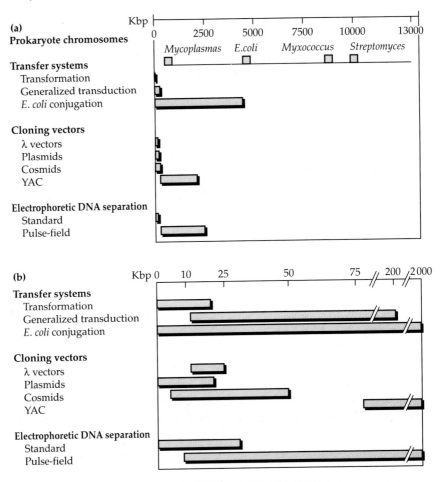

Fig. 14.7 Comparison of the DNA size scales involved in the different mapping techniques, as compared to the size range of prokaryote genomes. Two size scales are shown. (a) Range size of DNA fragments concerned. (b) Expanded scale.

to be tested are usually scarce enough (the rare-cut fragments) to be handled in one hybridization assay. More accurate mapping can be progressively pursued according to the refinement of the restriction map. It may be more efficient, although more expensive, to short-cut the *in vivo* genetic approach if dealing with a new species, especially if a very specific function is to be studied.

An improvement brought forth by the availability of physical mapping is the appearance of a map unit, the base (or kbp or Mbp), which is of universal usage and similar in all organisms. A comparison of the size scales at which the various mapping methods are accurate is shown in Fig. 14.7.

Construction of physical maps has provided evidence of several original genomic structures. While in most species circular maps of single chromosome structures have been arrived at, thus confirming the initial dogma as obtained from genetic data, cases which up to now appear to be exceptions have been described. The presence of large plasmids (≥50 kbp), as units distinct from the chromosome, is more easily and non-ambiguously established by PFGE physical mapping, while this is usually very difficult by conventional techniques, both molecular and genetic, since most of these molecules are still cryptic. Such large molecules appear fairly frequently. Maximal sizes observed reach the minimal chromosomal sizes known, indicating that they may contribute a relatively large amount of genetic information to the host (Table 14.3). The question then arises as to whether they are plasmids (molecules bearing dispensable information) or actual chromosomes, a situation forcing the conclusion that (some) prokaryotic karyotypes comprise several units ($n > 1$). Such a case has been non-ambiguously established. *Rhodobacter spheroides* has two large circular DNA molecules, between which the ribosomal RNA genes (*rrn*), considered as essential functions, are distributed as single copies (Table 14.3).

Linear plasmid and phage genomes (Chapters 3 and 4), as well as eukaryotic virus genomes, have already been described. PFGE techniques have now established that such structures exist also for much larger DNA molecules, classified as plasmids, e.g. in *Streptomyces* and *Borrelia* (Table 14.3). The unexpected discovery was the demonstration that the chromosome of one strain of *Borrelia burgdorferi*, which has both linear and circular plasmids, is itself linear (Table 14.3) (Chapter 2), while circular chromosomes have been described in other spirochaetes.

3.2 The *E. coli* K12 map

The case of *E. coli* K12, for which the large amount of information accumulated by conventional mapping procedures had already led to the elaboration of an accurate and largely documented map, typically enhances the advantage of available preexisting genetic data. It also demonstrates the complementarity of information obtained from the two approaches.

Since the first edition, in 1964, of a genetic linkage map showing 99 genes, seven further editions have been published, now positioning almost 1500 genes, representing 20–25% of the whole genetic information of the chromosome. About one-third of these genes have been cloned and sequenced, and their complete restriction maps are thus known.

In 1987, Kohara and coworkers, using classical cloning devices, obtained a complete physical map showing restriction sites for eight

(a)

(b)

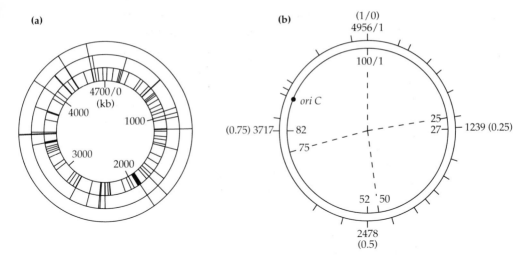

Fig. 14.8 Comparison of the genetic and physical maps of *E. coli* K12. (a) Construction of Kohara's map. The bars in the three concentric circles indicate (from innermost outwards) the regions that remained uncloned ('gaps') when the 1056, 2034 and lately, the 3400 library clones were analysed successfully. (b) Construction of Cantor's map aligned to the genetic map with the *thr* gene at 100/1 position. Inner circle: genetic scale; outer circle: physical scale, showing the *Not*I cleavage sites. Several regions of the genetic map have been distorted to match the physical map.

endonucleases (Fig. 14.8a). A genomic library, prepared in a modified λ phage used as cloning vector (Chapter 16), was contained in 1056 clones, each bearing 15–20-kbp-long inserts, the maximal size compatible with the vector and with the standard electrophoresis separation method. Procedures similar to those applied after PFGE, and the elaboration of adapted computer programs, led to the resolution of an almost complete map, reaching a size of 4700 kbp. Even then, a number of ambiguities or gaps remained, most of which were dealt with through analysis of another set of 2304 clones. Usage was then made of available partial sequences registered in gene banks, to confirm the already obtained map and, luckily, to elucidate ambiguities. One such example concerned the terminus region, for which a published sequence covering 470 kbp, the longest continuous sequenced stretch, allowed the corresponding gap to be filled. The final map, covered by a minimum of 381 selected clones from the λ library, still showed seven gaps, but the agreement of the physical map with the positions and published restriction patterns of cloned genes was excellent.

Simultaneously to this work, PFGE analysis performed by Cantor's group led to a 4600-kbp-long chromosome of which all ambiguities but one were solved by comparison with published partial restriction data. The whole map, covered by 22 *Not*I fragments, is mostly limited to information regarding the recognition sites for this enzyme. Correlation with the known linkage map was excellent as long as gene order was considered. Relative distances, however, differed slightly and alignment necessitated some distortions of the genetic map (Fig. 14.8b). One cause for these discrepancies probably lies in non-random crossing-over frequencies in certain regions of the chromosome.

Kohara's map is now used as a base into which all newly acquired sequence data are incorporated. An appropriate computer program prepared by Danchin's group allows the easing of the experimental work of mapping a new character by first working out its most probable location(s) through comparisons of restriction profiles. At present, more than one-sixth of the chromosome (comprising about 450 genes) has been sequenced and thus precisely mapped.

One accessory observation released by this work is the evidence for a very low level of polymorphism among the various *E coli* K12 isolates used by various laboratories from which restriction data are known. Most of it may be due to movement of mobile elements (ISs, transposons, phages). A consequence is the wider validity of the map than its unique use for the strain specimen from which it was constructed.

References

Kohara Y., Akiyama K. & Isono K. (1987) The physical map of the whole *E. coli* chromosome: application of a new strategy for rapid analysis and sorting of a large genomic library. *Cell*, 50, 495–508.

Smith C.L., Econome J.E., Schutt A., Klco S. & Cantor C.R. (1987) A physical map of the *E. coli* K12 genome. *Science*, 236, 1448–53.

15: Genetics as a Tool in Understanding Gene Expression and its Control: the Operon Model

Knowledge of genetic processes in prokaryotes and of their possible variations among species is *per se* an exciting field. Along with the development of bacterial genetics since the 1950s, another aspect of cellular function emerged, that of the means and pathways through which cells express their various metabolic and developmental capacities, and, moreover, how they control the level of expression of these functions, singly or as groups. In deciphering these processes, classical, and later molecular, genetics has played and still plays a key role. Other approaches, e.g. biochemistry and physiology, are essential complements which must never be underestimated. Molecular genetics, as will be demonstrated in this and the following chapter, is but a development, both fundamental and technological, of mostly genetic and biochemical knowledge.

Genetic tools available in the 1950s have largely contributed to the initial understanding of gene expression, studies which led to the concept of the operon, defining a unit of coordinately controlled genetic information; up to the present, this seems to be a general model for prokaryotes and their viruses.

1 The operon: early triumph of formal genetics

1.1 Precursors

When, in 1956, a first model was proposed to explain the biochemical events leading to expression of the hereditary characters inside a cell, it was built on very limited information. In brief, the model stated that each gene sequence specified the formation of an equivalent RNA sequence which, combined with proteins in structures called ribosomes, was 'translated' into a protein, usually an enzyme catalysing a biochemical transformation (Fig. 15.1a).

This model soon seemed to be in conflict with some of the experimental evidence becoming available. For instance, the similarity predicted by the model between the average base compositions of whole DNA and RNAs could never be observed. The existence of a new category of RNA molecules, characterized by their extreme instability, discovered at the same period, remained unexplained for some time as it did not fit into the proposed model. In parallel, in the late 1950s, Monod, studying the lactose catabolic pathway, demonstrated that a particular enzyme, β-galactosidase, was synthesized by the bacterium *E. coli* only when its substrate, lactose, was provided to the cells, and that its synthesis ceased when lactose was no longer available.

1.2 Analysis of mutants

Monod's complete working system comprised the early enzymes of lactose metabolism in *E. coli* K12, i.e. lactose permease and β-galactosidase, responsible for the cleavage of lactose into one glucose and one galactose residue. (In fact a third gene, for a transacetylase, is also present in this operon, but its function, still not clear, is not directly necessary for lactose metabolism and it was not considered in Monod's initial work.) Monod and his collaborators had established the strict and parallel dependence of *de novo* synthesis of the two enzymes on the presence of lactose, and the specificity of the phenomenon.

The deciphering of this 'adaptative process' was undertaken via a classical genetic approach, i.e. via the use of mutants. Two types (phenotypes) of mutants could be looked for, Lac⁻ ones, no longer capable of growing with lactose as unique carbon and energy source, and others which produced β-galactosidase even in the absence of the sugar in the growth medium. Selection took advantage of the previously demonstrated effects of lactose analogues, i.e. molecules showing particular effects equivalent to those of lactose but not useful to the cell. The discovery of these analogues provided assay conditions that made it possible to distinguish mutations in the different steps of the metabolic pathway. Thio-*o*-nitrophenyl-β-D-galactopyranoside, TONPG, a toxic competitor of lactose, allowed the isolation of permeation-deficient mutants through their resistance to the analogue. Mutants impaired in their β-galactosidase activity were detected by their incapacity to cleave *o*-nitrophenyl-β-D-galactopyranoside, ONPG, another competitor of lactose, the reaction yielding a yellow product. This reaction could also be used for quantitative activity assays. The induction of β-galactosidase by the addition of lactose could be mimicked by replacing this sugar by isopropylthio-β-D-galactoside, IPTG, said to be a gratituous inducer since it cannot be used by the cell; on solid medium, the presence of the enzyme could be visualized by addition of still another analogue, 5-bromo-4-chloro-3-indolyl-β-D-galactoside, XGal, a non-inducing chromogenic substance that is cleaved into a blue product, visible on the normally whitish colonies. Whilst, for instance, non-inducible colonies remained whitish, mutants producing the

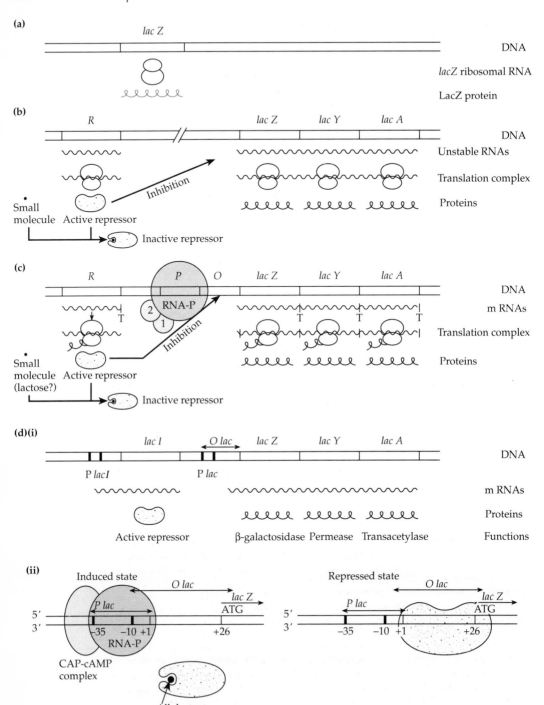

Fig. 15.1 (*See caption opposite.*)

Table 15.1 Phenotypic characteristics of various classes of Lac⁻ mutants in *E. coli* K12.

Class	Growth on IPTG + XGal mediumᵃ		Hydrolysis of ONPG		Toxicity of TONPG		Genetic locus affected
	+ lactose	+ lactose and glucose	+ lactose (or IPTG)	− lactose (or IPTG)	+ lactose (or IPTG)	− lactose (or IPTG)	
WT	Blue	White	+	−	+	−	−
I	−	White	−	−	−	−	*lacZ*
II	−	White	±	−	−	−	*lacY*
III-1	Blue	Blue	+	+	+	+	*lacO* or *lacI*
III-2	−	White	−	−	−	−	*lacI*

+ = growth; − = no growth. a, colour of the colonies on solid medium.

enzyme in the absence of inducer were blue, whether IPTG was added or not.

The mutants obtained could be classified in a finite number of categories (Table 1): (1) those affected only in β-galactosidase activity; (2) those affected only in lactose permease activity; and (3) those affected in both enzymic activities, which themselves could be separated into clones that synthesized the two enzymes continuously, i.e. even in the absence of lactose (constitutive mutants), and clones that never synthesized the enzymes at all (superrepressed mutants). The first two categories could be recognized as specifically mutated on the genes coding, respectively, for each of the two enzymes (structural genes for the corresponding proteins). Since mutants of the third class showed effects on both enzymes, it was considered that they were

Fig. 15.1. (*Opposite.*) The *E. coli* K12 lactose operon: the successive models. (a) The initial model of gene expression. 1950s. Gene *lacZ* allows the synthesis of a specific ribosomal RNA, which in turn leads to the synthesis of the LacZ protein. (b) The first model of an operon. The three genes *lacZ*, *lacY* and *lacA* are contiguous, and read into a single, intermediary, unstable RNA molecule which, with the ribosomes, leads to the synthesis of the three proteins. Gene R, by an unspecified process, allows synthesis of protein R, a repressor which inhibits reading of the *lac* unit. This inhibition is released if a small molecule (lactose?) inactivates it. (c) Further understanding of gene structure and expression. The three *lac* genes are contiguous, and preceeded on the chromosome by two 'genes', P (promoter) and O (operator) involved in their expression. They are transcribed into a single mRNA molecule by an RNA polymerase (RNA-P) assisted by two elements, a factor (1) insuring recognition of P and a small molecule (2) insuring efficiency of transcription. The mRNA possesses signals (T) indicating the end of each gene. It is translated, by a complex involving ribosomes, tRNAs and proteins (translational complex) into three independent proteins, LacZ, LacY and LacA. The repressor R, synthesized according to a similar scheme, prevents expression by binding to O, provoking steric hindrance for RNA-P fixation. The inducer is defined as above. (d) Present model of the *lac* operon, showing its structure (i) and regulation of expression (ii). See Section 3.2.2. for details.

affected outside the two structural genes. The possibility of double mutations, bearing deficiencies for both enzymes, could be excluded on the basis of their frequencies of appearance and the fact that reversions always occurred for both traits simultaneously.

1.3 Genetic localization and characterization of the mutations

During the same period, Wollman, Jacob and collaborators were studying the *E. coli* K12 conjugation system first described by Lederberg (Chapter 11). They had started to map genes on the chromosome and to isolate F-prime strains (Chapters 3 and 11). Amongst these were F' *lac* genes, in which the whole genetic *lac* system was inserted on the plasmid. Transferring such an F'*lac*$^+$ plasmid to an F$^-$Lac$^-$ recipient allowed the latter to recover a Lac$^+$ phenotype; both copies, *lac*$^+$ and *lac*$^-$, of the *lac* system remained stable in the cell. The structure was called a merodiploid (Chapter 11).

Conjugation was used to perform deletion and three-point-test mapping (Chapters 10 and 11) and thus to determine the number of loci implicated and their location and order on the chromosome. All mutants of interest here belonged to one of four loci, as illustrated in Table 15.1 and Fig. 15.1b.

Advantage was also taken of possible spontaneous transpositions of the *lac* region close to the *gal* locus or to the *tonB* gene (specifying sensitivity to phage T1 and colicins B and V) to prepare families of deletion mutants, since it had been observed that deletions were frequently obtained in the two latter regions. If *lac* was fused close to either, the deletion could extend over *lac*, and be visualized as either Lac$^-$Gal$^-$ or Lac$^-$T1^r. Also, it so happens that *tonB* and *gal* are transduced by the temperate phages ϕ80 and λ, respectively, a situation which enabled limited DNA transfer by specialized transduction rather than using conjugation.

Jacob and Monod (1961) took advantage of the F' system to test the interaction of various pairs of mutations. The rationale was that complementation (i.e. recovery of the wild-type phenotype) of one DNA by the other could be obtained only if the products of the genes involved were soluble. Different combinations of pairs of mutations, as shown in Table 15.2, allowed the positioning of mutations at different loci either in *cis* (on the same piece of DNA) or in *trans* (each on a separate DNA molecule). Genes *lacZ* and *lacY* were unambiguously shown to be the structural genes for β-galactosidase and permease, respectively, although the permease was not yet purified. The term cistron, previously proposed to refer to genetic units capable of complementing when organized in *trans* inside a cell, was then defined more precisely as a DNA unit coding for a protein, i.e. a structural gene. Ambiguity

Table 15.2 Merodiploid analysis of *E. coli* K12 Lac⁻ mutants.

Genotype[a]	β-galactosidase +IPTG	β-galactosidase −IPTG	Permease +IPTG	Permease −IPTG	Global phenotype	Interpretation
lacI⁺ lacZ⁺ lacY⁺	+	−	+	−	Lac⁺	
lacIᶜ lacZ⁺ lacY⁺	+	+	+	+	Lacᶜ	
lacI⁻ lacZ⁺ lacY⁺	−	−	−	−	Lac⁻	
lacI⁺ lacZ⁻ lacY⁺	−	−	+	−	Lac⁻	
lacI⁺ lacZ⁺ lacY⁻	+	−	−	−	Lac⁻	
lacOᶜ lacZ⁺ lacY⁺	+	+	+	+	Lacᶜ	
lacI⁺ lacZ⁻ lacY⁻ / *lacIᶜ lacZ⁺ lacY⁺*	+	−	+	−	Lac⁺	*lacI⁺* dominant over *lacIᶜ* in *trans* and in *cis*
lacI⁺ lacZ⁺ lacY⁺ / *lacIᶜ lacZ⁻ lacY⁻*	+	−	+	−	Lac⁺	
lacI⁻ lacZ⁺ lacY⁺ / *lacI⁺ lacZ⁺ lacY⁺*	−	−	−	−	Lac⁻	*lacI⁻* dominant over *lacI⁺* and *lacIᶜ* in *trans*
lacI⁻ lacZ⁺ lacY⁺ / *lacIᶜ lacZ⁺ lacY⁺*	−	−	−	−	Lac⁻	
lacOᶜ lacZ⁺ lacY⁺ / *lacO⁺ lacZ⁻ lacY⁻*	+	+	+	+	Lacᶜ	*lacOᶜ* dominant in *cis*
lacOᶜ lacZ⁻ lacY⁻ / *lacO⁺ lacZ⁺ lacY⁺*	+	−	+	−	Lac⁺	
lacOᶜ lacZ⁺ lacY⁻ / *lacO⁺ lacZ⁻ lacY⁺*	+	+	+	−	Lac⁺	
lacIᶜ lacO⁺ / *lacI⁻ lacOᶜ*	+	+	+	+	Lac⁺	*lacOᶜ* dominant in *cis* even in the presence of *lacI⁻*
lacI⁺ lacO⁺ / *lacI⁻ lacOᶜ*	+	+	+	+	Lacᶜ	

[a] All alleles not indicated are under the wild-type form.

later arose, however, concerning the difference between a gene and a cistron. The latter is nowadays seldom used, except when indicating polycistronic messenger RNAs (Section 3.2.1). In addition to dominance and *cis–trans* tests that delineated these two genes, biochemical analysis was important to show that, for instance, thermosensitive mutants in *lacZ* corresponded to a thermolabile β-galactosidase, or that material prepicitating with antiserum raised against purified β-galactosidase (cross-reacting material (Chapter 7)) was present in the cytoplasm of some of the *lacZ⁻* mutants.

The properties of mutants in the *lacI* and *lacO* loci (*I* and *O* stand for induction and operator, respectively) were altogether different. Constitutive mutants in *lacI* (*lacI^c*) were recessive with regard to *lacI$^+$*, whereas *lacI$^-$* mutants, which were unable to grow on lactose, produce β-galactosidase or allow lactose permeation, were dominant over both *lacI$^+$* and *lacI^c* mutants. This was interpreted as follows: *lacI^c* mutants lacked a molecule produced by the *lacI$^+$* allele that prevented in *trans* the synthesis of β-galactosidase (i.e. the expression of gene *lacZ*) when lactose, the inducer, was absent. *lacI$^-$* mutants behaved as superrepressed mutants, which would produce a molecule, the repressor I, that prevented the expression of gene *lacZ* even in the presence of the inducer (Table 15.2).

In contrast, *lacO^c* mutations were dominant in the *cis* position, that is, when they were on the same strand as the active *lacZ$^+$* or *lacY$^+$* alleles, even in the presence of a *lacI$^-$* allele, as if the repressor molecule needed a non-mutated *lacO$^+$* allele to be able to prevent expression of gene *lacZ$^+$* (Table 15.2). No soluble product which would have been capable of a *trans* effect was synthesized from *lacO* (or from *lacO^c*), meaning that the *lacO* DNA region itself must be implicated.

These results could be explained only if unstable molecules, intermediary between the genes and their protein products, were postulated. The existence of the mRNAs (Section 2.1) was thus predicted, yielding the first model of the operon, as depicted in Fig. 15.1b.

1.4 Experimental confirmation of the model

Besides the physiological and genetic data and the established reality of the genes, all other elements put forward to build the model were speculative. Thus the role suggested for the mRNA was proposed because the size and the instability of the small RNA molecules, which by then had been repeatedly demonstrated, correlated well with the requirements dictated from the kinetics of the enzymatic adaptation of the cells.

This model immediately seemed extremely fruitful in its potential. During the following decade, a large proportion of all research concerned with prokaryotic genetics was aimed at enriching it or at experimentally establishing its predictions in the few metabolic pathways which had been at the origin of the work. This proved to be a laborious task, since it was only in 1970 that repressors for two systems, the *lac* operon and the lysogenization system of phage λ, were purified. A year
later, the whole DNA region coding for the *lac* operon was isolated, and it was possible to have its function reproduced *in vitro*.

By then, new elements had been added to the model, yielding the scheme shown in Fig. 15.1c and d.

2 Gene expression, a rapid survey

Since it is not within the scope of this book to give a historical survey of the development of molecular biology, we shall 'jump' over 20–30 years, and come to the description of the present state of understanding of the organization and functioning of genes and operons. This knowledge has reached the level of fine molecular analysis, and has thus led to the definition of a large number of features (molecules, complexes, specific sequences or regions, etc.), which need to be defined in the general scheme of expression of genes and control of their expression. The steps and elements involved in this process will thus be rapidly surveyed before the actual description of the regulatory systems associated with the concept of operon is presented.

2.1 Functional organization of a gene

To be expressed, that is, to transform their informational content into functional molecules (RNAs or proteins), genes need several specific DNA regions in addition to those actually coding for the active molecules. The spatial organization of these regions is dictated by their respective roles (Fig. 15.2).

The stretch of DNA which actually codes for the active molecule is nowadays more often referred to as the coding region, or as the ORF (open reading frame) when it codes for a protein, although the word

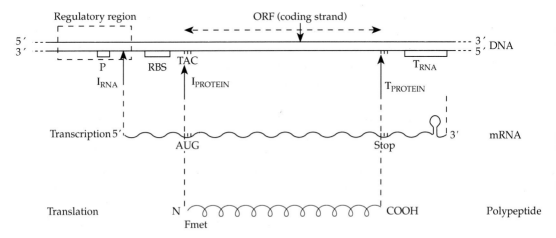

Fig. 15.2 Organization of a prokaryotic gene. I, initiation; P, promoter; RBS, ribosome binding site; T, termination; TAC and AUG, the transition initiation anti-triplet and codon (for Fmet).

gene is also ambiguously used to depict only this region of the unit. Expression of the gene includes two successive steps, transcription, the formation of an RNA, which is itself translated, if relevant, into a protein.

2.2 Transcription

In all cases, an RNA molecule must (first or only) be synthesized from the DNA coding region. The enzyme in charge, RNA polymerase, needs guidance as to where to start and end on the DNA molecule. The transcription initiation signal on the DNA molecule is called a promoter. The promoter works as a protein-binding region, recognized by the σ subunit of the RNA polymerase, thus promoting correct and efficient positioning of the core of the enzyme, which performs RNA polymerization in a process similar to polymerization of DNA by DNA polymerases (Chapter 2).

Besides this functional role, promoters are one of the key structures for the regulation of gene expression participating in the binding specificity of the σ factor. Different σ factors will recognize different promoter sequences, and a family of promoters will be recognized by only one σ factor. This allows differential expression of sets of genes, depending on the σ factor(s) available under given conditions. One well-known example is the development of spore formation in *B. subtilis*, a process that can be divided into a number of successive steps, each corresponding to the synthesis of new σ factors and thus to the expression of new sets of genes. Phages such as T4 use a slight variation of this scheme to discriminate with time the genes to be expressed (Chapter 4). Some promoters can bind RNA polymerase only if 'activated' by a specific protein, known as a positive activator (Section 3; Chapter 16).

A large number of promoter sequences corresponding, respectively, to the major σ factor or to minor ones, are now identified. This has led to the definition of consensus sequences composed of the most frequently occurring bases in each position. A frequently encountered one, initially recognized in *E. coli* K12, is a hyphenated region on the DNA including two elements of six bp each, starting respectively around −10 (this sequence is known as the Pribnow box from its discoverer) and −35 bp upstream from the first base of the transcript (negative backward numbering symbolizes the position of a base as oriented from the start and direction of reading) (Fig. 15.3a). It is present in a number of both Gram-positive and Gram-negative species, but not universally encountered. Some species belonging to the pseudomonads and the cyanobacteria possess genes for which the classical consensus promoter region can easily be detected and experimentally

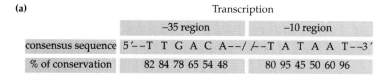

(a)

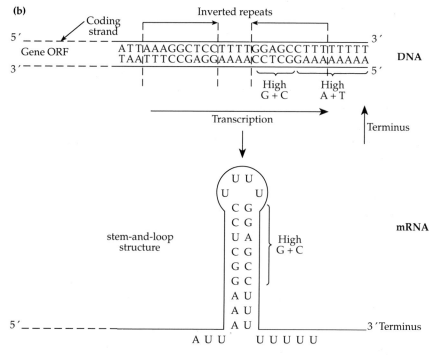

Fig. 15.3 Transcription signals in prokaryotic genes. (a) The consensus promoter. (b) Structure of a transcription termination region.

proved efficient, and others for which an equivalent sequence has not yet been defined. This is the case for about half of the more or less 200 genes that are now well known among several strains of cyanobacteria, with no apparent discrimination between regulated and constitutively expressed genes.

Termination of transcription is in many cases indicated by particular DNA sequences downstream of the coding region, such that a hairpin configuration will be formed on the extremity of the nascent RNA molecule (Fig. 15.3b). The general characteristic of these sequences is the presence of a G+C-rich stretch forming a palindromic inverted repeat, followed by a run of about six U residues (T–A pairs on the DNA) forming the very end of the RNA molecule. This poly-U sequence corresponds to A–U hydrogen bonds with the DNA, which have the weakest strength of all pairs. This feature, associated with a

pause introduced in the RNA polymerase movement by the presence
of the hairpin structure, contributes to the destabilization of the enzyme
from the DNA, causing transcription to stop. Two classes of terminators are known, although their structural differences are not clear.
Rho-dependent ones need a proteinic factor (a special helicase), ρ,
to efficiently separate the polymerase from its DNA substrate, while
others which are rho-independent do not. The RNA molecule is thus
liberated in the cytoplasm, and the DNA is ready for a new transcription process.

   The RNAs synthesized from protein-coding genes are thus always
longer than the actual ORF of the gene.

2.3 Translation

The large majority of known ORFs start with 5'AUG3', transcribed as a
translation-initiation codon (but GUG or UUG initiation codons are
encountered), leading to the positioning of a formyl-met (F-met) at the
N-terminal extremity of the protein to be synthesized. A stop signal
indicates the end of the ORF and corresponds on the mRNA to a
nonsense codon (not representing any amino acid), which induces
termination of translation (i.e. disruption of the ribosomal translation
complex).

   Translation itself is the synthesis of the protein specifically encoded
in the mRNA molecule to be translated. This process, the details of
which are outside the scope of this book, requires the participation
of a large number of constituents: ribosomes work as the stabilizing
machinery for the system; transfer RNAs do the actual translation since
they recognize the codons on the mRNA and associate to each codon
a specific amino acid; a peptidyltransferase covalently binds the successive amino acids to form the protein; many other proteins and cofactors
are required to control the various aspects of the process.

   A critical feature of translation is the correct recognition of the
initiation site on the mRNA sequence, and the in-frame, or in-phase,
reading of the ORF as given by the succession of three-letter codons.
This is achieved by the presence, upstream of the initiation codon,
of a particular base sequence on the mRNA molecule, specifically
recognized by the small ribosomal subunit, to form the initiation complex, allowing strict reciprocal positioning of the two elements. This
region, which is generally located seven bases upstream of the initial
AUG, is known as the ribosome-binding site (RBS), or the Shine –
Dalgarno (SD) region, from the researchers who first identified it in
E. coli. This sequence is a hexameric stretch of bases, 5'CCUCCU3', on
the mRNA, of which three to five are complementary to a region on
the ribosomal RNA of the small subunit of the ribosome.

Genes whose final product is an RNA (tRNAs, rRNAs) and which therefore do not need to be translated do not have such requirements.

3 Regulation of expression: operons and regulons

It has now been widely established that all organisms, except very small phages, have several means of modulating the expression of at least part of their genes, both in choice and level, as a response to specific needs dictated by environmental conditions. It is not the aim here to describe the various molecular devices available, since they are largely explained in many specialized books. Suffice it to recall that all successive steps of the expression process, transcription, translation and, if relevant, protein maturation, are possible targets for regulatory processes.

As we are concerned with genome organization, we shall focus on the regulatory process that has led to the concept of the operon. It is a widely used system among probably all prokaryotes and many bacteriophages. No such organization has yet been described in eukaryotes.

3.1 Principle of the operon-type regulatory process

As stated above, expression of a gene is initiated by the recognition by the RNA polymerase σ factor of the promoter of that gene. The most economical way of modulating the level of expression consists in modifying the accessibility or the affinity of σ for the promoter, either negatively or positively, so as to synthesize the corresponding protein only when appropriate. This is achieved through a reversible modification of the region immediately adjacent to and/or including the promoter (Figs 15.1b and 15.2). A specific regulatory protein, known as an activator or a repressor depending on its effect, binds to this regulatory region, an event resulting in either enhancement (activator) or hindrance (repressor) in the capacity for the RNA polymerase to detect the promoter.

In the simplest cases, specific responses to environmental changes and reversibility of the regulating process are ensured through the effect of an effector, inducer or corepressor, a small molecule directly involved in the environmental change, which triggers, by allosteric modification of the regulatory protein conformation (i.e. modification of the structural configuration of the protein as a result of its association with the effector), its binding to or release from the promoter region. Studies concerned with the structural characteristics of regulator proteins and with their conformational modifications allowing these functions are presently well developed, but are beyond the scope of this book.

Both mechanisms can be involved in either of the two opposing types of control, positive and negative, known as repression (prevention of gene expression due to the presence of the corepressor) and induction (trigger of gene expression due to the presence of the inducer). There is no correlation between the mechanism of control at the gene level (positive or negative) and the type of control (positive or negative) at the metabolic level. The effector is usually either a substrate or an end-product of the reaction involved in the regulated function, depending on whether the reaction is catabolic or anabolic. Thus the initial step(s) of utilization of a sugar by a heterotrophic bacterium is often regulated, positively, by the sugar itself. In contrast, synthesis of metabolites (amino acids, bases, vitamins, etc.) is negatively regulated by the end-product, e.g. the amino acid, if present in excess.

3.2 The regulatory units: operons and regulons

Thus the minimal requirements to ensure regulation of expression of a gene include specific DNA regions in addition to the actual 'reading unit', and one or several proteins. If these elements were utilized for the control of a genetic unit limited to a single gene, in spite of its high effectiveness, such a situation would have two important disadvantages. It would be costly to the cell, since one regulatory protein would be needed for the regulated synthesis of each metabolic protein. Secondly, it would be difficult for the cell to ensure coordinated expression of genes intimately associated in a metabolic pathway. Prokaryotes have developed two supraorganizations which allow these drawbacks to be overcome.

3.2.1 *The operon*

The most evident supraorganization is the association of a set of genes into a single regulatory unit, the operon, on which a single copy of the regulatory protein can simultaneously (and thus coordinately) control the different components (Fig. 15.4a). This is the structure of the *lac* operon, the first to be described, which includes the three enzymes involved in the early steps of lactose catabolism (Sections 1 and 3.2.2). In such a system, several coding regions are located adjacent to each other, and a single set of expression signals (promoter, regulatory region, transcription termination) controls transcription of the whole set, which is also designated a transcriptional unit. A single mRNA is synthesized for the whole unit. It is called polycistronic, the name

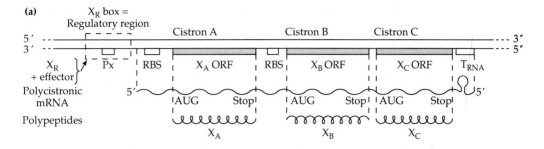

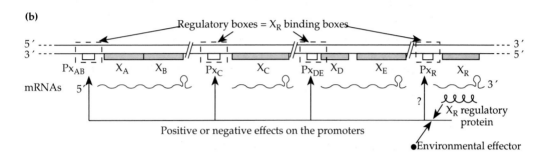

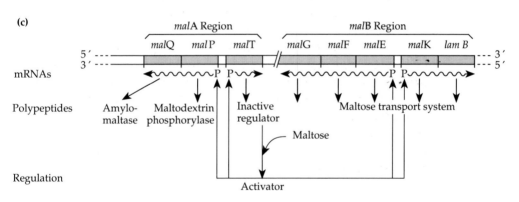

Fig. 15.4 Organization of complex genetic units. The lengths of the different elements have not been drawn to scale. (a) Organization of an operon. (b) Organization of a regulon. (c) The *E. coli* K12 maltose regulon.

evoking the individual coding units, or cistrons (Section 1.3), composing the structure. Depending (often) on the length of the intergenic region separating two adjacent coding regions, RBS sites may or may not be present upstream of each coding region.

Besides the above-cited advantages, this organization tends to superimpose yet another level of regulation. Indeed, it follows from the orientated occurrence of both transcription and translation along the unit that possible fall-off of either RNA polymerase or ribosomes may lead to a decreasing gradient of expression from initial to down-stream cistrons. This difference, known as the polar effect, may be an important part of the regulatory process. In fact, it is the only known one functioning in RNA bacteriophage development (Chapter 4).

Another consequence of such an organization concerns the effects of mutations. Point mutations and small in-frame deletions or additions in each coding region will generally affect only that cistron. In contrast, a nonsense mutation in one of the genes of an operon will lead to the premature termination of translation, i.e. fall-off of the ribosomes at the site of the mutation, thus hindering translation of the distal (downstream) part of the gene itself but also of the unit. Translation of downstream gene(s) is more or less affected depending on the position of the nonsense mutation with respect to the initiation codon of the next gene, the level of translation decreasing with this distance, thus justifying the notion of polar effect given to this phenomenon. A short distance favours maintenance of the ribosomes, which will 'jump' to the next initiation site and translate it. Deliberate constructions of mutations aimed at elucidating the transcriptional organization of a group of functionally related genes will be described in Chapter 16.

3.2.2 *The lactose operon model: present state of the art*

Interpretation of the observations presented in Section 1 and of others not reviewed here led to the present model depicting expression of the *lac* operon as presented in Fig. 15.1c. *lacI*, expressed independently of the rest of the system, codes for a soluble repressor protein, I, which is active on the regulatory DNA region, or operator, *lacO*, upstream of the coding regions for *lacZ* and *lacY*. A single mRNA, here tricistronic (including the *lacA* cistron also), which has been transcribed from promoter *lacP* located inside the *lacO* region, is impeded by the fixation of repressor I on *lacO*. When lactose is present, its binding (or rather the binding of an isomeric form of the sugar, allolactose) to a specific site on the repressor protein decreases the affinity of the repressor for *lacO*. The allolactose – repressor I complex can then be displaced by the σ factor, and transcription of the whole unit can take place. Depletion of lactose will again allow the repressor to bind to *lacO* and stop expression of the operon and hence synthesis of enzymes then useless to the cell.

lacI⁻ mutations can be of two sorts, either preventing synthesis of the repressor or producing synthesis of a repressor with a lower affinity for *lacO*. *lacIˢ* mutations mostly correspond to repressor molecules on which the allolactose-binding site is inoperable, thus making it impossible to induce expression. The modified operator in *lacOᶜ* mutants does not allow stable binding of I, and hence lactose is no longer necessary to trigger expression.

The system is in fact more complex, since a second regulatory process is superimposed on that controlled by lactose. The σ factor of *E. coli* has a low affinity for *lacP*, and efficient transcription requires, in addition, that a positive regulator, in the form of a protein specifically associated with cyclic AMP, the cAMP – CAP complex, is bound to the region upstream of the operon. This peculiarity is at the origin of the well-known effect of catabolic repression, or glucose effect, which consists in the inhibition of the utilization of lactose (or other sugars) as long as glucose is present. Glucose, by preventing the synthesis of cAMP, limits the possibility of making cAMP – CAP complexes, and hence of having efficient expression of the *lac* operon.

3.2.3 *The regulon*

A regulon (Fig. 15.4b) covers a group of operons and/or single genes related by function and submitted to the same regulatory process, but dispersed over the genomic map. All units, otherwise complete for their transcription and translation signals, possess similar regulatory DNA regions, on which the same regulator protein can specifically bind. The first example described was the maltose regulon of *E. coli* K12 in the 1970s, which comprises four units of one to three cistrons each, located at 74 and 90 min on the chromosome (Fig. 15.4c).

The regulon is the simplest of all global regulatory systems. Responses to modifications of nutrient supply (catabolic repression, stringent response to nutrient shift-down, etc.) or to stresses (heat, salt, etc.) are examples. One such system, the understanding of which has made use of many genetic and molecular advances, will be described in Chapter 16.

References

Jacob F. & Monod J. (1961) Genetic regulatory mechanism in the synthesis of proteins. *Journal of Molecular Biology*, 3, 318–56.

16: Genetics as a Tool in Unravelling the Molecular Organization of Genetic Functions

As in any field of human activity, each advance acquired in genetics can in turn be exploited for further unravelling of genetic knowledge. A direct demonstration of the extent of these applications would be tedious, as repeating previous chapters, and most probably incomplete. We shall thus choose to illustrate this process through the discussion of a single example, typical in the large span of both classical and molecular genetic tools that have been used to promote its understanding. The description of the relevant methodological principles and the rationale for the construction of the corresponding tools are presented alongside the development of the example as appendices given at the end of the chapter.

The function chosen will be nitrogen fixation. Although nitrogen gas, N_2, is the most abundant source of nitrogen, its direct utilization is exclusively limited to certain eubacteria and archaebacteria – hence the frequent dependence of plants, the best known being legumes, especially under natural conditions – on N_2-fixing bacteria for optimal supply of assimilable nitrogenous nutrients. Lichens, which also fix nitrogen, do so thanks to the cyanobacterium that lives in symbiosis with the fungus.

Nitrogen fixation consists of the enzymic conversion of N_2 to NH_3. Most bacteria capable of nitrogen fixation, also called diazotrophs, can be found in the soil, in fresh or salted waters or in the hindgut of termites. Many diazotrophs fix N_2 as free organisms, but the most efficient do so in symbiotic associations with plants: they are the Gram-negative *Rhizobium*, *Bradyrhizobium* and some cyanobacteria, and the actinomycete *Franckia*. Some archaebacteria also possess this capacity. None of them, however, is an obligate diazotroph. Other possible nitrogen sources are the NO_3^- and NH_4 radicals (referred to as combined nitrogen) and organic molecules such as amino acids.

The property shared by all diazotrophs is that they contain a complex Fe–Mo enzyme, called nitrogenase, which catalyses the conversion of $N{=}N$ to NH_3. The nitrogenases from all these organisms are very similar. An important common feature is their inactivation by molecular oxygen, a somewhat remarkable trait, since many of these organisms are obligate aerobes while cyanobacteria actually draw their energy from O_2-evolving photosynthesis.

The N$_2$-fixing function is symbolized by the abbreviation Nif and the genes by *nif*.

1 Functional identification of the *nif* genetic system of *Klebsiella pneumoniae*

K. pneumoniae is a facultative anaerobe, belonging to the enteric bacteria and closely related to *E. coli*, which is able to perform nitrogen fixation as a free living organism under anaerobic conditions.

1.1 Determination of the genes involved in nitrogen fixation

1.1.1 *Nif⁻ mutants*

As in any other genetic analysis before the recombinant DNA era, the availability of Nif⁻ mutants was the only means of, and still is a key tool in, deciphering the genetic organization of this function. The first Nif⁻ mutants, defined as unable to grow on a medium devoid of combined nitrogen, were obtained through classical mutagenesis and penicillin enrichment (Chapter 7). Using gene transfer techniques (conjugation and transduction had been described in *K. pneumoniae* in the early 1970s), it had been shown that all mutations were clustered in a region located next to the operon for histidine biosynthesis.

Later, this situation was used to isolate a conjugative R' plasmid (Chapter 11), pRD1, bearing a fragment of the *K. pneumoniae* genome encompassing the *his* operon and the *nif* cluster, pRD1*his,nif*. (More reproducible and more efficient methods of transferring genes on an accessory vector are now widely developed (Appendix 1)). pRD1 can replicate in both *E. coli* K12 and *K. pneumoniae*. When a wild-type copy of this DNA region is transferred into a His⁻ *E. coli*, it can complement the His⁻ mutation (this allows isolation of the exconjugants), but does not allow N$_2$ fixation in this bacterium. However, it confers a specific phenotype on this host strain, i.e. the ability to form large, opaque colonies on a medium supplemented with aspartate as sole nitrogen source, whereas non-recombinant cells grow very poorly and yield small translucent colonies. Nif⁻ mutations generated on pRD1 in *E. coli*, while maintaining the His⁺ phenotype, yielded colonies of wild-type morphology, susceptible to easy screening.

This property was widely used to make Nif⁻ mutants by transposon mutagenesis. Since no convenient suicide vector (Chapter 13) was available in *K. pneumoniae*, this was achieved by preparing a collection of transposon insertions in the *nif* cluster present on pRD1 in *E. coli* K12. Transposons Tn5, Tn7, Tn10 and Mu were thus utilized.

Another property of transposons, mostly of Mu and Tn10, was also

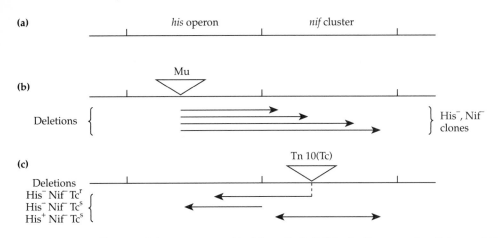

Fig. 16.1 Deletions generated by Mu or Tn10 (Tc^r) in the *K. pneumoniae nif* cluster. (a) The *nif-his* region. (b) Deletions generated by a Mu inserted in the *his* operon. A set of clones bearing deletions of increasing sizes encompassing part of both *his* and *nif* genes (i.e. Nif$^-$ His$^-$) are prepared. (c) Deletions generated by a Tn10 inserted in the *nif* cluster.

taken advantage of, i.e. the ability to generate deletions (Chapter 13). For instance, if Mu transposes into a *K. pneumoniae his* gene on a pRD1*his,nif* plasmid (Fig. 16.1a) present in *E. coli*, a series of deletions encompassing variable lengths of the *nif* cluster can be recovered by isolating colonies that show a His$^-$ Nif$^-$ phenotype in a His$^-$ *E. coli* (Fig. 16.1). All deletions start in the *his* gene, at the site of insertion of Mu (Fig. 16.1b). Conversely, *his* deletions have been screened from Nif$^-$ mutants resulting from Tn10 insertions at different locations in the *nif* cluster. Some deletions generated by Tn10 are non-adjacent to the transposon insertion site, while others may have also lost the transposon (Fig. 16.1c). The mutated genes thus created were subsequently integrated by homologous recombination (Chapter 9) into the chromosome of *K. pneumoniae*. This was achieved by transduction (Chapter 12) from *E. coli* to *K. pneumoniae* by bacteriophage P1, which can infect both strains, and selection on both the resistance borne by the transposon and the Nif$^-$ phenotype. This mutagenic process has been termed inverse, or reverse, genetics (Chapter 13) (Appendix 2).

1.2 Numbering and mapping *nif* genes

Determination of the number of genes involved in the Nif function was performed by complementation (Chapter 15) and recombination analysis of crosses between pairs of Nif$^-$ mutants. Regular mapping procedures, after transfer by transduction or conjugation and analysis of recombinants, enabled the ordering of the genes. However, rapid

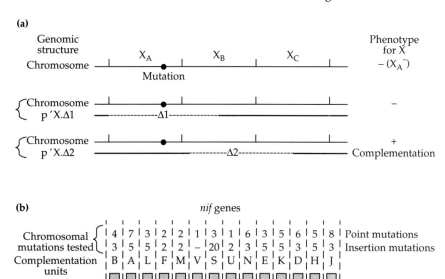

Fig. 16.2 Deletion mapping by complementation. (a) Principle of the method. (b) Map of *nif* genes obtained by ordering Mu deletions. Deletions (bottom part) were mapped against point and insertion mutations (upper part) present in the *K. pneumoniae* chromosome. Figures indicate the number of mutants tested.

progress in mapping was performed by the deletion mapping method (Appendix 3). pRD1*his*,*nif* derivatives on which deletions had been generated were transferred into a series of recipient *K. pneumoniae* clones, each carrying a Nif⁻ point mutation or an insertional mutation. Complementation, detected as growth of the exconjugants in the absence of combined nitrogen, occurred only when the deletion on the plasmid did not overlap the gene in which the chromosomal mutation was located (Fig. 16.2a). A very large number of mutations and deletions were ordered by this method (Fig. 16.2b). A less rapid but more precise method consisted in transferring by transduction the DNA bearing the point or insertional mutations into cells of various clones with deleted *nif* regions. The capacity/incapacity to obtain a wild-type *nif* cluster after recombination between the two DNAs allowed the precise positioning of the point or insertional mutations versus the deletion ones. First 11 and then 13 genes were thus detected in 1978 and 1980, respectively.

1.3 Functional organization of the *nif* cluster

1.3.1 *Determination of transcriptional units*

Transcriptional units can be determined by taking advantage of the polar effect of an insertional mutation in a given cistron on the expression of other cistrons organized in the same operon and located downstream from the mutated one (Chapter 15). Standard complementation analysis is performed between clones bearing point mutations on the one hand and insertion mutations on the other. Besides allowing detection of the different genes when assayed against appropriate point mutations, as seen above, the polar effect induced by the insertion will also highlight the possible presence of multiple transcriptional units (Chapter 15) and their orientation of reading. Figure 16.3a illustrates this strategy. Each of a series of plasmids bearing a transposon in one of the genes it carries, X_A to X_D, is introduced into a series of recipient clones, each having a point mutation in one of the same genes. If genes X_A, X_B and X_C belong to a single transcriptional unit read in the indicated order, insertion of a transposon in X_A, proximal to the promoter, prevents transcription of the downstream cistrons, and no complementation will be possible by any of the corresponding alleles, X_A^-, X_B^- or X_C^-, bearing point mutations. An insertional mutation present in a gene further downstream of the transcriptional unit, X_B, will similarly prevent complementation of all distal cistrons on this unit, e.g. X_C, but will permit complementation of the cistrons located between the promoter and this gene, e.g. X_A.

As shown in Fig. 16.3b, this analysis, performed in the 1970s, revealed a complex transcriptional organization for the *nif* cluster, comprising both genes read individually and operons. Obviously, the limit of resolution of the method depends on the availability of mutants. Thus genes *nifB*, *A*, *M* and *N*, considered then as monocistronic units (Fig. 16.3b), do in fact each belong to operons, the other cistrons of which were not determined at that time.

Such tests do not yield information as to the orientation of transcription of the (seemingly) monocistronic units. This was obtained by investigating whether cross-reacting material, i.e. non-active protein material (CRM (Chapter 7)), was made by a series of clones carrying frameshift (nonsense or insertion) mutations located at different sites inside the gene; those located toward the upstream extremity (close to the promoter) were recognized in that they abolish all peptide synthesis, as indicated by the absence of CRM. Direct analysis of DNA sequences (Section 1.3.2) and the use of gene fusions (Section 2.1), made available by molecular techniques, are now more conveniently used to gather such information.

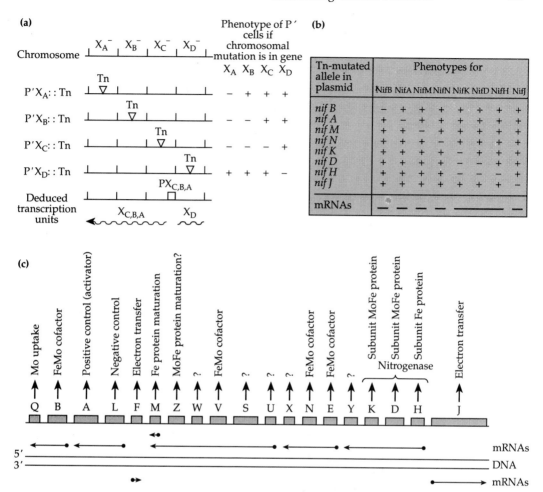

Fig. 16.3 Use of transposon insertions to determine transcriptional organization (units, orientation) by complementation. (a) Principle of the method. Genomic structure and phenotype of different plasmid prime derivatives of chromosomal mutations of genes *X* involved in the same metabolic process. (b) Determination of the transcriptional organization of some *nif* cistrons of *K. pneumoniae*. (c) Transcriptional organization of the whole *nif* cluster of *K. pneumoniae*.

1.3.2 *Exhaustive detection of ORFs*

As mentioned, some genes could not be detected by mutant analysis. Their existence could be revealed only by the determination of potential ORFs (Chapter 15), detected by sequence analysis of the different parts of the *nif* cluster (Appendix 4). In particular, continuous open reading frames, flanked by proper expression signals (Chapter 15), were good candidates for potential genes.

Thus, in 1986, 17 genes organized in eight transcriptional units were known, and two more were added to the map in 1987 (Fig. 16.3c).

The three subunits of nitrogenase are produced by three cistrons belonging to a *nifH,D,K* operon.

The actual functions of some of the gene products were determined by the biochemical comparison, e.g. enzyme activities, of mutants. However, most functions were not detectable by this method. Different genetic methods used to elucidate some functions will be described below.

2 Genes involved in nitrogen fixation in other diazotrophs

2.1 *Azotobacter vinelandii*

A. vinelandii is not only a free-living diazotroph, but is also an obligate aerobe, a case which raises the question, still unanswered, of how its nitrogenase can be protected against molecular oxygen.

The method used to study the *nif* genetic system in this strain took advantage of the homologies of all previously known nitrogenases. A library of chromosomal DNA from *A. vinelandii*, i.e. a family of plasmids bearing random fragments of the whole chromosomal DNA in a number sufficient to statistically encompass the whole genome, was constituted (Appendix 5). This library was used to select the DNA region carrying the genes looked for by probing it, i.e. each individual plasmid present in independent clones of an *E. coli* strain, with a fragment of DNA corresponding to the *nifH*, *D* or *K* genes from *K. pneumoniae* (Appendix 6).

Not only did the three probes hybridize with some clones from the library, but a clone was found which hybridized with all of them, suggesting both the presence of three genes homologous to each of the *K. pneumoniae* ones and thus of a similar multimeric enzyme in *A. vinelandii*, and a close linkage of the three genes, since they could be borne by a single fragment of DNA. Detailed analysis (DNA sequencing (Appendix 4) and interspecies complementation) confirmed that the three genes of *A. vinelandii* were homologous to the *K. pneumoniae* ones, and showed that they were linked in the same order.

Having established the restriction map (Chapters 6 and 14) of the DNA region (Appendix 7), it was possible to make a controlled deletion encompassing all three genes (Fig. 16.4): cleavage of the vector-borne fragment by appropriately chosen restriction enzymes allowed a defined part of the cloned DNA fragment to be cut out. The deleted fragment was replaced by a gene for an antibiotic resistance, here Km^r. This gene, referred to as a cassette (Appendices 2 and 8), served as a direct selection marker. This modified plasmid was then used to transform *A. vinelandii* (this strain is naturally transformable). Homologous recombination allowed the replacement of the wild-type region by the

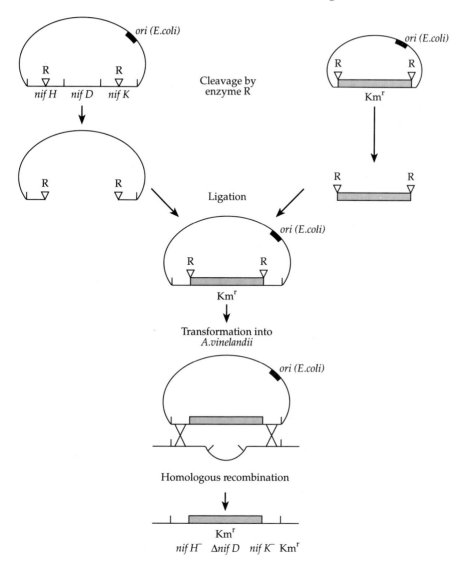

Fig. 16.4 Construction of a deletion mutation in the *nifH,D,K* operon of *Azotobacter vinelandii*.

deleted one (recombinants being selected through their acquisition of the Km^r marker), resulting in the absence of production of the Fe–Mo nitrogenase in the transformants. However, this mutated strain remained able to fix nitrogen, provided Mo was omitted from the medium. Further mutagenesis of this strain led to the discovery of a new nitrogenase with a vanadium cofactor. Recently a third one, with an Fe cofactor, was detected (Fig. 16.5).

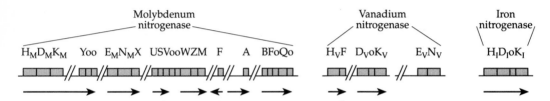

Fig. 16.5 Genetic organization of the genes for the three nitrogenases of *A. vinelandii*.
o, ORF of unknown function; Capital letters, *nif* genes.

2.2 Heterocystous cyanobacteria

All heterocystous cyanobacterial strains are diazotrophs, with an external morphology of filaments, composed of a large number (up to several hundred) of individual cells held together by a polysaccharidic sheath. These strains solve the problem of the sensitivity of nitrogen fixation to oxygen, which these cells permanently produce as a result of their oxygenic photosynthetic activity, by compartmentalizing these two incompatible processes. A lack of combined nitrogen in their growth medium induces the differentiation of a proportion of the cells along the filament into heterocysts, in which both nitrogenase synthesis is induced and oxygenic photosynthesis is blocked. The latter effect, besides depriving heterocysts of their energy supply (in fact only partially, thanks to a particular, non-oxygenic functioning of the photosynthetic apparatus), results in arrest of their CO_2 assimilation capacity, and thus in exhaustion of their endogenous carbon reserves. Exchanges of carbohydrates, synthesized from CO_2 by the vegetative cells, and of NH_2-containing molecules, made in heterocysts from N_2 reduction, ensure proper distribution of nutrients along the filament.

The *nifH, D* and *K* genes from one such cyanobacterium, *Anabaena variabilis*, were identified by hybridization with those of *K. pneumoniae* used as probes, confirming the homology of these genes among different species. However, cloning the genes from a genomic library constructed with DNA extracted from a suspension of vegetative cells, i.e. grown in a medium containing combined nitrogen, indicated that the organization on the chomosome was different from the previously known cases. While all three *nif* probes could hybridize with a single clone, indicating a cluster, another clone hybridized only with the *nifD* probe, suggesting the presence, elsewhere, of a second copy of this gene. When the same analysis was performed on DNA extracted from heterocysts, this second copy could no longer be found.

Restriction mapping (Appendix 7) and sequence analyses (Appendices 2 and 8) of the regions encompassing all the detected *nif* genes provided the clue to this unexpected situation: the *nifD* gene is

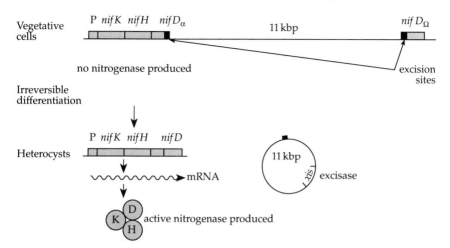

Fig. 16.6 Chromosomal re-organization of the *nifKHD* operon of *Anabaena variabilis* leading to efficient transcription in heterocysts.

interrupted, in the chromosomes of vegetative cells, by a stretch of 11 kbp, which is excised in the differentiated, functional heterocysts (Fig. 16.6). The size of this intervening region resulted in the two halves of the *nifD* gene being on two separate fragments in the library. This explains why two clones could be recognized by the *nifD* probe used, which covered the whole gene.

It has since been shown that the 11-kbp segment excises itself in the manner of a mobile element (Chapter 5), since the process does not require homologous recombination and is performed by an 'excisase', a specific endonuclease, whose gene is located inside the fragment. This complex process is used as a regulatory pathway by *Anabaena variabilis*.

All this information was obtained after analyses performed either in *E. coli* or *in vitro*, on DNA isolated from both types of cells and integrated into cloning vectors (Appendix 1). No DNA transfer system was then available in *Anabaena*, thus preventing any easy use of mutants, classical genetic analysis or reverse genetics (Section 1.1) as could be applied for *A. vinelandii* or *Rhizobium*.

2.3 The symbiotic Rhizobiaceae

Rhizobium and *Bradyrhizobium* are two different genera of Gram-negative bacteria that fix nitrogen upon development of symbiotic nodules in the roots of specific legumes. Their economic importance (they provide 70% of the total amount of reduction of nitrogen gas from the air) has prompted numerous studies.

Applications of methodologies similar to those described above (hybridization of genomic libraries with the *K. pneumoniae* probes, restriction and sequence analysis, gene inactivation by insertion of cassettes) enabled the cloning of some *nif* genes, and their location and mode of expression. In the fast growing *Rhizobium* species, the *nifH,D,K* genes are contiguous, cotranscribed (Chapter 15) and located on a large plasmid (pSym) (Chapter 3). Other *nif* genes are scattered on the same plasmid. In the slow-growing. *Bradyrhizobium japonicum*, which nodulates soybean, *nifH* and *nifD,K* are located on the chromosome, separated by 17 kb, with *nifE,N* and *nifB* in between (Fig. 16.7).

Reverse genetics was applied to the cloned *nif* genes (Appendix 2) and allowed the isolation of Nif⁻ mutants, obtained after transfer by conjugation and stabilization by recombination into the *Rhizobium* genome of the mutated genes. Some mutants nodulated their host although they did not reduce nitrogen, indicating that the two functions, symbiosis and N₂ fixation, were contained in different genetic systems and could be uncoupled.

Genes involved in nodulation, denoted *nod*, have been detected by transposon mutagenesis (Chapter 13), using a strategy that highlights the extreme usefulness of transposons in deciphering complex systems. To detect a mutant impaired in any step of symbiosis, sterilized germinating seeds must be individually inoculated by a single clone each from a mutagenized *Rhizobium* culture and incubated in a medium devoid of combined nitrogen. The screening criterion looked for is the limited development of germinating seeds infected with non-nodulating mutants from the *Rhizobium* culture, a characteristic which does not allow any selective advantage. Due to this absence of direct phenotypic selection, the screening process would have implied analysis of at least thousands of clones, even after mutagenesis of the bacteria, a colossal undertaking when considering the complexity of the detection system. Transposon mutagenesis provided a very efficient tool in yielding both a high enough proportion of mutants and a preliminary screening. The bacterial culture was mutagenized with a transposon. The transposon-

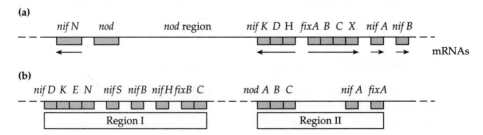

Fig. 16.7 Organization of some *nif* genes involved in nitrogen fixation in two genera of Rhizobiacae. (a) *Rhizobium meliloti*. Vegetative cells. (b) *Bradyrhizobium japonicum*. Heterocyst.

borne marker was used to select the mutagenized cells from the whole bacterial population (approximately $1/10^4$ to $1/10^5$ cells) and these were then used to inoculate the plant seeds. Only the problem of finding the mutants specifically impaired in the nodulation genetic system, which could be expected to range between 0.1 and 1%, remained, since the absence of plant development could also result from seeds infected with mutants impaired in any other step of the nitrogen fixation process, e.g. *nif⁻* clones.

Appropriate suicide vectors (Chapter 13) had first to be constructed. The first one, pJB4JI, was an IncP::Mu::Tn5 plasmid, which is not maintained stably in *Rhizobium* cells. The presence of Mu seems to be responsible for this, as yet unclear, phenomenon. Another more recent and widely used suicide vector is a small mobilizable one, pSUP1011, which contains the *oriV* of an *E. coli*-specific plasmid (a colE1 type) that is not recognized by the *Rhizobium* replicating system, the *oriT* (Chapter 3) from RP4 and a Tn5 copy. The plasmid is maintained in a strain of *E. coli* which harbours, integrated in its chromosome, a copy of RP4 deprived of its resistance genes but with intact *tra* functions. Plasmid pSUP1011 can thus be mobilized by the *tra* functions of RP4 and transferred to a *Rhizobium* recipient during a heterologous mating. Using either of the transposon-bearing plasmids, transposition of the Tn5 on the *Rhizobium* genome (chromosome or endogenous plasmids) will result in the stabilization of its marker in the cell, allowing selection of mutagenized clones that have acquired this antibiotic resistance character. The two methods have yielded mutants of two different phenotypes, Nod⁻, which is unable to nodulate the legume, and Fix⁻, which is deficient in nitrogen fixation. The latter led to the discovery of a set of genes that are not homologous to the *K. pneumoniae nif* genes (Fig. 16.7).

3 Regulation of nitrogen fixation in *K. pneumoniae*: from Nif-specific to general nitrogenous metabolism regulation

As mentioned, nitrogenase is an enzymic complex that is irreversibly inactivated by oxygen. In addition, the presence of NH_3 and O_2 in the culture medium of *K. pneumoniae* completely represses N_2 fixation. Elucidation of the mechanism of this regulation, i.e. determination as to whether it takes place at the level of transcription of *nif* genes, has been greatly facilitated by the utilization of gene fusion (Appendix 9). This technique, which consists of the introduction of a gene (called a reporter gene) whose expression is easily detected inside the genetic system under study, is used every time direct analysis of expression of the genes under study is experimentally difficult. *lacZ* is often used for this purpose.

3.1 Negative regulation of the *nif* cluster by NH₃ and O₂

A *lacZ* reporter gene (i.e. *lacZ* devoid of its promoter) was fused to the *nif* operon by inserting it in the same orientation, either between the *nif* promoter and the *nifH* ORF (the first cistron of the operon) without preventing expression of the *nif* genes, or inside *nifH* (Fig. 16.8a). The modified operon was integrated in place of the chromosomal one by transfer into wild-type *K. pneumoniae* and homologous recombination between both *nif* regions (a marker allowing direct selection, not shown here, was also introduced). The *lacZ* gene could thus be read from the *nif* promoter, its regulation of transcription being that of the *nif* operon. The *K. pneumoniae* cells were incubated under various conditions known to influence N₂ fixation activity, and expression of *nifH,D,K* was assayed by measuring the β-galactosidase activity.

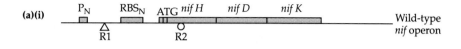

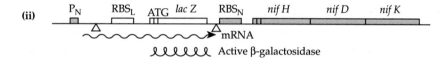

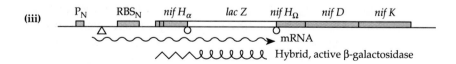

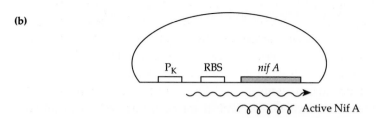

Fig. 16.8 Different fusions used to study *nifH,D,K* regulation of expression in *K. pneumoniae*. (a) Insertion of a promoter-less *lacZ* upstream of (ii) or inside (iii) *nifH* of the *nifHDK* operon (i). (b) Insertion of a promoter-less *nifA* downstream of the constitutive promoter, P_K, of a Km^r gene. The Km^r gene originated from Tn5 or Tn903. P, promoter; RBS, ribosome binding site; R1, R2, cleavage sites for restriction enzymes R1 and R2; ATG, the translation initiation triplet. L and N subscripts refer to *lac* and *nif* systems, respectively.

Table 16.1 Regulation of the *nifH,D,K* operon by NH_3 and O_2 in *K. pneumoniae*. β-galactosidase activities were measured in strains harbouring different *lacZ* fusions incubated in the presence or the absence of NH_3 and/or O_2.

	Medium			
	---	---	---	---
Strain	$+NH_3$ $+O_2$	$+NH_3$ $-O_2$	$-NH_3$ $+O_2$	$-NH_3$ $-O_2$
A WT/*nifH–lacZ*	−	−	−	+
B WT/*nifH–lacZ* (P$_k$–*nifA*)	+	++	+	++
C WT(*nifL–lacZ*)	−	−	+	+
D *nifA*⁻(*nifL–lacZ*)	−	−	+	+

A, *nifH,D,K* regulation, measured by a *nifH–lacZ* fusion; B, positive regulation by NifA; C and D, regulation of *nifL,A* measured by a *nifL–lacZ* fusions in *nifA*⁺ and *nifA*⁻ strains.

The results, given in Table 16.1 (line A), show that the synthesis of β-galactosidase, a measure of the expression of *nifH* (and thus of the *nifH,D,K* operon), was repressed by both NH_3 and O_2. However, when a copy of another *nif* gene, *nifA*, fused to a non-regulated promoter (Fig. 16.8b), i.e. allowing its constitutive expression, was introduced in the cells, this almost completely lifted the repression of *nifH,D,K* under all conditions (Table 16.1, line B). This can be interpreted as indicating that repression of *nifH,D,K* does not take place directly at the level of the promoter of this operon, but at the level of *nifA*. The product of this gene, synthesized only in the absence of both NH_3 and O_2, would act as a positive regulatory protein, necessary to promote expression of *nifH,D,K*. When provided constitutively, as in line B of Table 16.1, no repression of *nifH,D,K* can occur. It was established separately that *nifA* belongs to a *nifL,A* operon. Line C of Table 16.1 demonstrates that expression of this operon is repressed by NH_3 (but only slightly by O_2), but the NifA protein itself is not involved in this regulation, since the same pattern of expression is observed in both *nifA*⁻ and *nifA*⁺ strains (lines C and D of Table 16.1). Yet another process must control expression of *nifL,A*.

The regulatory system was indeed found to be quite complex and involves genes that regulate glutamine synthesis as well. The initial regulatory genes are similar in all enteric bacteria and have been studied extensively in *E. coli* and *S. typhimurium*. The different names given to these homologous genes, *glnL* and *glnG* in *E. coli* and *S. typhimurium*, and *ntrB* and *ntrC* in *K. pneumoniae*, reflect the nature of the regulated pathway in which they were first described in each strain, before their common identity was established.

3.2 Determination of the DNA and protein regulatory components

3.2.1 *DNA signals (sequences upstream of the genes)*

Availability, by gene cloning and sequencing, of the complete sequences lying upstream of both the *nif* and the *ntr* genes has allowed the elucidation of a new kind of gene regulation mechanism. Comparison of the nucleotide sequences (Appendix 8) upstream of the *nif* genes and of other genes, such as those involved in glutamine metabolism in both *K. pneumoniae* and *E. coli*, brought important information.

No typical 'consensus' promoter, as defined in *E. coli* and several Gram-negative bacteria (Chapter 15), can be found upstream of the *nif* genes. Instead, a new consensus structure, including bases around positions −12 and −24 (Fig. 16.9a), is found. A similar sequence is also detected upstream of the *ntr* operon, although in this case both the standard consensus promoter and the −12 and −24 regions are present. These sequences, which were good candidates as promoters, have been shown to be recognized by RNA polymerase provided it functions with a different σ factor, the σ^{54} coded by the *ntrA* (*rpoN* in *E. coli*) gene. However, this σ factor itself is constitutively produced,

(a)

K. pneumoniae	*nifH*	—CTGG T A T G————CTGCA——
	nifE	—CTGG A G C G————TTGCA——
	nifU	—CTGG T A T C————TTGCT——
	nifB	—CTGG T A C A————TTGCA——
	nifM	—CTGG C C G G————TTGCA——
	nifF	—CTGG C A C A————TCGCA——
	nifL	—AGGG C G C A————TTGCA——
R. meliloti	*nifH*	—CTGG C A C C————TTGCA——
	P2	—CTGG C A C G————TTGCA——
R. japonicum	*nifH*	—TTGG C A C G————TTGCT——
	nifD	—CTGG C A T G————TTGCA——
R. parasponiae	*nifH*	—TTGG C A T G————TTGCT——
consensus sequence		—CTGGPyАPyPu————TTGCA——

(b)

$$\text{T G T N}_{10}\,\text{A C A} ——————\text{G G N}_{10}\,\text{G C}$$

. −130 region   −24   −12 region

nif promoter

NifA box   σ^{54} (NtrA) binding site

Fig. 16.9 Organization of *nif* genes upstream sequences. (a) Sequences of the −24, −12 promoters of the *nif* genes in four strains. (b) Organization of a *nif* gene upstream region.

indicating that yet another mechanism is responsible for *nif* regulation by NH_3.

A comparison of all *K. pneumoniae* regions upstream of *nif* genes led to the discovery of a sequence common to all of them, except *nifL,A*, and which is located 100–200 bp upstream from their transcriptional start site. This sequence, $TGTN_{10}ACA$ (Fig. 16.9b), was shown to be specifically recognized by the NifA protein, and has been called the upstream activator sequence (UAS) or NifA box. Another upstream activator sequence was found in the regions upstream of *nifL,A* and of *ntrL,B,C* (the NtrC box), the promoters of which are both recognized by σ^{54}.

Other such regulatory boxes have now been discovered in many instances and, in some cases, the σ factors involved can be very numerous and diverse. The best and most complex example known is the cascade of σ factors involved in the control of the sporulation process of *B. subtilis*. The finding of either a particular σ-type promoter or a UAS box in the sequence upstream of an as yet unknown or poorly known gene or operon can indeed be taken as a good indication of a regulation common to other systems using the same signals. The *E. coli* chromosome sequencing programme might help in the discovery of further functions sharing common regulations.

3.2.2 *Protein functional domains (protein-sequence comparison)*

Similarly, comparisons of protein sequences, which are nowadays more easily obtained by deduction from the nucleotide sequences of the ORFs than they were by classical protein sequencing, have led to tremendous progress in the understanding of functions and of the relationships between function and structure (Appendix 8). A single but significant example will be described for the case of nitrogen metabolism. The pair of proteins NtrB and NtrC were shown to share homologies (i.e. identical or similar amino acid sequences) with numerous couples of proteins (Fig. 16.10), most of which are involved in adaptation to modifications of the growth conditions and which are known as signal-transducing proteins, or two-component regulators (Table 16.2).

All members of the family in which NtrB fits show a relatively high homologous C-terminal domain. They are histidine protein kinases, i.e. proteins whose function is to transfer a phosphate from a histidine residue of their protein to an aspartate residue of other specific, regulatory proteins. Their N-terminal domains, in contrast, show large differences in their primary structure (i.e. amino acid sequence), but many display a strong hydrophobicity, suggestive of their being located in the membrane (Fig. 16.10). These proteins are often called

Table 16.2 Functions known to be regulated by the joint activities of a histidine–protein kinase and a response regulator protein.

Organisms	Functions
Agrobacterium tumefaciens	Virulence
Bradyrhizobium parasponiae	Nitrogen regulation
Bacillus subtilis	Phosphate regulation; sporulation and competence; synthesis of extracellular proteins
Bordetella pertussis	Virulence
Enterobacter aerogenes	Chemotaxis
Escherichia coli	Chemotaxis; nitrate reductase; porin expression; hydrogenase; nitrogen, phosphate and oxygen regulation
Klebsiella aerogenes	Nitrogen regulation
Myxococcus xanthus	Motility; development
Pseudomonas aeruginosa	Alginate production
Rhizobium leguminosarum	Dicarboxylic transport
Rhizobium meliloti	Nitrogen fixation
Staphylococcus aureus	Exoprotein synthesis
Salmonella typhimurium	Chemotaxis; nitrogen regulation; porin expression; virulence; phosphoglycerate and tricarboxylic acid transport systems

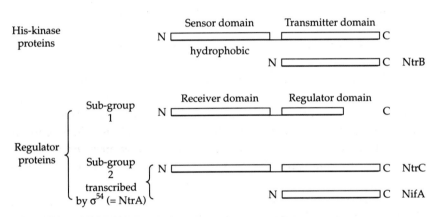

Fig. 16.10 Conserved domains in signal-transducing proteins.

'sensors' because they are activated by a specific change in the growth conditions.

The members of another family, called response regulators, to which NtrC belongs, are similar to each other in their N-terminal domains, which have been shown to include the target for phosphorylation by the corresponding protein kinases. This domain has been called the 'receiver' (Fig. 16.10). Their C-terminal domains can be grouped into several families. The sequence of the NtrC protein C-terminal domain is reminiscent of that of DNA-binding proteins. It shares significant homology with all the activators implicated in enhancing recognition by the σ^{54} RNA polymerase of its promoters. Protein NifA belongs to this category in this regard, although it lacks the common N-terminal domain of NtrC.

3.2.3 *Conclusions*

The interest of sequence comparisons is obvious. Studies performed on one protein of a family can be extrapolated to all others, thus considerably speeding up the understanding of functions. Furthermore, accumulated data on a protein family greatly facilitate understanding of the functional level of action of a newly described one. Chromosome sequencing can also bring a wealth of new information, extending further than the mere functional classification of an unkown protein, since comparisons of accumulated sequences favours the discovery of classes of proteins.

A striking feature that arises from these comparisons is that they tend to minimize the apparent diversity of biological molecules. Families of protein domains are highlighted, covering both prokaryote and eukaryote proteins. General functions, such as DNA binding, GTP binding and ATP hydrolysis, have been shown to correspond to protein domains recognized by their high degree of homology throughout the living world.

Although still not unambiguously interpreted, the secondary and tertiary protein structures that can be hypothesized from their amino acid sequence yield even more information as to the relationships between particular protein structures and functions. Such correlations allow predictions to be made. These can obviously be tested by analysing mutants, and the great progress brought about by DNA technologies means that mutants can now be engineered *in vitro* to the exact requirements of the researcher, both in nature and location, a process known as site-directed, or site-specific, mutagenesis. This consists in replacing a local sequence of the gene studied by a different sequence bearing the base changes required to bring the amino acid change looked for (Appendix 10).

4 Perspectives

This rapid survey highlights how genetic knowledge and the techniques consequently set up can be exploited for the unravelling of a given function, here shown through the unique example of dinitrogen assimilation, by utilizing a large number of genetic systems. Transposons are used for mutagenesis, mapping (by formation of deletions) and determination of transcriptional organization; fusions are the best suited tools for investigation of regulatory pathways; sequence analyses allow a deep insight into genetic organization, signals for gene expression and its regulation at the level of both the DNA and the proteins implicated.

Once a new bacterium is given proper genetic tools, that is, at least a way to promote gene transfer and recombination, there is now virtually no limit to the possibilities of dissection and description of its functions. Construction of mutants is a critical step, in which transposons are almost unavoidable. These elements, however, as plasmids, are not universal, and the most time-consuming part of a study is often its initial steps, the very finding of the convenient tools: shuttle vectors, transposons, reporter genes, vectors for tests of 'gene expression domains (promoters, ORFs, etc.). Once all this is done, the remaining work can be amazingly rapid.

It must be recalled that, as a result of the possibilities opened by the development of molecular genetic technologies, rules have been made which aim at minimizing the risk of artificial species spreading in an uncontrolled way. One aspect of these regulations concerns the strains used as intermediary hosts, in particular *E. coli*. Besides the specifications required for their use as hosts, all the usual strains bear a number of other mutations, such as auxotrophies, which render their proliferation outside laboratories more difficult. In addition, the required characters (e.g. conjugative capacities, regulatory signals, etc.) are often distributed, together with necessary alleles, over several plasmids, which, in the absence of appropriate selection pressure, would have a good chance of being counterselected and thus lost, leading to an incapacity of the cell to divide.

Appendix 1: cloning vectors and hosts

Cloning vectors are usually artificially constructed replicons, of small size, arranged so that foreign segments of DNA can be stably and reversibly inserted into their molecule. As a replicon, a cloning vector must possess an origin of replication, which defines its host-range and the incompatibility group to which it belongs (Chapter 3). Its cognate information for replication regulation may either have been maintained, giving a controlled copy number, or deleted or mutated, usually yielding high-copy-number molecules appropriate for amplification. Amplification, i.e. the multiplication of the vector, occurs after its transfer into its host. Depending on the host, different modes of transfer, natural or artificial, are available (Chapters 10–12), but clearly this step is critical, which explains the efforts made, when starting on a new strain, to set up a transfer process.

For obvious reasons, *E. coli* is the prokaryotic host most frequently used to achieve all intermediary constructions or molecular analyses. Physiological analyses, e.g. regulation of expression, must, however, be performed in the original host, thus often requiring the construction of vectors capable of replication in both *E. coli* and the other host. This is achieved by introducing into standard *E. coli* vectors an origin of replication specific for this host, usually obtained from an endogenous plasmid. Such molecules are referred to as shuttle vectors.

In present-day vectors, the site for insertion of the exogenous DNA (insert) is usually an artificial sequence carrying restriction sites for a series of restriction enzymes (Chapter 6) for which no other site exists on the rest of the molecule. This region is called a polylinker, or multiple cloning site (MCS). The DNA fragment to be inserted must be prepared by cutting with a restriction enzyme, yielding extremities which fit the possibilities offered by the MCS (Chapter 6). This procedure allows precise and reversible integration, without destruction of either the vector or the DNA.

Beside these compulsory requirements, cloning vectors are rendered much easier to work with when additional features are added, a critical one being a marker allowing direct detection of their presence in a host cell. Indeed, these vectors are manipulated *in vitro*, and the resulting constructions are amplified by autonomous replication after their transfer to an appropriate host. The transformed bacterial cells are isolated through the expression of the selection marker. The most frequently used markers are the numerous antibiotic resistance genes, which are normally present in transposons and which are isolated from these elements and stably integrated into the vectors. These genes, for unexplained reasons, appear to be expressed in many different bacterial hosts although with apparently different efficiencies (i.e. acquired level of resistance).

Another convenient construction is the positioning of the MCS at the beginning of a *lacZ* gene, which is itself stably integrated into the vector. Expression of *lacZ*, i.e. synthesis of β-galactosidase, although not necessary for the host cell (another carbon source is provided), can nevertheless be induced, and is revealed by the appearance of a blue colour on the colonies when grown on special media (Chapter 15). The most widely used system (and that which is constructed, in particular, in the artificial pUC vectors) consists of only the N-terminal part of the *lacZ* gene (known as *lacZα*) being on the vector and yielding an inactive LacZα peptide (if no insert prevents its synthesis), while the C-terminal part, *lacZΩ*, which is cloned downstream of a convenient promoter, is borne in the recipient strain by either the chromosome or another resident plasmid. β-galactosidase activity is restored by association (known as α-complementation) of the two peptides, a particularity of this enzyme. A large variety of such vectors are now available, allowing cloning for almost any purpose.

The presence of the MCS between the promoter and the ribosome-binding site (Chapter 15) does not interfere with translation of *lacZ*, unless a larger stretch of DNA has been ligated into it, as is the case when an exogenous insert has been integrated. This device allows easy recognition of clones which carry a vector bearing an insert (white colonies) as against plain vectors (blue colonies). This is an important trick for the selection of the appropriate clones since the ligation of the insert into the vector may be poorly efficient, depending on the conditions, thus yielding mostly non-modified vectors.

Other constructions exist. Thus, omitting the promoter of the *lacZ* ORF creates conditions for the identification of exogenous promoters or for detection of controlled expression of exogenous genes (promoter-probe vectors), by insertion of a promoter-bearing insert upstream of the *lacZ* sequence. The use of some of these is described in this chapter.

Several artificial vectors very frequently used are derived from the *E. coli* colE1 plasmid (Chapter 3), because several characteristics of this plasmid and of related ones are particularly favourable: they possess an origin of transfer that is mobilizable by the F and RP4 transfer systems; they are replicated by PolI (a very abundant enzyme in *E. coli*) and not by PolIII; and the regulation of their copy number, which requires a negatively active peptide, can be easily abolished by blocking protein synthesis. The two latter properties allow for large amplification, since up to 1000 copies may be produced per cell. The first vector thus constructed was pBR322 and more recent ones include the pUC family, in which the gene for the copy number regulatory peptide has been deleted.

Most vectors are mere plasmids. However, being replicons, phages can be adapted for use as cloning vectors. Thus, deleting the region bearing the lysogenic control of λ and replacing it with exogenous DNA yields a molecule which reproduces as a virulent phage and can be stored as mature particles. Up to approximately 20 kb, that is, a size larger than those usually inserted in standard vectors, can be inserted without interfering with the packaging of the DNA into the phage heads. The F-pili-specific *E. coli* single-stranded DNA phages, e.g. mostly M13, have been adapted for other purposes, since their DNA can be isolated either in its mature form in the particle or as double-stranded circles when isolated during the reproductive cycle. A major application is encountered in the setting up of DNA sequencing methods (Appendix 4).

The vector, if a plasmid, is purified from the cells using methods identical to those described for natural, endogenous plasmids (Chapter 3). If a phage, it is similarly purified from the mature particles.

Appendix 2: preparation of transposon mutants by inverse genetics

Mutations are generated in a gene or group of genes that is precloned on a plasmid vector (Appendix 1) specific for a convenient secondary host (usually *E. coli*) carrying the desired transposon. After transposition has had the opportunity to occur in this strain, the DNA segments are transferred back into their original host. The carrying plasmid is chosen so as not to be able to replicate in this strain (suicide vector). The mutants with the phenotype looked for are screened among the clones that have acquired the transposon marker, by their specific phenotypic modification. This method is now widely used.

Appendix 3: comparison of the diverse fine-mapping methods

While the deletion mapping method is rapid and gives non-ambiguous results (one looks for a + or − answer), it does not yield information about distances between genes (for an example, see Section 1.2). Recombination analysis (possibly through three-point tests), in contrast, provides relative distances between the mutations involved in the cross. Only complementation analysis demonstrates whether two mutations belong to the same gene or not.

Neither type of approach can ensure that all the genes involved in the function studied have been identified. Since the methods consist of analyses of all possible point and insertion mutations, only those for which such a mutant has been isolated can be mapped. The larger the number of mutants analysed (all mutants having the same phenotype), the higher the probability of having hit all the genes involved. However, genes for which mutations result in a lethal phenotype under the selective conditions used will remain hidden by this method, but can be revealed by direct DNA analysis (Section 3.2.1) (Appendix 9).

Appendix 4: DNA sequencing

The method most frequently used to determine the base sequence of a molecule of DNA was set up by Sanger *et al.* (1977) and is referred to as the chain-termination, or dideoxy, method (Fig. 16.11). The DNA fragment to be sequenced is inserted into the double-stranded, RF, form of a modified M13 phage genome. The RFs are obtained from a lysate of M13-infected, F-bearing *E. coli* cells (Chapter 4). The M13 genome has been modified so as to include a multiple cloning site (MCS). As in other cloning vectors, markers have been introduced to facilitate selection (Fig. 16.11a) (Appendix 1). A mutation

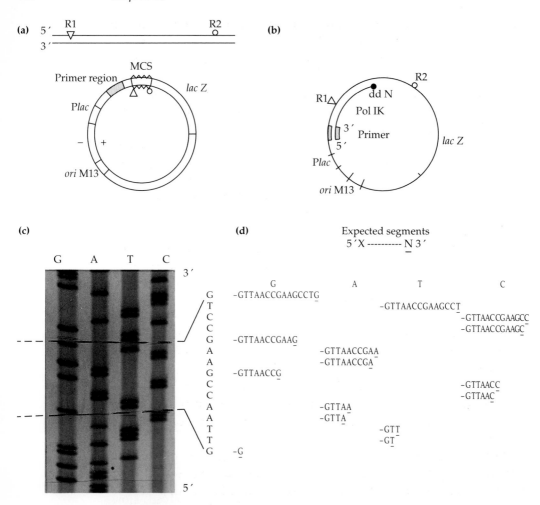

Fig. 16.11 Schematic representation of the Sanger DNA sequencing method. (a) The M13 modified genome is drawn as an RF. (b) *In vitro* integration of the segment to be sequenced into the M13 modified DNA. (c) Autoradiograph of an electrophoretic gel of radioactively-labelled dideoxy fragments, and the deduced reading of the DNA sequence. (d) The expected dideoxy fragments synthesized during the four labelling, chain-terminating reactions, and their expected position on the electrophoretic gel. Underlined bases symbolize the inserted dideoxyribonucleotides.

in the gene coding for protein II (involved in the phage DNA replication) prevents the autonomous reproduction of the phage in the absence of a helper M13 (Chapter 4), which, when simultaneously introduced in the host, provides the missing protein. A short known sequence inserted close to the MCS will later serve as a primer recognition site for *in vitro* DNA polymerization by PolI (Chapter 2). The fragment to be sequenced, prepared as a digest by one of the restriction enzymes that have a site in the MCS, is inserted by annealing with and ligation to the M13 genome, previously linearized by the same enzyme (Fig. 16.11b) (Chapter 6). The circular, double-stranded molecule, including the DNA insert, can be introduced into *E. coli* by transformation (Chapter 12) and reobtained as either double-stranded molecules or single-stranded circles coated into mature particles. The increase in length of the DNA of the M13 vector does not hinder encapsidation, since this is achieved by a DNA-coating procedure (Chapter 4). Only the insert strand ligated to the + strand of M13 will be recovered in the phage particles (Fig. 16.11b). Inverting the orientation of integration of the cloned fragment into the M13 RF provides the other strand in the mature phage DNA.

The single-stranded molecules, extracted from the phage particles, are used to synthesize, *in vitro*, a labelled, complementary strand in the presence of PolIK, the *E. coli* PolI enzyme devoid of its exonuclease activities (also known as the Klenow enzyme after its discoverer (Chapter 2)) and of all necessary factors and precursors. Since DNA polymerases require a primer, an oligonucleotide fragment complementary to that inserted in the phage DNA (see above) is used for this purpose, thus defining the start site for synthesis. One of the four precursor triphosphate-nucleotides (each in turn, in four different reaction tubes) is partially replaced by a modified, dideoxyribonucleotide, a molecule lacking the 3'-OH group on the sugar. Integration of such a dideoxyribose in place of the normal deoxyribose results in prevention of further chain elongation (Fig. 16.11b). By adjusting the relative proportion of normal/modified nucleotides, interruptions will occur at various positions, and statistically all possible lengths ending with the corresponding base are obtained (Fig. 16.11d). Simultaneous labelling is obtained by introducing one or several radioactively labelled normal precursors. The newly made chains are separated from the template by heat denaturation, and the four reaction mixtures are then submitted to electrophoresis on a polyacrylamide gel, which allows precise resolution of fragments differing in lengths by one nucleotide. The positions of the fragments are revealed on an autoradiogram (Fig. 16.11c). Each lane gives the pattern of fragments ending with a particular base. Sequential reading from the bottom (the faster-migrating and thus the smaller molecules) on all four lanes directly provides the base sequence in the 5' → 3' orientation (that of synthesis).

Since a maximum of about 300 bases can be sequenced on one standard gel, a series of fragments covering (with overlaps) the entire fragment to be sequenced are inserted into M13 vectors and treated as described.

Appendix 5: preparation of and screening from a genomic library

A genomic library (or bank) is, ideally, a collection of vectors (plasmids or phages), each carrying a fragment of a given genome, sufficient to cover the entirety of the genome. Any region of the genome looked for (i.e. to be cloned) can thus be found, using appropriate detection criteria.

Without going into details as to sizes of fragments and strategies for restriction-enzyme digestion, both the genomic DNA and the vector are cut by the same (or compatible (Chapter 6)) restriction enzyme and ligated *in vitro*. The constructs are then amplified, as described above (Appendix 1).

Several methods are available for selective screening of the clone(s) carrying the gene(s) under study. Direct screening from the library is possible in three cases:

1 If a homologous DNA sequence (e.g. the same gene from another strain) or an oligonucleotide synthesized from a known sequence of the protein is available, the corresponding DNA region is looked for by hybridization (this chapter).

2 If an antibody against the protein under study is available, an immunoscreening is carried out. This implies that the gene coding for the protein can be expressed, and it may require a particular construction of the library, referred to as an expression library.

3 If the gene can either complement a deficiency in or provide a particular phenotype for the intermediary *E. coli* cells, this can obviously be used (an example is given in this chapter).

Indirect screening consists in the transfer of the library into an appropriate mutant of the original host, in which either complementation by or recombination with the cloned gene leads to a detectable modification of the phenotype.

The clone carrying the gene can thus be detected, the vector purified, and the DNA insert analysed by various molecular techniques, e.g. restriction mapping (Chapter 6) or sequencing (Appendix 4).

Appendix 6: hybridization from colonies or plaques

The principle of this method is similar to the standard hybridization technique performed on purified DNA and usually resolved on agarose gels, which is known as Southern blotting. The cells or phages containing the plasmids are grown as colonies or plaques on solid medium. By applying a nitrocellulose or nylon sheet on the surface of the plate, a fraction of each colony or plaque is transferred to the filter. These are then fixed and lysed by alkali to liberate their DNA content, the latter denatured *in situ*. The fitter is incubated in the presence of the labelled DNA fragment, also denatured, used as a probe.

Labelling of the probe is often radioactive. It is achieved by nick translation (Chapter 2): the double-stranded fragment is treated simultaneously with a non-specific endonuclease (e.g. from phage T4) and with PolI of *E. coli*, which repairs the nicks made by the endonuclease and incorporates the radioactive precursor provided. (Other methods use covalent linkage of easily detectable molecules, e.g. luminescent dyes, as labelling.)

After drying, the filters are analysed by autoradiography (or other appropriate method) and the positive clone(s) recovered from a replica of the original plate.

Appendix 7: elaboration of restriction maps

The principle of the methodology yielding the restriction map of a cloned DNA region, i.e. the pattern of distribution of possible cleavage sites for a number of restriction enzymes, has been described in Chapter 6. Experimentally, the corresponding digestions are performed on copies of the DNA regions prepared as inserts in a vector, which is amplified in a host, usually *E. coli*. The R–M genotype of the host is thus critical, since possible methylations might hide cleavage sites to restriction enzymes recognizing the same or compatible sequences. Standard *E. coli* strains used for DNA work are thus methylation-deficient, which requires that they are also restriction-deficient (Chapter 6).

The restriction-deficient character is thus critical for the strains used as host for library preparation, since no information is usually available as to the modification pattern of the exogenous DNA, and its maintenance is best ensured in a host totally devoid of restriction capacities. The same *E. coli* hosts are usually used for all intermediary analyses and preparations.

Experimental constructions of restriction maps are usually limited to up to 5–10 different enzymes, since the aim of this process is mostly to provide working facilities for further subcloning, deleting, etc. Complete restriction maps, including any required enzyme, can be deduced from the base sequence when sequencing has been performed. The presence of possible restriction sites does not mean that cleavage would automatically be obtained in the genuine DNA, since this depends on the overall methylation pattern of the strain.

Appendix 8: deciphering DNA sequences: sequence comparisons

The base-sequence data obtained by DNA sequencing are interpreted, usually with the aid of computer programs, so as to determine a number of important features. The first one is the search for genes, either protein, tRNA- or rRNA-encoding regions. The former are known as open reading frames, or ORFs. These are recognized as long stretches of significant in-frame triplets, starting with an initiation one (usually ATG) and ending with a triplet corresponding to a nonsense codon. ORFs must be looked for in the three possible reading frames, and in both chains (and thus orientations) of the DNA.

When a potential ORF is detected, its transcriptional expression signals should be looked for. This is now often facilitated by comparing the upstream and downstream regions of the ORF with those of well-known genes, and looking for the possible presence of consensus regions (both in sequence and position) for promoters, ribosome-binding sites and transcription-termination structures (Chapter 15). A number of other criteria, e.g. adequation with the strain-specific codon usage, search for strain-specific expression signals, are indicative of the ORF being actually expressed. The non-observation of such sequences does not mean that the potential ORF is a fake, but may indicate that expression signals are particular to that gene (or strain). Non-ambiguous determination of the role of this region as an actual ORF can be obtained either by detection of its transcriptional product (mRNA) or by the observation of a modified phenotype when an inactivated ORF replaces the wild-type one. Such mutated ORFs can be constructed *in vitro*, the most frequent strategies being either to delete (part of) the ORF or to interrupt it by the insertion of another gene, used as marker and referred to as a cassette or cartridge, or interposon.

Comparisons of long segments, the size of genes or longer, which are now possible by a search in gene banks (which are usually performed on the deduced amino acid sequences for protein-coding regions) also constitute an important argument for the identification of the function of a DNA sequence.

One can also compare proteins either as to their primary structure (DNA sequence-deduced amino acid sequence) or for other characteristics, such as distribution of hydrophilic amino acids (hydropathy profile), presence of specific patterns of amino acid organization and probable secondary and tertiary structures.

Appendix 9: the use of gene fusion for the analysis of regulation

Gene fusion is the covalent ligation of two genes, the one to be studied and a second one used as selection marker and referred to as the reporter gene. The reporter gene, deleted of part of its expression signals, is thus ligated downstream of the expression signals and/or part of the gene to be studied, so that a unique transcriptional unit is created (Fig. 16.8). Expression of the reporter gene is thus dependent on the regulatory pathway of the gene studied. Fusions are used whenever the activity of the gene is difficult to assay, whatever the reason. The reporter gene used is very often the *lacZ* gene from *E. coli*, because of the easiness and flexibility of detection of β-galactosidase activity both in liquid cultures and on plates. But the *cat* (*cam* acetyl *t*ransferase, yielding chloramphenicol resistance) originating from transposon Tn9 and the luciferase gene from *Vibrio fischeri*, *luxAB*, the expression of which gives the cells a readily detectable luminescence, are also often used.

The fusion product can be either borne by a plasmid transferred to a wild-type cell, in which case both the intact gene and the fused one are present, or integrated in replacement of the wild-type gene. The two constructions will be denoted WT(*geneX-lacZ*) and WT/*geneX-lacZ*, respectively.

Two main kinds of fusion can be envisaged (Fig. 16.8):

1 The reporter gene, devoid of its transcription signals but having retained its translational signals, is inserted anywhere downstream of the promoter of a transcription unit, yielding a transcriptional, or operon, fusion. The insertion site may be chosen so as not to prevent expression of the genuine gene(s), for instance, between the promoter and the RBS sequence of the first cistron of the transcriptional unit.

2 If deprived of all its upstream expression signals, including the initiation codon, it must be inserted in-frame inside an ORF, usually as close as possible to the translational signals, forming a gene fusion, also called protein or translational fusion. The product synthesized will be a chimeric protein, with an N-terminal extremity belonging to the original peptide and the C-terminal one to the reporter gene. Conversely, the fusion may consist in changing the promoter of a gene for a constitutive one, or for one whose regulation is well known. This will be denoted (P_A-*geneX*) (Fig. 16.8b).

Gene fusions can be obtained by specially engineered transposons or by *in vitro* gene manipulation.

Appendix 10: site-directed mutagenesis

When a specific mutation, i.e. the specific replacement of a base or a small deletion, is required, a small DNA molecule (approximately 20–30 bases long), known as an oligonucleotide, identical to the region encompassing the modification looked for except for the modification introduced, is synthesized *in vitro*. Used as a primer for DNA polymerization, it is hybridized to the original gene sequence carried on a plasmid, often an M13 derivative, which allows the use of single-stranded DNA molecules as template. Polymerization is carried out in the presence of all DNA precursors. The plasmid also possesses a number of restriction sites for a given enzyme, used in a later step (see below). Protection of these sites on the newly made strand is achieved by replacing one of the precursor nucleotides by a modified derivative. When the whole double-stranded molecule is formed and ligated into a covalent double-stranded circle, it is treated by the appropriate restriction enzyme, which will cut the template strand. The single-strand nicks thus formed serve as start sites for an exonuclease which will digest the nicked fragments and yield the newly made strand as a single-stranded circular molecule. This in turn is used as template for *in vitro* synthesis of its complementary strand, as above, thus creating a homoduplex mutated molecule, which is transferred into an appropriate *E. coli* host to be amplified. The resulting vectors can then be purified from the *E. coli* host and used as convenient (for the integration of the mutated gene into another plasmid, transfer to a desired strain, etc.).

Reference

Sanger F., Nicklin S. & Coulson A.R. (1977) DNA sequencing with chain-terminating inhibitors. *Proceedings of the National Academy of Science (USA)*, 74, 5463–7.

Glossary

Alkylating agent an organic compound able to transfer an alkyl group to nucleotides.

Allele one of alternative forms of a gene.

Allosteric refers to the stereospecific modification of a protein by an effector to influence the activity of another site of the protein.

Amber codon the UAG nonsense codon, not coding for any amino acid in the universal code, the presence of which results in termination of synthesis of a proteinic chain.

Amber mutation any mutation that results in an amber codon (UAG) being created from a different codon.

Amber suppressors mutated tRNA genes that now code for a tRNA that delivers an amino acid to the UAG codon.

Annealing the process by which single-stranded nucleic acid pairs with its complementary strand to form a duplex.

Antibiotic a substance interfering with a particular step of cellular metabolism, causing either bactericidal (irreversible) or bacteriostatic (reversible) inhibition; a substance that can be used as a therapeutic antibacterial agent, a term sometimes restricted to those having a natural biological origin.

Anticodon the three bases in a tRNA that are complementary to those in the codon of the mRNA.

Antiterminator a protein allowing RNA polymerase to transcribe through a termination site.

Attenuation a means of regulating the level of transcription of certain operons responsible for synthesis of amino acids by interfering with mRNA elongation. (During transcription, annealing of particular sequences in the mRNA molecule, depending on the rate of concomitant translation from the mRNA, itself a function of the availability of the amino acid concerned, results in destabilization of the RNA polymerase–DNA template structure. Can exist only in prokaryotes since it requires concomitant translation, the regulatory trigger being the end-product (the amino acid) of the biosynthetic pathway.)

Autolysin an enzyme capable of partially degrading the cell wall of the bacterium which synthesizes it as part of its cell-wall synthesis machinery.

Auxotroph an organism requiring a specified organic supplement for growth, referred to as growth factor.

Bacteriophage (phage) a virus that infects bacteria.

Base pairs the pair formed by one purine or pyrimidine base in one nucleic acid strand hydrogen-bonded to the complementary pyrimidine or purine base in the other strand of the duplex.

Bidirectional replication occurs when two replication forks proceed in opposite directions from the same origin of replication.

Blunt ends extremities of a DNA duplex produced when both strands end at the same position, with no single-strand extension.

Capsduction a DNA transfer system occurring in *Rhodobacter capsulatus* species whereby the donor DNA is carried in a phage-like structure known as a gene transfer agent (GTA), which injects the donor DNA derived from the GTA-producing cells.

Capsid the protein layer surrounding a phage chromosome in the free particle.

Catabolite repression the decreased expression of several bacterial operons due to the addition of glucose.

Centromere the region of a eukaryotic chromosome responsible for attachment to the mitotic or meiotic spindle leading to controlled partitioning of chromosomes during nuclear division.

Chi site a sequence of eight nucleotides providing a recognition site for genetic recombination to occur in *E. coli* K12.

cis describes the position of two genetic determinants on the same DNA molecule, as opposed to *trans*.

Cistron rather obsolete term for gene. Mostly used for genes organized inside a single polycistronic transcription unit.

Clone a group of cells derived from a single cell and thus supposed all genetically identical. Deviations from genetic identity stem from either random spontaneous mutations or from chromosome segregation of polyploid, genetically heterogenous cells, arising during formation of the clone.

Cloning vector a DNA molecule capable of replication in a suitable host cell and possessing suitable recognition site(s) for restriction endonuclease(s) allowing for the insertion of DNA fragments by recombinant DNA techniques and selective markers.

Codon the sequence in the mRNA of three purine and/or pyrimidine bases which code for an amino acid or punctuation mark. (Corresponds to triplets in the DNA molecule.)

Cointegrate a structure resulting from the fusion of two replicons, one of which possesses a transposon. (The resulting structure has two copies of the transposon, one at each junction point, orientated in direct repeats.)

Competence the physiological state of a bacterial cell allowing the natural uptake of transforming DNA into the cell.

Complementation the ability of non-allelic genes present in *trans* in a (partial) heterozygote formed from two mutants showing the same deficient phenotype to restore a proficient phenotype by providing diffusible products.

Concatemer a polymer of repeated unit genomes.

Conjugation a means of gene transfer by contact between responsive bacterial cells, the donor possessing a transfer device (e.g. the *E. coli* F factor).

Constitutive enzyme an enzyme which is continuously synthesized by the cell without modification of its level of synthesis depending on environmental conditions or developmental state.

Contig (i) a segment of chromosome from a set produced by cleavage with a frequent-cutting restriction endonuclease sharing part of its sequence with two adjacent fragments from a set produced by cleavage with a rare-cutting restriction endonuclease; (ii) two partially overlapping fragments used in DNA sequencing to ensure proper reading of the sequence.

Crossing-over (or crossover) the site where two homologous DNA strands originating from homologous chromosomes are resealed to form the recombinant chromosome.

Deletion the loss of a segment (one to many bases) inside a genomic region.

Deletion mapping the use of deletion mutations overlapping point mutations to acheive fine mapping of a restricted chromosomal region.

Denaturation of DNA separation of the two strands of a duplex molecule, by melting (usually by heat or alkali) the hydrogen bonds joining the complementary bases from the two opposite strands.

Density shift the change of growth medium containing either a heavy or a light isotope in a DNA constituent to a medium containing the opposite density. (5-bromouracil is often used to give a 'heavy' precursor.)

D-loop the structure in a duplex DNA due to the invasion of single-stranded DNA in recombination.

Duplex DNA a double-stranded DNA molecule.

Electroporation the technique for making cell membranes permeable to DNA by the application of an electric field.

Endonuclease a nuclease which cleaves phosphate-deoxyribose bonds within a nucleotide sequence – specific for either single- or double-stranded DNA, or RNA.

Engineered (molecule) an artificially modified molecule, protein or DNA, obtained by use of molecular genetic technology.

Episome former name given to a genetic element that can exist either as an autonomous, usually circular, particle in the cytoplasm or integrated into the chromosome.

Exonuclease a nuclease which cleaves phosphate-deoxyribose bonds one at a time from one end of a nucleotide sequence – specific for single- or double-stranded DNA, or RNA, and specific for 3′ or 5′ extremities.

Expression vector an artificially constructed plasmid typically possessing a strong and regulatable promoter, downstream of which appropriate restriction sites allow the insertion of an ORF under study so that it will be read from this promoter.

F⁺ cell a cell possessing an F factor as an autonomous plasmid, which enables the cell acting as a donor (male) to transfer it to a recipient (female) cell.

F⁻ cell a cell devoid of an F factor, and hence able to act as a recipient (female) in a conjugative DNA transfer in matings with F⁺ or Hfr strains.

F factor an *E. coli* plasmid coding for conjugative transfer (fertility) functions, which may exist as an autonomous plasmid in the cytoplasm or integrated at specific sites in the chromosome.

F-prime (F′) plasmid an extrachromosomal plasmid consisting of a fragment of chromosomal DNA attached to an F plasmid.

Frameshift shift in a reading frame due to the deletion or addition of one or two (or a non-multiple of 3) nucleotides in a coding DNA sequence.

Gene the genetic (functional) unit, formerly equivalent to cistron, a term now rarely used; a part of a nucleic acid molecule carrying the genetic information coding for either (i) a polypeptide chain, and the regions necessary for this synthesis, i.e. transcriptional signals (promoter, regulatory regions, termination region) and translational signals (ribosome-binding site, initiation and termination triplets), or (ii) a stabler RNA (i.e. transfer or ribosomal RNAs) or (iii) a non-expressed unit directly implicated in metabolism (e.g. *ori*).

Gene library a collection of fragments of the genome of an organism, usually obtained by restriction cleavage, cloned into a cloning vector introduced into a suitable host organism.

Generation time the lapse of time necessary under strictly defined conditions for one cell to reproduce, forming two daughter cells, when in exponential growth.

Genetic code the system comprising all the triplet codons of mRNA assigned to amino acids and punctuations.

Genetic determinant the region of a genome encoding a given genetic trait.

Genetic recombination the process by which a fragment of DNA from one molecule (chromosome, plasmid, phage genome) is exchanged with and/or integrated into another molecule to give (two) recombinant molecule(s).

Gene transfer agent (GTA) the phage-like structure produced by donor cells of *Rhodobacter capsulatus* species which transfers host DNA to recipient cells.

Genome the total content of gene-bearing molecules of an organism, or the total genetic information of an organism.

Genotype the genetic composition of an organism as coded on its genome, i.e. defined by the allelic form present for each gene – usually, only departures from the wild type are described.

Genus (genera pl.) a taxon grouping related species.

Growth factor the metabolite needed by an auxotrophic strain to grow.

Growth rate the rate of increase of cellular density or concentration during exponential growth, per unit time.

Gyrase an enzyme classed as a type II topoisomerase which unwinds coiled DNA during, for instance, replication.

Haploid the state in which only one copy of each chromosome exists in the cell. (Prokaryotes are haploid, although transient states exist during which more than one copy may be present simultaneously in the cell, depending on growth rate.)

Helicase an enzyme which unwinds duplex DNA molecules by releasing interstrand hydrogen bonds.

Helix–turn–helix a configuration of a DNA-binding protein, comprising two α-helices at an angle with each other, allowing specific recognition and fixation to the DNA.

Heteroduplex a double-stranded DNA molecule in which the two strands have different sources and are therefore likely to be imperfectly matched.

Heteroimmune phage a phage not sensitive to the inhibition of lytic development of another phage stabilized as a prophage in a given cell.

Heterologous (i) DNA regions coding for the same function but having differences in their nucleotide sequences; (ii) DNAs which originate from different species.

Heterotrophic the capacity for an organism to use an organic molecule as carbon source.

Heterozygote a diploid (usually partial in prokaryotes) organism having different alleles of (a) particular gene(s).

Hfr (high frequency of recombination) cell a cell in which the F factor has integrated into a specific location in the chromosome, causing it to act as a high-frequency donor of chromosomal genes in crosses with F⁻ cells.

Histone a eukaryotic basic protein which binds to DNA, participating in the formation of nucleosomes, the structural units of chromatin.

Holliday junction the cruciform structure postulated as an intermediate in the process of homologous genetic recombination.

Homologous (i) nucleic acid molecules having the same base sequence; (ii) nucleic acid molecules originating from strains of the same species, thus having at least long stretches of identical DNA base sequences; (iii) gene or protein families having a recognizable common evolutionary origin.

Homozygote a diploid (usually partial in prokaryotes) organism having the same allele of (a) particular gene(s) in both copies of its genetic material.

Host-range the strain or species specificity of a cell-dependent entity (bacteriophage, plasmid).

HU a histone-like protein of *E. coli*.

Hybridization the technique whereby a denatured (single-stranded) nucleotide chain (DNA or RNA) is allowed to associate with another single-stranded nucleotide chain − if sufficiently related, the two chains will pair by their complementary bases.

Hyphae multicellular filaments formed during the vegetative reproduction of fungi or actinomycetes.

Immunity (conferred by a bacteriophage to a bacterium) the capacity for a bacterium lysogenic for a given phage (stabilized as a prophage) to survive infection (referred to as superinfection) by a second phage related to the prophage, as a result of inhibition of reproduction of the superinfecting phage by the repressor synthesized by the prophage.

Immunoassay the test by which a protein is recognized by its response to a specific antibody.

Inducible enzyme an enzyme the synthesis of which requires the presence of a specific substance, the inducer.

Inoculum the initial sample of a microorganism or phage suspension added to a culture medium used to start a new culture.

Insert a fragment of DNA, usually obtained after cutting a larger molecule with a restriction enzyme, integrated into another one (usually a vector).

Insertion sequence (IS) a self-contained DNA sequence capable of transposition or integration at specific or more or less random sites (targets) of a DNA molecule by non-homologous recombination. (Does not specify other phenotypic properties.)

Intercalation the positioning of flat polycyclic molecules, such as the acridines, between nucleotides in a DNA duplex, thus potentially causing frameshift mutations.

Intron a naturally occurring intervening stretch of bases inside an ORF, defined by characteristic features, which is transcribed and excised (by a process known as mRNA maturation or splicing) before translation takes place. In prokaryotes, known only in Archaebacteria and in a few *E. coli* phages and bacterial species.

Inverted (terminal) repeats two copies of the same sequence of nucleotides repeated in opposite directions in the same DNA molecule, often present at the extremities of insertion sequences.

kb (kbp) kilobase (pairs) 1000 bases of single-stranded DNA or RNA, or 1000 base pairs of a double-stranded DNA.

Leaky (mutation, phenotype) a modification in the amino acid sequence of a protein resulting in only partial loss of its activity.

Leucine-zipper motif a stretch of amino acids in a protein including a leucine as every seventh residue, and lying adjacent to a stretch of positively charged amino acids responsible for DNA binding.

Ligase an enzyme joining a 3'-OH residue of a deoxyribonucleotide to the 5'-phosphate residue of another deoxyribonucleotide.

Ligation the action of a ligase.

Linkage describes the possibility for two or more genetic determinants to be simultaneously transmitted to a recombinant organism.

Locus (loci) the position on a chromosome where a particular genetic trait resides.

LPS lipopolysaccharides, the main constituents of the outer layer of the outer membrane of Gram-negative bacteria.

Lysate the phage output resulting from the development of a lytic cycle of a bacteriophage.

Lysogen a bacterium harbouring the prophage form of a temperate phage.

Lysogeny the ability of a temperate bacteriophage to maintain itself as a quiescent prophage until induced into the lytic cycle.

Lytic cycle the development of a bacteriophage, either after infection of a host bacterium or after induction of a lysogenic one, resulting in production and release of free progeny phage particles, and often in the lysis of the host cell.

Map unit an arbitrary unit representing the distance between genes, usually derived from the percentage of recombination, but in conjugation by the time at which the gene is transferred.

Marker a genetic trait of which one allelic form is used as a selection criterion during DNA transfer processes.

Merodiploid a partially diploid organism – in prokaryotes results from the introduction of a homologous fragment of chromosome by a transducing particle or by a plasmid.

Mineral medium a salt solution comprising only minerals, sufficient to allow growth of a given organism when supplemented with carbon and energy sources.

Mismatch a defect in the pairing of two complementary DNA sequences where a base(s) in one strand is different from that expected according to complementarity with the other.

Missense mutation a mutation consisting in the change of a codon for a different one, resulting in an amino acid change in the encoded polypeptide.

Mobile element an insertion sequence or a transposon, able to promote its own transposition.

Modification (of DNA) the specific addition of secondary residues (methyl, hydroxymethyl, glucosyl) to deoxyribonucleotides by an organism to differentiate its DNA from any other, or to use as regulatory purposes.

mRNA an RNA molecule synthesized from a gene (or an operon) encompassing the coding region(s) and the translation signals, serving as intermediate to specify the amino acid sequence of the encoded polypeptide(s) for the translation process.

Multiplicity of infection (moi) the ratio of the number of bacteriophage particles to the number of host bacterial cells in an infection mixture.

Mutagen an agent (physical or chemical) capable of increasing the frequency of mutation.

Mutagenesis the process by which mutation occurs.

Mutant (i) an organism with an altered base sequence in one or several genes; (ii) an organism in which a mutation engenders a phenotypic difference from the wild type.

Mutation an alteration in the base sequence on the genetic material, does not necessarily give rise to the formation of a functionally impaired clone.

Nick the cutting of a phosphodiester bond between adjacent nucleotides in one strand of a DNA duplex.

Nick translation a method whereby DNA polymerase I is used to first produce a nick in a DNA duplex, then to degrade a length of several nucleotides by its 5'-exonuclease activity, and replace them by labelled nucleotides, thus producing a probe.

Non-permissive conditions growth conditions not allowing a conditionally lethal mutant to survive.

Nonsense codon a codon which does not code for any amino acid, but signals a termination of translation, or punctuation.

Nuclease an enzyme which cleaves phosphate-deoxyribose bonds within (endonuclease) or at the extremity of (exonuclease) a nucleotide sequence – usually recognizes a specific substrate, e.g. single- or double-stranded DNA or RNA.

Nucleoid the folded, dense structure of a prokaryote chromosome inside the cell.

Ochre mutation a change in a codon to create the nonsense codon UAA (the ochre codon).

Okasaki fragments lengths of some 1000–2000 nucleotides long formed in the daughter strand of a DNA duplex newly synthesized from the lagging matrix by discontinuous replication of DNA, later joined together by ligation.

Oligonucleotides small lengths of nucleotides usually created by *in vitro* synthesis.

Open reading frame (ORF) the stretch of DNA in a gene which actually codes for the active molecule (the protein), also often referred to as a coding region. (The word gene is also ambiguously used to depict only this region of the unit.)

Operon a sequence of adjacent genes read as a single mRNA, and thus, if relevant, coordinately regulated, usually by a regulator protein in association with an external cofactor.

Origin of replication (*ori*) the base sequence on a replicon from which replication is initiated.

Outer membrane a lipopolysaccharide-containing layer, containing some proteins, which form the (generally) outermost layer of a Gram-negative bacterium.

Palindrome a sequence of base pairs in which the order on the complementary strands of the DNA duplex reads the same starting from a central point of symmetry.

Permease an enzyme system concerned with the transport of specific substances, usually nutrients, through the cytoplasm membrane.

Permissive conditions growth conditions allowing a conditionally lethal mutant to survive.

Phage see Bacteriophage.

Phenotype the observable characteristics of an organism as expressed depending on the environmental conditions.

Phylogeny classification of organisms according to evolutionary criteria.

Pilus a filamentous protein polymer protruding from the cell surface. e.g. F pili, used as recognition structures between donor and recipient cells during F-mediated conjugation in *E. coli*.

Plaque a clear area in a lawn of bacterial cells due to the lytic action of bacteriophages in infected cells that had been developing in the cleared area.

Plasmid a molecule of DNA existing as an autonomous replicon in the cytoplasm. (Most plasmids are covalently closed circles (CCC), which under certain circumstances may have a strand nicked, when they are referred to as open circles (OC), although cases of linear plasmids are known.)

Pleiotropic gene a gene which affects several apparently non-related characteristics of the phenotype.

Polar mutation a mutation which affects not only the gene in which it is located but other(s) located immediately downstream on the DNA molecule.

Porin a protein complex (usually a homopolymer) located in the outer membrane of a Gram-negative bacterium, endowed with more or less specific permeation properties.

Positive supercoiling additional coiling of the circular duplex DNA molecule in the same direction as the winding of the double helix.

Primer a short length of ribo- or deoxyribonucleotides complementary to the strand to be synthesized, providing a 3'-OH end used as an anchor for DNA polymerases to start synthesis.

Probe a fragment of DNA labelled and used to visualize, by hybridization, a homologous region looked for on another DNA molecule.

Promiscuous plasmid a plasmid displaying a broad host-range.

Promoter (site) a sequence on DNA serving as the specific site of attachment of RNA polymerase, and defining the transcription start site. (Most known prokaryotic promoters contain two short conserved (consensus) base sequences, known as the '−10 and −35 boxes'.)

Prophage the DNA of a temperate phage which is quiescent and not in the lytic state – the prophage is usually integrated into the host chromosome, but may remain (e.g. phage P1 of *E. coli*) as a circular autonomous particle (a plasmid) in the cytoplasm.

Protease an enzyme which breaks down proteins.

Protoplast the structure of a Gram-positive bacterium maintained in an isotonic environment when its cell wall has been digested by appropriate enzymes.

Prototroph an organism not requiring a specified organic supplement (growth factor) to grow.

Reading frame one of the six possible ways of reading a nucleotide sequence or its complementary strand as a series of triplets.

Read-through abnormal lack of arrest of transcription between two adjacent genes.

RecA protein the protein product of gene *recA*, essential for homologous recombination, and also involved (in *E. coli*) in SOS DNA repair and in the induction of the lytic cycle of prophage λ.

Recombinant organism resulting from an exchange of gene(s) between different DNA molecules.

Recombinant DNA a molecule of DNA in which a region from a different source has been inserted after cutting with a restriction endonuclease and subsequent ligation.

Recombination see Genetic recombination.

Regulon a group of genes and/or operons responding to a common regulatory process, but dispersed over the whole chromosome, and thus transcribed as several mRNA molecules.

Repair (of DNA) the replacement, by different specialized processes, of DNA regions of a double-stranded molecule conferring defects in the normal structure of the double helix.

Replication the act of duplicating a DNA molecule forming a genome (bacterial or viral).

Replication fork the point, on a replicating double-stranded DNA, where synthesis of DNA is taking place.

Replicon a unit of replication, i.e. a molecule of DNA or RNA containing an origin of replication and, most often, the information for its regulation of replication, and therefore capable of autonomy of regulation for its replication.

Repressor a protein, coded by a repressor gene, which decreases the transcription level of other specific genes, often by coordinated action with a cofactor on the regulatory region (operator) of a gene or operon; also used to designate the cofactor which interacts with a repressor or activator protein (the proper term should be corepressor).

Restriction the cleavage of a DNA duplex by an endonuclease (restriction enzyme) recognizing it as foreign by the absence of protection of specific sites by secondary modification (e.g. methylation). (Cleavage takes place at specific or random sites depending on the class of the endonuclease, although the enzymes always have (rather) specific recognition sites.)

Restriction endonuclease an enzyme which cuts double-stranded DNA by recognizing specific sites, in many cases arranged in palindromes.

Reversion the alteration (by mutation) of the mutated genome so as to return to the original (wild-type) phenotype.

R factor a plasmid carrying genes coding for resistance to antibacterial substances.

Rho factor a protein which assists the transcription-termination region of some genes to terminate transcription and detach the RNA polymerase.

Ribosome an RNA–protein complex responsible for the correct positioning of mRNA and charged tRNAs allowing proper alignment of amino acids during polypeptide synthesis.

RNA polymerase (DNA-dependent) an enzyme responsible for the synthesis of RNA from its constituent ribonucleotides, using one strand of DNA as template.

rRNA one of the RNA molecules forming a ribosome.

Septation the synthesis of a septum.

Septum the invagination produced by neosynthesis, directed toward the inside of the cell, of the envelope layers, forming the physical separation of the two bacterial daughter cells at division.

Sigma (σ) factor a protein acting as a subunit of bacterial RNA polymerases responsible for specificity of recognition of promoters.

Single-strand DNA-binding (Ssb) protein basic protein showing high affinity for single-stranded DNA.

Site-specific recombination a recombination event requiring only a small sequence of homology between the two molecules implicated, and independent of the Rec systems.

SOS (response) the coordinate induction of several processes, including DNA repair, in response to the presence of damage on the DNA – initiated, in *E. coli*, by degradation of the regulon-repressor LexA by the activated RecA protein.

Southern hybridization (blotting) annealing of a labelled probe (single-stranded DNA) with fragments of denatured DNA resolved by electrophoresis and fixed on a solid support (nitrocellulose or nylon membranes).

Spheroplast the structure of a Gram-negative bacterium maintained in an isotonic environment when its cell wall has been digested by appropriate enzymes.

Start (initiation) codon the AUG (sometimes also GUG or occasionally UUG) codon on mRNA at which polypeptide synthesis is initiated.

Sticky (cohesive) ends the two complementary single-stranded ends of a DNA duplex, e.g. resulting from cuts occurring at different positions by a Class II restriction endonuclease.

Stop (nonsense) codon one of the UAG, UAA or UGA codons, not representing any amino acid, at which polypeptide synthesis stops.

Stringent response the ability of a bacterium to cut down its synthesis of tRNAs and ribosomes when in a poor growth medium.

Structural gene originally, a gene coding for a polypeptide actually involved in a metabolic pathway (as opposed to a regulatory gene); often ambiguously used to designate any region of a genome coding for a polypeptide.

Suicide vector a modified plasmid or phage genome used as gene-carrying vector in molecular genetic technology that cannot replicate in a given host, used to force the integration into host replicons of genetic determinants that it carries and on which selection can be made.

Supercoiled DNA double-stranded circular DNA in which either overwinding or underwinding of the duplex makes the circle twist.

Superinfection the secondary infection by a phage in a cell already harbouring the same phage.

Suppressor gene a mutated gene which produces a product which reverses the effect of a previous specific mutation without actually correcting the original mutation in the DNA – can be either intergenic or intragenic.

Synapsis the pairing of homologous chromosomes or of homologous chromosomal regions.

Tandem repeats multiple copies of the same sequence lying in contiguous series.

Taxonomy the classification of organisms according to various criteria.

Telomere the terminal part of a eukaryotic chromosome, with a DNA sequence consisting of a single-stranded end that may fold into a hairpin structure, conferring stability on the molecule.

Temperate phage a bacteriophage that is capable of becoming a prophage in the bacterial host, i.e. maintain itself as a quiescent genome. Improperly but frequently called a lysogenic phage.

Terminal redundancy presence of identical DNA sequences repeated at the two ends of a linear DNA molecule.

Three-point-test mapping the determination of the relative localizations of three genetic determinants on a genome using as criterion their relative frequency of cotransfer during a cross between two strains bearing complementary allelic forms of these determinants.

Topoisomerase an enzyme which introduces or removes over- or underwinding of the DNA circular duplex by causing a nick, repositioning the strands and then ligating them.

trans describes the position of two genetic elements on different DNA molecules present in the same cell, as opposed to *cis*.

Transcription synthesis of RNA from a DNA template.

Transcription unit the length of DNA (a gene or an operon) transcribed as a single RNA.

Transduction a means of gene transfer between bacteria in which the bacterial donor DNA is carried by a bacteriophage.

Transformation a means of gene transfer between bacteria in which the donor DNA is 'naked' and enters the cell with no exogenous vector.

Transition the substitution of one type of base of a base pair by the same type, e.g. purine by purine or pyrimidine by pyrimidine.

Translation synthesis of a protein from an mRNA template.

Translocation the shift of a DNA region from one part of a chromosome to another (or another chromosome in eukaryotes).

Transposon a mobile element possessing accessory characterizable genetic traits, often resistance to antibiotic(s), that are not involved in the transposition capacity.

Transversion the substitution of one type of base of a base pair by the other type, i.e. purine by pyrimidine and vice versa.

Triplet the sequence in the genomic molecule of three bases, similar to the mRNA triplets except for thymine replacing uracil, which code for an amino acid or punctuation mark when transcribed into mRNA codons.

tRNA the translational key between the nucleic acid code and the amino acid language; a small RNA molecule bearing specific correlated anticodon- and amino acid-binding specificity.

tRNA synthetase the enzyme which charges a specific amino acid to a specific tRNA.

Vector a replicon used in molecular genetic technology to carry DNA fragments so that they can be transferred to other cells.

Virulent phage a bacteriophage which can only enter the lytic cycle of infection in a bacterial host.

Wild type an organism, often the naturally occurring strain, whose genetic traits are considered as reference, often either not requiring a specified growth factor(s) or sensitive to a given antibacterial substance(s).

Zinc-finger motif a region of a protein which, in association with zinc, comprises a specific DNA-binding domain.

Zygotic induction the lytic development of a prophage upon its transfer, usually by conjugation, into a recipient cell devoid of the same prophage, and thus of the specific repressor stabilizing the prophage state.

Further Reading

The prokaryotes

Newton A. & Ohta N. (1990) Regulation of the cell division cycle and differentiation in bacteria. *Annual Review of Microbiology*, 44, 689–719.

Vining L.C. (1990) Functions of secondary metabolites. *Annual Review of Microbiology*, 44, 395–427.

Bacterial chromosomes

Brock T.D. (1988) The bacterial nucleus: a history. *Microbiological Reviews*, 52, 397–411.

Chase J.W. (1986) Single-stranded DNA binding proteins required for DNA replication. *Annual Review of Biochemistry*, 55, 103–36.

D'Ari R. & Bouloc P. (1990) Logic of the *Escherichia coli* cell cycle. *Trends in Biochemical Sciences*, 15, 197–8.

de Boer P.A.J., Cook W.R. & Rothfield L.I. (1990) Bacterial cell division. *Annual Review of Genetics*, 24, 249–74.

Drlica K. & Rouviere-Yaniv J. (1987) Histone-like proteins of bacteria. *Microbiological Reviews*, 51, 301–19.

Firshein W. (1989) Role of the DNA/membrane complex in prokaryotic DNA replication. *Annual Review of Microbiology*, 43, 89–120.

Gilson E., Clément M., Perrin D. & Hofnung M. (1987) Palindromic units: a case of highly repetitive DNA sequences in bacteria. *Trends in Genetics*, 3, 226–30.

Holland I.B. (1987) Genetic analysis of *E. coli* division clock. *Cell*, 48, 361–2.

Kornberg A. (1987) Enzyme systems initiating replication at the origin of the *Escherichia coli* chromosome. *Journal of Cell Science*, (suppl.) 7, 1–43.

Krawiec S. & Riley M. (1990) Organization of the bacterial chromosome. *Microbiological Reviews*, 54, 502–39.

Kuempel P.L., Pelletier A.J. & Hill T.M. (1989) Tus and the terminators: the arrest of replication in prokaryotes. *Cell*, 59, 581–3.

Lohman T.M., Bujalowski W. & Overman L.B. (1988) *E. coli* single strand binding protein: a new look at helix-destabilizing proteins. *Trends in Biochemical Sciences*, 13, 250–4.

Lutkenhaus J. (1990) Regulation of cell division in *E. coli*. *Trends in Genetics*, 6, 22–5.

McHenry C.S. (1988) DNA polymerase III holoenzyme of *Escherichia coli*. *Annual Review of Biochemistry*, 57, 519–50.

Meyer R.R. & Laine P.S. (1990) The single-stranded DNA-binding protein of *Escherichia coli*. *Microbiological Reviews*, 54, 342–80.

Moses R.E. & Summers W.C. (1988) *DNA replication and mutagenesis*. American Society of Microbiology, Washington DC.

Nossal N.G. (1983) Prokaryotic DNA replication systems. *Annual Review of Biochemistry*, 53, 581–615.

Schmid M.B. (1988) Structure and function of the bacterial chromosome. *Trends in Biochemical Sciences*, 13, 131–6.

Schmid M.B. (1990) More than just 'histone-like' proteins. *Cell*, 63, 451–3.

Zyskind J.W. & Smith D.W. (1986) The bacterial origin of replication, *oriC*. *Cell*, 46, 489–90.

Plasmids

Clewell D.B. (1981) Plasmids, drug resistance, and gene transfer in the genus *Streptococcus*. *Microbiological Reviews*, 45, 409–36.

Couturier M., Bex F., Bergquist P.L. & Maas W.K. (1988) Identification and classification of bacterial plasmids. *Microbiological Reviews*, 52, 375–95.

Dunny G.M. (1990) Genetic functions and cell–cell interactions in the pheromone-inducible plasmid transfer system of *Enterococcus faecalis*. *Molecular Microbiology*, 4, 689–96.

Foster T.J. (1983) Plasmid-determined resistance to antimicrobial drugs and toxic metal ions in bacteria. *Microbiological Reviews*, 47, 361–409.

Hardy H.G. (ed.) (1987) *Plasmids: a practical approach*. IRL Press at Oxford University Press, Oxford.

Ippen-Ihler K.A. & Minkley E.G. Jr (1986) The conjugation system of F, the fertility factor of *Escherichia coli*. *Annual Review of Genetics*, 20, 593–624.

Krishnapillai V. (1988) Molecular genetic analysis of bacterial plasmid promiscuity. *FEMS Microbiology Reviews*, 54, 223–38.

Kues U. & Stahl U. (1989) Replication of plasmids in Gram-negative bacteria. *Microbiological Reviews*, 53, 491–516.

Nordström K. (1989) Mechanisms that contribute to the stable segregation of plasmids. *Annual Review of Genetics*, 23, 37–69.

Nordström K. (1990) Control of plasmid replication – how do DNA iterons set the replication frequency? *Cell*, 63, 1121–4.

Novick R.P. (1987) Plasmid incompatibility. *Microbiological Reviews*, 51, 381–95.

Rothman Scott J. (1984) Regulation of plasmid replication. *Microbiological Reviews*, 48, 1–23.

Silver S. & Misra T.K. (1988) Plasmid-mediated heavy metal resistances. *Annual Review of Microbiology*, 42, 717–43.

Stephens C.R.C. (1986) Pheromones among the prokaryotes. *Critical Reviews in Microbiology*, 13, 309–34.

Thomas C. (ed.) (1989) *Promiscuous plasmids of Gram-negative bacteria*. Academic Press, New York.

Thomas C.M. & Smith C.A. (1987) Incompatibility group P plasmids: genetics, evolution, and use in genetic manipulation. *Annual Review of Microbiology*, 41, 77–101.

Wiedemann B., Bennett P.M., Linton A.H., Sköld O. & Speller D.C.E. (eds) (1986) *Evolution, ecology and epidemiology of antibiotic resistance*. Proceedings of a joint meeting held in Bad Honnef (Bonn) by the Paul-Ehrlich-Gesellschaft für Chemotherapie and the British Society for Antimicrobial Chemistry.

Bacteriophages

Black L.W. (1989) DNA packaging in dsDNA bacteriophages. *Annual Review of Microbiology*, 43, 267–92.

Calendar R. (1988) *The Bacteriophages*. Plenum Press, New York.

Echols H. (1986) Bacteriophage λ development: temporal switches and the choice of lysis or lysogeny. *Trends in Genetics*, 2, 26–30.

Feiss M. (1986) Terminase and the recognition, cutting and packaging of λ chromosomes. *Trends in Genetics*, 2, 100–4.

Hirth L. & Girard M. (1989) *Virologie Moléculaire*. Doin, Paris.

Mathews C.K., Kutter E.M., Mosig G. & Berget P.B. (1983) *Bacteriophage T4*. American Society of Microbiology, Washington DC.

Mosig G. (1987) The essential role of recombination in phage T4 growth. *Annual Review of Genetics*, 21, 347–71.

Rasched I. & Oberer E. (1986) Ff coliphages: structural and functional relationships. *Microbiological Reviews*, 50, 401–27.

Symonds N., Toussaint A., van de Putte P. & Howe M.M. (eds) (1987) *Phage Mu*. Cold Spring Harbor Symposium Quantitative Biology, Cold Spring Harbor Laboratory, New York, pp. 375–6.

Warren R.A.J. (1980) Modified bases in bacteriophage DNAs. *Annual Review of Microbiology*, 34, 137–58.
Zinder N.D. & Horiuchi K. (1985) Multiregulatory element of filamentous bacteriophages. *Microbiological Reviews*, 49, 101–6.

Transposable elements

Berg D.E. & Howe M.M. (eds) (1989) *Mobile DNA*. American Society of Microbiology, Washington DC.
Clewell D.B. & Gawron-Burke C. (1986) Conjugative transposons and the dissemination of antibiotic resistance in *Streptococci*. *Annual Review of Microbiology*, 40, 635–59.
Craig N.L. (1988) The mechanism of conservative site-specific recombination. *Annual Review of Genetics*, 22, 77–105.
Derbyshire K.M. & Grindley N.D.F. (1986) Replicative and conservative transposition in bacteria. *Cell*, 47, 325–7.
Kingsman A.J., Chater K.F. & Kingsman S.M. (eds) (1988) *Transposition*. Symposium 43. SGM Cambridge University Press, Cambridge.
Mizuuchi K. & Craigie R. (1986) Mechanism of bacteriophage Mu transposition. *Annual Review of Genetics*, 20, 385–429.
Shapiro J.A. (ed.) (1983) *Mobile genetic elements*. Academic Press, New York.

DNA methylation, modification/restriction

Marinus M.G. (1987) DNA methylation in *Escherichia coli*. *Annual Review of Genetics*, 21, 113–31.
Messer W. & Noyer-Weidner M. (1988) Timing and targeting: the biological functions of Dam methylation in *E. coli*. *Cell*, 54, 735–7.
Price C. & Bickle T.A. (1986) A possible role for DNA restriction in bacterial evolution. *Microbiological Sciences*, 3, 296–9.
Wilson G.G. (1988) Type II restriction-modification systems. *Trends in Genetics*, 4, 314–19.
Yuan R. (1981) Structure and mechanism of multifunctional restriction endonucleases. *Annual Review of Biochemistry*, 50, 285–315.

Mutations

Boe L. (1990) Mechanism for induction of adaptive mutations in *Escherichia coli*. *Molecular Microbiology*, 4, 597–601.
Cairns J. (1990) The origin of mutants. *Nature*, 335, 192–4.
Cox E.C. (1976) Mutator genes. *Annual Review of Genetics*, 10, 135–56.
Demerec M., Adelburg F.A., Clark A.J. & Hartman P.E. (1966) A proposal for a uniform nomenclature in bacterial genetics. *Genetics*, 54, 61–76.
Luria S.E. & Delbruck M. (1943) Mutations in bacteria from virus sensitivity to virus resistance. *Genetics*, 28, 491–511.
Ripley L.S. (1990) Frameshift mutation: determinants of specificity. *Annual Review of Genetics*, 24, 189–213.

DNA repair

Claverys J.-P. & Lacks S.A. (1986) Heteroduplex deoxyribonucleic acid base mismatch repair in bacteria. *Microbiological Reviews*, 50, 133–65.
Echols H. & Goodman F. (1990) Mutation induced by DNA damage: a many protein affair. *Mutation Research*, 236, 301–11.
Friedberg E.C. (1985) *DNA repair*. W.H. Freeman & Company, New York.
Grilley M., Holmes J., Yashar B. & Modrich P. Mechanisms of DNA-mismatch correction. *Mutation Research*, 236, 253–67.
Lindahl T. (1990) Repair of intrinsic DNA lesions. *Mutation Research*, 238, 305–11.
Lindahl T., Sedgwick B., Sekiguchi M. & Nakabeppu Y. (1988) Regulation and expression of the adaptive response to alkylating agents. *Annual Review of Biochemistry*, 57, 133–57.
Modrich P. (1987) DNA mismatch correction. *Annual Review of Biochemistry*, 56, 435–66.

Radman M. & Wagner R. (1986) Mismatch repair in *Escherichia coli*. *Annual Review of Genetics*, 20, 523–38.

Sakumi K. & Sekiguchi M. (1990) Structures and functions of DNA glycosylases. *Mutation Research*, 236, 161–72.

Sancar A. & Sancar G.B. (1988) DNA repair enzymes. *Annual Review of Biochemistry*, 57, 29–67.

Sancar G.B. (1990) DNA photolyases: physical properties, action mechanisms and roles in dark repair. *Mutation Research*, 236, 147–60.

Van Houten B. (1990) Nucleotide excision repair in *Escherichia coli*. *Microbiological Reviews*, 54, 18–51.

Walker G.C. (1985) Inducible DNA repair systems. *Annual Review of Biochemistry*, 54, 425–57.

Walker G.C., Marsh L. & Dodson L.A. (1985) Genetic analyses of DNA repair: inference and extrapolation. *Annual Review of Genetics*, 19, 103–26.

Genetic recombination

Holliday R. (1990) History of heteroduplex. *Bioessays*, 12, 133–42.

Low B.K. (ed.) *The Recombination of Genetic Material*. Academic Press, London.

Miller R. & Kokjohn T.A. (1990) General microbiology of *recA*: environmental and evolutionary significance. *Annual Review of Microbiology*, 44, 365–94.

Smith G.R. (1983) Chi hotspots of generalized recombination. *Cell*, 34, 709–10.

Smith G.R. (1987) Mechanism and control of homologous recombination in *Escherichia coli*. *Annual Review of Genetics*, 21, 179–201.

Smith G.R. (1988) Homologous recombination in procaryotes. *Microbiological Reviews*, 52, 1–28.

Thaler D.S. & Stahl F.W. (1988) DNA double-chain breaks in recombination of phage λ and of yeast. *Annual Review of Genetics*, 22, 169–97.

Wang J.C., Caron P.R. & Kim R.A. (1990) The role of DNA topoisomerases in recombination and genome stability: a double-edged sword. *Cell*, 62, 403–6.

West S. (1990) Processing of recombination intermediates *in vitro*. *Bioessays*, 12, 151–4.

Genetic transformation

Butler L.O., Harwood C. & Moseley B.E.B. (1989) *Genetic transformation and expression*. Intercept, Andover, Hants.

Dubnau D. (1991) The regulation of genetic competence in *Bacillus subtilis*. *Molecular Microbiology*, 5, 11–18.

Stewart G.J. & Carlson C.A. (1986) The biology of natural transformation. *Annual Review of Microbiology*, 40, 211–35.

Conjugation

Lederberg J. (1986) Forty years of genetic recombination in bacteria. *Nature*, 324, 627–31.

Lederberg J. (1987) Genetic recombination in bacteria: a discovery account. *Annual Review of Genetics*, 21, 23–46.

Smith G.R. (1991) Conjugational recombination in *E. coli*: myths and mechanisms. *Cell*, 64, 19–27.

Artificial transfer systems

Chassy B., Mercenier A. & Flickinger J. (1988) Transformation of bacteria by electroporation. *Trends in Biotechnology*, 6, 303–9.

Solioz M. & Bienz D. (1990) Bacterial genetics by electric shock. *Trends in Biochemical Sciences*, 15, 175–7.

Zimmerman U. (1983) Electrofusion of cells: principles and industrial potential. *Trends in Biotechnology*, 1, 149–55.

Genetic Maps

Bachmann B.J. (1990) Linkage map of *Escherichia coli* K-12 edition 8. *Microbiological Reviews*, 54, 130–97.

Burke D.T., Carle G.F. & Olson M.V. (1987) Cloning of large segments of exogenous DNA into yeast by means of artificial chromosome vectors. *Science*, 236, 806–10.

Medique C., Bouché J.-P., Ménaut A. & Danchin A. (1990) Mapping of sequenced genes (700 kbp) in the restriction map of the *E. coli* chromosome. *Molecular Microbiology*, 4, 196–7.

O'Brien S.J. (1988) *Genetic maps.* Cold Spring Harbor Symposium Quantitative Biology, Cold Spring Harbor Laboratory, New York.

Piggot P.J. & Hoch J.A. (1985) Revised genetic linkage map of *Bacillus subtilis*. *Microbiological Reviews*, 49, 158–79.

Sanderson K.E. & Roth J.R. (1988) Linkage map of *Salmonella typhimurium*, edition VII. *Microbiological Reviews*, 52, 485–532.

Schlessinger D. (1990) Yeast artificial chromosomes: tools for mapping and analysis of complex genomes. *Trends in Genetics*, 6, 248–58.

The operon model

Booth I.R. & Higgins C.F. (1986) *Regulation of gene expression, 25 years.* Society of Genetic Microbiology, Cambridge University Press, Cambridge.

Miller J.H. & Reznikoff W.S. (eds) (1978) *The operon.* Cold Spring Harbor Symposium Quantitative Biology, Cold Spring Harbor Laboratory, New York.

Reznikoff W.S. (1992) The lactose operon-controlling elements – a complex paradigm. *Molecular Microbiology*, 6, 219–22.

Genetics as molecular tools

Albright L.M., Huala E. & Ausubel F.M. (1989) Transduction mediated by prokaryotic signal sensor and regulator protein pairs. *Annual Review of Genetics*, 23, 311–36.

Gussin G.N., Ronson C.W. & Ausubel F.M. (1986) Regulation of nitrogen fixation genes. *Annual Review of Genetics*, 20, 567–91.

Old R.W. & Primrose S.B. (1985) *Principles of gene manipulation – an introduction to genetic engineering.* Blackwell Scientific Publications, Oxford.

Postgate J. (1989) Trends and perspectives in nitrogen fixation research. *Advances in Microbial Physiology*, 30, 1–22.

Saunders V.A. & Saunders J.R. (1987) *Microbial genetics applied to biotechnology.* Croom Helm Ltd., London.

Silhavy T.J. & Beckwith J.R. (1985) Uses of *lac* fusions for the study of biological problems. *Microbiological Reviews*, 49, 398–418.

Smith I., Slepecky A. & Setlow P. (eds) (1988) *Regulation of prokaryotic development: structural and functional analysis of bacterial sporulation and germination.* American Society of Microbiology, Washington DC.

Smith M. (1985) *In vitro* mutagenesis. *Annual Review of Genetics*, 19, 423–62.

Stock J.B., Ninfa A.J. & Stock A.M. (1989) Protein phosphorylation and regulation of adaptive responses in bacteria. *Microbiological Reviews*, 53, 450–90.

Genetics of bacterial species

Baltz R.H. & Seno E.T. (1988) Genetics of *Streptomyces fradiae* and tylosin biosynthesis. *Annual Review of Microbiology*, 42, 547–74.

Chatterjee A.K. & Starr M.P. (1980) Genetics of *Erwinia* species. *Annual Review of Microbiology*, 34, 645–76.

de Vries G.E., Kües U. & Stahl U. (1990) Physiology and genetics of methylotrophic bacteria. *FEMS Microbiology Reviews*, 75, 57–102.

Dybvig K. (1990) Mycoplasmal genetics. *Annual Review of Microbiology*, 44, 81–104.

Forterre P. (1986) Les archaebactéries: origine et évolution des lignées cellulaires procaryotes et eucaryotes. *Année Biologique*, 25, 273–306.

Gasson M.J. (1990) *In vivo* genetic systems in lactic acid bacteria. *FEMS Microbiology Reviews*, 87, 43–60.

Guidolin A. & Manning P.A. (1987) Genetics of *Vibrio cholerae* and its bacteriophages. *Microbiological Reviews*, 51, 285–98.

Holloway B.W. & Morgan A.F. (1986) Genome organization in *Pseudomonas*. *Annual Review of Microbiology*, 40, 79–105.

Jones W.J., Nagle D.P. Jr & Whitman W.B. (1987) Methanogens and the diversity of archaebacteria. *Microbiological Reviews*, 51, 135–77.

Kennedy C. and Toukdarian A. (1987) Genetics of *Azotobacters*: applications to nitrogen fixation and related aspects of metabolism. *Annual Review of Microbiology*, 41, 227–58.

Leblond P., Demuyter P., Simonet J.M. & Decaris B. (1990) Genetic instability and hypervariability in *Streptomyces ambofaciens*: towards an understanding of a mechanism of genome plasticity. *Molecular Microbiology*, 4, 707–14.

Lidstrom M.E. & Stirling D.I. (1990) Methylotrophs: genetics and commercial applications. *Annual Review of Microbiology*, 44, 27–58.

Long S.R. (1989) *Rhizobium* genetics. *Annual Review of Genetics*, 23, 483–506.

Neidhardt F.C. (ed.) (1987) Salmonella *and* Escherichia coli: *cellular and molecular biology*. American Society of Microbiology, Washington.

Novick R.P. (1989) Staphylococcal plasmids and their replication. *Annual Review of Microbiology*, 43, 537–65.

Razin S. (1985) Molecular biology and genetics of mycoplasmas (mollicutes). *Microbiological Reviews*, 49, 419–55.

Winkler H.H. (1990) *Rickettsia* species (as organisms). *Annual Review of Microbiology*, 44, 131–53.

Young M., Minton N.P. & Staudenbauer W.L. (1989) Recent advances in the genetics of the clostridia. *FEMS Microbiology Reviews*, 63, 301–26.

Evolution

Beacham I.R. (1987) Silent genes in prokaryotes. *FEMS Microbiology Reviews*, 46, 409–17.

Belfort M. (1989) Bacteriophage introns: parasites within parasites? *Trends in Genetics*, 5, 209–12.

Belfort M. (1991) Self-splicing introns in prokaryotes: migrant fossils? *Cell*, 64, 9–11.

Hopwood D.A. & Chater K.F. (eds) (1989) *Genetics of bacterial diversity*. Academic Press, New York.

Inouye M. & Inouye S. (1991) Retroelements in bacteria. *Trends in Biochemical Sciences*, 16, 18–21.

Kuhsel M.G., Strickland R. & Palmer J.D. (1990) An ancient group I intron shared by eubacteria and chloroplasts. *Science*, 250, 1570–3.

Riley M. & Anilionis A. (1978) Evolution of the bacterial genome. *Annual Review of Microbiology*, 32, 519–60.

Schleifer R.H. & Stachebrandt E. (1983) Molecular systematics of prokaryotes. *Annual Review of Microbiology*, 37, 143–87.

Woese C.R. (1987) Bacterial evolution. *Microbiological Reviews*, 51, 221–71.

Xu, M.-Q., Kathe S.D., Goodrich-Blair H., Nierzwicki-Bauer S.A. & Shub D.A. (1990) Bacterial origin of a chloroplast intron: conserved self-splicing group I introns in cyanobacteria. *Science*, 250, 1566–70.

General books

Birge E.A. (1988) *Bacterial and Bacteriophage Genetics*, 2nd edn. Springer-Verlag, New York.

Glass R.E. (1982) *Gene Function*. Croom Helm Ltd., London.

Hayes W. (1968) *The Genetics of Bacteria and their Viruses* 2nd edn. Blackwell Scientific Publications, Oxford.

Scaife J. & Galizzi A. (eds) (1985) *Genetics of Bacteria*. Academic Press, London.

Index

Page numbers in *italics* refer to tables/figures